The International Karakoram Project

THE INTERNATIONAL

KARAKORAM

PROJECT

Volume 1

Proceedings of the International Conference
held at
Quaid-i-Azam University
Islamabad
Pakistan

EDITED BY K. J. MILLER

Cambridge University Press

Cambridge

London New York New Rochelle

Melbourne Sydney

CAMBRIDGE UNIVERSITY PRESS
Cambridge, New York, Melbourne, Madrid, Cape Town, Singapore,
São Paulo, Delhi, Dubai, Tokyo

Cambridge University Press
The Edinburgh Building, Cambridge CB2 8RU, UK

Published in the United States of America by Cambridge University Press, New York

www.cambridge.org
Information on this title: www.cambridge.org/9780521129749

First published 1984
This digitally printed version 2009

A catalogue record for this publication is available from the British Library

ISBN 978-0-521-26339-9 Hardback
ISBN 978-0-521-12974-9 Paperback

These proceedings are dedicated to the memory of
J. F. BISHOP
who died on Kurkun (4730 m) on 14 July 1980
during the course of
The International Karakoram Project

Contents of Volume 1

viii

HOUSING AND NATURAL HAZARDS

GEOMORPHOLOGY

Contents of Volume 2

GLACIOLOGY

KARAKORAM RESEARCH CELL

Preface

These volumes detail a comprehensive series of inter-related studies carried out by seventy-three scientists from China, Pakistan and Britain who visited the Karakoram mountains in the summer of 1980. These men and women were members of the Royal Geographical Society Project to the world's highest mountain range as part of the Society's 150th anniversary celebrations.

The proceedings are divided into two volumes; the first volume being reserved for papers presented and discussed, in conference, in Islamabad, Pakistan prior to the field work. The second volume is reserved for papers based on the field work conducted during the course of the International Karakoram Project, or IKP for short, and read at a conference at the Royal Geographical Society, London.

The two volumes present a comprehensive cover of the earth sciences, ranging in scale from surveys to measure the effects of impacts between the Eurasian and Indian tectonic plates, to electron microscope studies on weathering processes. The six general subject areas are: geology, glaciology, geomorphology, seismology and topography (survey). The sixth subject is concerned with the indigenous population namely social, economic and health aspects of villagers, the architectural form and structural stability of their homes and the order of priorities they have to accept when combating the greatest number of natural hazards known to man which include earthquakes, drought, fire, floods, famine, landslides, avalanches and disease.

These proceedings therefore bring together a collection of papers to provide a datuum for future scientific research in the Karakoram. The IKP also proved to be a catalyst for the founding of the Karakoram Research Cell (KRC) by the University Grants Commission of Pakistan. The KRC will be responsible for co-ordinating the research work conducted in the Karakoram by future researchers; see Volume 2.

The varied scientific programmes undertaken were supported by many national and international bodies, listed elsewhere, but a special mention must be made of Academia Sinica, Beijing and the Ministry of Science and Technology, Islamabad who joined the Royal Geographical Society in sponsoring the venture. These three together with The Overseas Development Administration of the UK Government and the British Council initially made the project a reality and finally an international success in scientific collaboration.

K. J. Miller
Leader of the
International Karakoram Project

Sheffield 1982

Other publications related to the International Karakoram Project: Continents in Collision, 212 pp, K.J. Miller, ISBN 0 540 010669. George Philip and Son Ltd. London.

Preface to Volume 1

This volume contains a unique set of inter-related technical papers. The text, supported by 55 figures, 62 maps, 44 photographs, 58 graphs and 36 tables of data, is a record of a conference at which, for the first time, scientists from many countries gathered together to discuss the multi-disciplinary problems created within, and by, the Karakoram and neighbouring high mountain ranges. Here are the world's largest glaciers outside the polar regions, the greatest precipices on the land-surface of the earth, the most rapidly changing topography known to man, and the greatest concentration of 7000 m peaks.

Furthermore, although the conference concerned itself predominantly with the earth sciences and the conditions of human life amidst an unimaginable chaotic landscape, a most important aspect of the meeting was to discuss recent scientific and technological developments that could assist the many programmes of research to be conducted during the course of the International Karakoram Project (IKP).

As the conference progressed a third and most important feature developed, namely the interaction between members of various teams. Usually an expedition to the Karakoram is limited in time and manpower because of finance, weather and logistical problems, but this large scale venture permitted, for example, geomorphologists to discuss transient landforms with the housing and natural hazards team examining the optimal siting of buildings. This second team in turn necessarily involved itself with earthquake studies whilst the seismology team interacted with the surveyors measuring surface deflections of tectonic plates. Similarly the glaciologists discussed modern survey systems to measure the dangerous movements of the immense glaciers that threaten the day-to-day pattern of existence of the indigenous population. Thus it can be seen that the conference allowed a full circle of debate to take place between all the scientists involved with the project and the many other experts who attended the conference as delegates. There were many varied forms of interactive discussions, some based on the usefulness of specific sets of data to the different groups of investigators, some based on the requirement to develop similar and collaborative, data-collecting techniques, while some discussions were concerned with methods of data analysis and their comparison.

Another important thread which passes through these proceedings is one related to the history of the Karakoram and many papers present a review and bibliography of man's knowledge of the Karakoram Range, thereby forming a datum for future work. Papers such as the analysis of the 1913 trigonometrical survey linking the grids of the Indian subcontinent to that of the USSR to form the greatest land survey network on earth is of note, particularly since the 1913 link is now known to have crossed the suture zone between the Eurasian and Indian tectonic plates. Other examples are the papers by the Chinese members who have briefly and succinctly presented many of their own long-awaited results from extensive research programmes in and around the Karakoram between 1974 and 1976.

This volume also presents details on modern systems, equipment and techniques for investigating difficult problems in the earth sciences and which were to be employed in the Karakoram, many for the first time. Additional information is presented that relates to other parts of the world but which have an important bearing on the initiation of specific Karakoram studies, e.g., the first ever recording and presentation of depth profiles of wet ice of temperate glaciers and ice caps. A few papers discuss problems of great concern to all mankind, problems that, once solved, will bring benefits to the peoples of the Karakoram.

The importance of the conference and these proceedings can perhaps best be judged from the opening speechs made by the President of Pakistan, General Zia-ul-Haq, and Lord Shackleton to whom we are indebted for their assistance in the entire Project as well as their attendance at the conference. Lord and Lady Hunt together with the British Ambassador, Oliver Forster and the staff of the British Council also gave valuable assistance throughout the period of the IKP. Finally a special mention and our grateful thanks must be awarded to Dr S. A. R. Jafree of Quaid-i-Azam University who organised the conference on behalf of the International Karakoram Project.

K. J. Miller

Sheffield 1982

Acknowledgements

The authors of these proceedings and all members of the IKP acknowledge the assistance given, in terms of time, effort and finance, by persons far too numerous to list here. Without the help given by these countless men and women, and those listed below, the IKP would never have completed the numerous studies reported in these volumes. We are forever in their debt.

IKP SPONSORS

Government of Pakistan
Academia Sinica, Beijing
Royal Geographical Society

INTERNATIONAL PROJECT COMMITTEE

Dr John B. Auden (Chairman)
Dr G. Colin L. Bertram
Professor Eric H. Brown
Sir Douglas Busk
Dr Andrew Goudie (Deputy Leader)
George Greenfield
Lt. Col. David Hall
Brigadier George Hardy
John Hemming (Director RGS)
Lord Hunt of Llanfair Waterdine
   (President RGS)
Professor Keith J. Miller (Leader)
Mr Michael Ward
Nigel de N. Winser (Deputy Leader
   and Secretary)

PAKISTAN LIAISON

David Latter, British Council
   Representative, Islamabad

DISTINGUISHED VISITORS

H. E. The British Ambassador
   Oliver Forster
Major General Mushtaq Ahmad Gill
Lord and Lady Hunt
Lord Shackleton

CONSULTANTS

GEOMORPHOLOGICAL RESEARCH

Professor E. H. Brown
Professor K. J. Gregory
Professor N. J. Stevens
Dr R. J. Price
Dr J. M. Grove
Dr D. Q. Bowen

RADAR ICE-DEPTH SOUNDING RESEARCH

Sir Vivian Fuchs
Dr S. Evans
Dr D. Drewry
Dr G. de Q. Robin
Dr C. W. M. Swithinbank
Dr C. M. Doake

cont'd...

HOUSING AND NATURAL HAZARDS RESEARCH

Professor N. Ambraseys
Professor W. G. V. Balchin
Professor O. Koenigsberger
D. J. Dowrick

GEOLOGICAL RESEARCH

Asrarullah (Geological Survey of Pakistan)
Dr B. F. Windley

SURVEY RESEARCH

P. G. Mott
J. G. Oliver
A. L. Allen
Mianmohammed Sharif (Survey of Pakistan)

FINANCIAL SUPPORT

Barclays Bank International
Binnie & Partners
British Council
William A. Cadbury Charitable Trust
Carnegie Trust for the Universities of Scotland
Winston Churchill Memorial Trust
The Drapers' Company
The Goldsmiths' Company
Harrods Ltd
Longdin & Browning
Manchester Geographical Society
Mount Everest Foundation
National Geographic Society

Natural Environment Research Council
Overseas Development Administration
Albert Reckitt Charitable Trust
Rio Tinto Zinc Corporation
Royal Geographical Society
Royal Institution of Chartered Surveyors
Royal Scottish Geographical Society
The Royal Society
Science Research Council
Wexas International
George Wimpey & Co. Ltd
United Nations

EXPEDITION EQUIPMENT

Berec International
Brillo Manufacturing Co
Briton Chadwick Ltd
Bryant & May
Burgess Rucksacs
Camping Gaz (GB) Ltd
Damart Thermal Underwear
Estercol Sales Ltd
Flatetec Ltd
Hamish Hamilton
Jen Shoes & Boots
Karrimor International
Kitbin
Laughton & Sons
Lacrinoid Ltd
Racal Radios
Mountain Equipment
Mitsui Machinery Sales
Northcape Textiles

Nevisport
Nikwax
A B Optimus Ltd
Penguin Books
Phillips of Axminster
Portacel Waterfilters
Pindisports
Rohan
Spacecoat Garments
Stanley Tools Ltd
Supreme Plastics
Swains Packaging
Thor Hammer Co
Thermos Ltd
Tog 24
Tri Wall Containers
Troll Safety Equipment
Vango Scotland Ltd

SURVEY EQUIPMENT

Aga Geotronics
British Aluminium
Casio Electronics
Decca Surveys
Hunting Surveys
Mapping and Charting Establishment RE

School of Military Survey
SLD Sitelink Services
Survey & General Instruments
Tellurometer (UK) Ltd
Wild Heerbrugg

OTHER SCIENTIFIC EQUIPMENT

Avon Rubber Company
Chloride Batteries (Pakistan) Ltd

Hewlett Packard Ltd (Geneva)
Perex Ltd

TRANSPORT

Anchor Line Ship Management Ltd
Associated Tyre Specialists
Automobile Association (Stockport)
British Airways
Brown Jenkinson (Liverpool) Ltd
Burmah Castrol Ltd
Caltex Oil (Pakistan) Ltd
Fairey Engineering Ltd
Land Rover Ltd

Lifting Gear Hire Ltd
Lloyds Industries
Manor National Group Motors
Michelin Tyre Company Ltd
Pakistan National Shipping Corp
Pakistan State Oil Co Ltd
Protofram Ltd
Trailvan Ltd
Unipart Ltd

FOODSTUFFS

British Food Export Council
Coca Cola Corporation
Colmans Foods
Drinkmaster Ltd
Frank Cooper
H J Heinz
The Honey Bureau & Gales Honey
John West Foods
Nabisco Ltd

Raven Food
Ryvita
Sabatani & Taylor Associated
    for Schwartz Spices
St Ivel Ltd
Tate & Lyle
Weetabix Ltd
Whitworths Holdings
Unilever Export Ltd

MEDICAL SUPPLIES

Abbott Laboratories Ltd
A D International Ltd
Allen & Hanburys Ltd
Astra Chemicals Ltd
Bayer (UK) Ltd
Beecham Research Laboratories
Bencard
Boehringer Ingleheim Ltd
Boots Company Ltd
Calmic Medical Division of the
    Wellcome Foundation
Ciba Laboratories
Davis & Geck
Dista Products
Dome Laboratories
Downs Surgical Ltd
Duncan Flockhart & Co Ltd

Merck Sharp & Dohme Ltd
Miles Laboratories
Montedison Pharmaceuticals Ltd
Nicholas Laboratories
Novo Laboratories
Lakeland Plastics
Parke Davis & Co Ltd
Pfizer Ltd
Pharmacia (GB) Ltd
Pharmax Ltd
A H Robins Co Ltd
Roche Products Ltd
Roussel
Reckitt & Colman Pharmaceuticals
Scholl (UK) Ltd
Serle Laboratories
E R Squibb & Sons Ltd

Eli Lily & Co Ltd
Fair Laboratories
Farley Health Products
Geigy Pharmaceuticals
Glaxo Laboratories
Hoechst Pharmaceuticals
Imperial Chemical Industries
Janseen Pharmaceuticals Ltd
Johnson & Johnson Ltd
Kirby Warwick Ltd
Laboratories for Applied Biology
Lederle Laboratories
Leo Laboratories
May & Baker Ltd

Smith & Nephew Pharmaceuticals
Smith Kline & French Laboratories
Strentex Fabrics Ltd
Chas. Thackery Ltd
3M UK Ltd
Upjohn Ltd
Vernon Carus Ltd
Wander Pharmaceuticals Ltd
Wlm Warner Ltd
Warner-Lambert Ltd
WB Pharmaceuticals
The Wellcome Foundation
Winthrop Laboratories
Wyeth Laboratories

PHOTOGRAPHIC WORK

Classic Vases
C & J Clark Ltd
Geographical Magazine
Japanese Cameras (Minolta)
Kodak Ltd

Laptech Studies
National Geographic
Quest Vest
Three Arrows Films
Tenba Bags

INSURANCE SERVICES

Sedgewick International

International Conference

Recent Technological Advances
in Earth Sciences

22–25 June 1980

ORGANIZED BY

>The Department of Earth Sciences
>Quaid-i-Azam University Islamabad, Pakistan

IN COOPERATION WITH

>The Royal Geographical Society, London
>Academia Sinica, Beijing
>Ministry of Science and Technology, Islamabad
>Ministry of Education, Islamabad
>The University Grants Commission, Islamabad
>The British Council, Islamabad
>Pakistan Science Foundation
>Pakistan Atomic Energy Commission
>Geological Survey of Pakistan
>UNESCO - UNDP
>Attock Refinery Ltd, Rawalpindi

CONFERENCE COMMITTEE

Dr Ahmed Mohiuddin (Chairman),
Vice Chancellor,
Quaid-i-Azam University.

Dr S. Arif Raza Jafree (Secretary),
Chairman, Dept. of Earth Sciences,
Quaid-i-Azam University.

Dr M. D. Shami,
Chairman, Pakistan Science Foundation.

Professor Ismail Sethi,
University Grants Commission.

Dr Nisar Ahmad,
Ministry of Science & Technology.

Dr Tahir Hussain,
Ministry of Education.

Dr Ashafq Ahmad,
Pakistan Atomic Energy Commission.

Mr A. M. Khan,
Ministry of Petroleum &
  Natural Resources.

Professor Keith Miller, Leader,
International Karakoram Project.

Mr David Latter,
Representative, British Council.

Mr Liu Shi Wei,
Chinese Embassy.

Dr R. A. Khan Tahirkheli,
Director, Dept. of Geology,
(now Vice Chancellor),
Peshawar University.

Dr Saif-ul-Islam Saif,
Quaid-i-Azam University.

Mr S. N. Quraishi,
Quaid-i-Azam University.

Mr Shabbir Ahmad,
Quaid-i-Azam University

xxii

Further assistance was provided by:

Dr Raziuddin Siddiqi,
Chairman,
Pakistan Academy of Sciences.

Dr M. S. H. Siddiqi,
Chairman,
Hydrocarbon Development Institute of
    Pakistan.

Dr Aslam Khan,
Chief Scientist, D.E.S.T.O.

Mr Asrarullah,
Director General,
Geological Survey of Pakistan.

Dr Qayyum Qazi,
Deputy Scientific Adviser,
Ministry of Science & Technology.

Mr M. H. Rizvi,
General Manager,
Oil' & Gas Development Corporation.

Dr S. H. Hashmi,
Quaid-i-Azam University.

Dr Fayyazuddin,
Quaid-i-Azam University.

Dr M. S. K. Razmi,
Quaid-i-Azam University.

Dr M. Hafeez,
Quaid-i-Azam University.

Dr C. M. Hussain,
Quaid-i-Azam University.

Dr Mahboob Mohammad,
Quaid-i-Azam University.

Dr Misbahul Islam,
Quaid-i-Azam University.

Mr Azhar Kaleem,
Quaid-i-Azam University.

and

Dr   Zubarr Ahmad Malik
Mr   Ali Jafar Zaidi
Dr   Masood Pervaiz
Dr   Tahira Chauhan
Mr   Jabir Khan
Mr   M. Ashraf
Mr   Abdullah Khan
Mr   Karim
Mr   Shakil Ahmad
Mr   Fateh Mohammad Malik
Mr   Saeed Shafqat
all of Quaid-i-Azam University

Authors and
Members of the Expedition

Members of the IKP

Profession/Team

Abbas, S. Ghazanfar

Deputy Director
Geological Survey of Pakistan
Geology Team

Ahmad, Shabbir

Lecturer
Quaid-i-Azam University
Seismology Team

Akbar, Khurshid

Assistant Geophysicist
Geological Survey of Pakistan
Seismology Team

Allen, John

Postgraduate research student
Sheffield University
Survey Team

Aman, Ashraf

Mountaineer
Rawalpindi Tourist Industries
Survey Team

Atkinson, Nigel

Surveyor
Hunting Surveys and Consultants Ltd
Survey Team

Awan, Abdul Razzaq

Surveyor
Survey of Pakistan
Survey Team

Bilham, Roger

Geophysicist
Lamont-Doherty Geological Observatory
Survey Team

Bishop, James

Civil Engineer
Sir Alexander Gibb and Partners
Survey Team

Brunsden, Denys

University Reader
King's College, University of London
Geomorphology Team

Charlesworth, Ronald

Freelance Photographer
Three Arrows Limited
Film Unit

Chen, Jianming

Engineer and Surveyor
Institute of Glaciology and Cryopedology,
    Lanzhou
Survey Team

Collins, David Nigel

University Lecturer
Manchester University
Geomorphology Team

Colvill, Alan John

Chartered Accountant/Land Surveyor
University of Colorado
Survey Team

Crompton, Thomas Oliver

University Lecturer
University College, University of London
Survey Team, Deputy Director

Davis, Ian Robert

Principal Lecturer
Oxford Polytechnic
Director, Housing and Natural Hazards
    Team

Davison, Ian

Geologist
British National Oil Corporation
Seismology Team

Derbyshire, Edward

University Reader
University of Keele
Geomorphology Team

Dong, Zhi Bin

University Lecturer
Lanzhou University
Radar Ice-Depth Sounding Team

Durrani, Nasir Ali

Professor of Geology
University of Baluchistan, Quetta
Geology Team

Farooq, Mohamed

Surveyor
Survey of Pakistan
Survey Team

Ferguson, Robert Ian

University Senior Lecturer
Stirling University
Geomorphology Team

Ferrari, Ronald Leslie

University Lecturer
Cambridge University
Radar Ice-Depth Sounding Team

Francis, Marcus

Research Engineer
British Hydraulics Research Association,
    Cranfield
Co-Director, Radar Ice-Depth Sounding
    Team

Ghauri, Arif Ali Khan

Associate Professor
Peshawar University
Geology Team

Giles, David Peter Vaughan Lindsey	Doctor Medical Practice, Bude Director, Medical Team
Goudie, Andrew Shaw	University Lecturer Oxford University Deputy Leader and Director, Geomorphology Team
Holmes, Robert Edward	Photographer (Stills) Freelance Photographer Film Unit
Hughes, Richard Edward	Geotechnical Archaeologist Ove Arup & Partners Housing and Natural Hazards Team
Illi, Dieter	University Lecturer Zurich University Housing and Natural Hazards Team
Islam, Shaukat	University Lecturer Peshawar University Botanist
Israr-ud-Din	Associate Professor Peshawar University Housing and Natural Hazards Team
Jan, Qasim M.	Associate Professor Peshawar University Geology Team
Jackson, James Anthony	Research Fellow Queen's College, Cambridge University Co-Director, Seismology Team
Jones, David Keith Crozier	University Lecturer London School of Economics, University of London Geomorphology Team
Khan, Islam M.	University Lecturer Peshawar University Botanist
Khan, Muhammad Zakir	Major, Pakistan Army Liaison Officer
Khattak, Rehman	Electrical Engineer Pakistan Atomic Energy Commission Seismology Team
King, Geoffrey Charles Plume	Senior Research Assistant Cambridge University Director, Seismology Team

Li, Jijun — Associate Professor
Lanzhou University
Geomorphology Team

Lin, Ban Zuo — Engineer
Institute of Geophysics, Beijing
Seismology Team and Co-Leader of
Chinese Team

Massil, Helen — Doctor
Medical Practitioner
Medical Team

Miller, Keith John — Professor of Mechanical Engineering
Sheffield University
Leader of Expedition

Moughtin, James Clifford — Professor of Planning
Nottingham University
Housing and Natural Hazards Team

Moughtin, Timothy — Assistant
Housing and Natural Hazards Team

Muir Wood, Robert — Journalist/Geologist
Writer for New Scientist

Musil, George Jiri — Postgraduate Research Student
British Antarctic Survey
Radar Ice-Depth Sounding Team

Nash, David Francis Tyris — University Lecturer
Bristol University
Housing and Natural Hazards Team
(Mrs Nash also gave valuable service
to the Project)

Nunn, Paul James — Mountaineer/Lecturer
Sheffield Polytechnic
Film Unit/Radar Ice-Depth Sounding Team

Oswald, Gordon Kenneth
Andrew — Electronics Research Engineer
Cambridge Consultants Ltd
Co-Director, Radar Ice-Depth Sounding Team

Perrott, Frances Alayne — Lecturer
Oxford University
Geomorphology Team

Rajab, Ali — Sirdar of Porters
Survey Team

Rana, Javaid Akhtar — Major, Pakistan Army
Liaison Officer

Redhead, Charles Stephen

Driver
Transport Team

Rehman, Mohammad Abdul

Geologist
Pakistan Atomic Energy Commission
Geology Team

Rendell, Helen

Lecturer
Sussex University
Geomorphology Team

Riley, Anthony

Freelance Cameraman
Three Arrows Limited
Film Unit

Said, Mohammed

Professor of Geography
Peshawar University
Geomorphology Team

Smith, Edward Whittaker

Lecturer
University of Manchester
    Institute of Science and Technology
Survey Team

D'Souza, Frances

Director
International Disaster Institute
Housing and Natural Hazards Team

Spence, Robin John

Lecturer
Cambridge University
Deputy Director, Housing and Natural
    Hazards Team
(Dr Spence had his research student,
    Andrew Coburn, assist for a short period)

Stoodley, Robert Arthur

Chairman and Managing Director
Manor National Group Motors Ltd
Transport Team Manager

Tahir, Iqbal

Major, Pakistan Army
Liaison Officer

Tahirkheli, Rashid Ahmad
    Khan

Professor of Geology
Peshawar University
Director, Geology Team and
    Leader of Pakistani Scientists

Walton, Jonathan Lancelot
    William

Land Surveyor
University College, London University
Director, Survey Team

Waters, Ronald Sidney

Professor of Geography
Sheffield University
Geomorphology Team

Wesley-Smith, Shane
Administrator
Royal Geographical Society
Administrative Team

Whalley, Brian
Lecturer
Queen's University, Belfast
Geomorphology Team

Winser, Nigel de Northop
Administrator
Royal Geographical Society
Deputy Leader and Director,
    Administrative Team

Xu, Shuying
Associate Professor
Lanzhou University
Geomorphology Team

Yielding, Graham
Postgraduate research student
Cambridge University
Seismology Team

Zafar, Hashmat
Hydrogeologist
Water and Power Development Authority
Geomorphology Team

Zhang, Xiangsong
Associate Professor
Institute of Glaciology and Cryopedology,
    Lanzhou
Radar Ice-Depth Sounding Team and
    Co-Leader of Chinese Team

OTHER AUTHORS OF CONFERENCE PAPERS WHO WERE NOT MEMBERS
OF THE IKP

Andrews, R. M.
Sheffield University, UK

Bull, P. A.
Christ College, Oxford, UK

Coburn, A. W.
University of Cambridge, UK

Cumming, A. D. G.
University of Cambridge, UK

Huang Maohuan
Inst. of Glaciology & Cryopedology, Lanzhou
People's Republic of China

McGregor, A. R.
Sheffield University, UK

Owen, G.
Trinity College, Cambridge, UK

Shi Yafeng
Inst. of Glaciology & Cryopedology, Lanzhou
People's Republic of China

Simpson, D.
Lamont-Doherty, Geological Observatory of
    Columbia University, New York, USA

Teng Ji Wen   Inst. of Geophysics, Academia Sinica,
Beijing, People's Republic of China

Wang Wenying   Inst. of Glaciology & Cryopedology, Lanzhou
People's Republic of China

Conversion Units

To convert from	to	multiply by
inch	metre (m)	2.54×10^{-2}
foot	metre (m)	30.48×10^{-2}
mile	kilometre (km)	1.609
pound force	newton (N)	4.448
kilogram force	newton (N)	9.807
kilogram force/metre^2	pascal (Pa)	9.807
pound mass	kilogram mass (kg)	4.536×10^{-1}
pound force/in^2	pascal (Pa)	6.895×10^{3}
torr	pascal (Pa)	1.333×10^{2}
bar	pascal (Pa)	1×10^{5}
angstrom	metre (m)	1×10^{-10}
calorie	joule (J)	4.184
foot-pound	joule (J)	1.356
degree Celsius	kelvin (K)	$T_K = T_C + 273.15$

Multiplication factor	Prefix	Symbol
10^{-12}	pico	p
10^{-9}	nano	n
10^{-6}	micro	μ
10^{-3}	milli	m
10^{3}	kilo	k
10^{6}	mega	M
10^{9}	giga	G

Inauguration ceremony

Inaugural address
by
The President of Pakistan
General Mohammad Zia-ul-Haq

Federal Education Minister, Mr Muhammad Ali Hoti;

Vice-Chancellor of the Quaid-i-Azam University, Dr Ahmed Mohiuddin,

Former President of the Royal Geographical Society, Lord Shackleton;

Leader of the Chinese delegation, Professor Zhang;

Honourable delegates; and

Distinguished Guests;

Assalam-o-Alaikum

I am thankful to you for giving me the honour to inaugurate this international conference on Earth Sciences. I am also thankful to the Royal Geographical Society that it selected the great Karakoram Range as its subject for scientific research to celebrate its 150th anniversary. I welcome all participants of this important conference, particularly foreign delegates, and I hope necessary facilities will be available to them to make the expedition a success.

A welcome aspect of this Conference is that besides Pakistan, prominent experts from our great neighbour, the People's Republic of China, and from a developed country of the West, Britain, are taking part in the Conference. In modern times, the task of scientific research has become so complicated and expensive that countries instead of monopolising science and technology should undertake such work with one another's association. It is necessary that the developed and the developing countries should co-operate with one another in this field so that the entire human race should benefit simultaneously from scientific knowledge. This association and co-operation will help accelerate the speed of scientific advancement.

Pakistan is not only desirous of research promotion in the field of science and technology but is making all necessary effort in this respect within its resources. Besides other sectors we are concentrating on two aspects (1) research for increase in agricultural productivity; and (2) discovery of the country's resources of energy and their development.

In the first sector, the Pakistan Agricultural Research Council is endeavouring to bring about a revolution with the help of modern research, so that we should not only be self sufficient in agricultural requirements, but should also meet the needs of other countries. In the field of development of energy we are trying to discover all resources of energy and develop existing ones so that our energy requirements are met locally. In this connection the sources of energy which are being given special attention include, besides the search for oil, solar wind, hydro, bio-gas, geo-thermal energy resources etc. We hope that by developing these sources, we would attain self-sufficiency in the field of energy.

In the context of Pakistan, I wish to emphasise that we have reached a stage in the field of science and technology, including Earth Sciences, where we have left behind the dark ages of the past. But we have to go a long way to achieve our objective. By the grace of God we have among us noted and competent scientists who are working with devotion in their respective fields. They need further opportunities and facilities to advance their work. This is being looked after. The Government has high hopes that these experts who are working in the fields of agriculture, water and mineral resources, will achieve results. In many sectors of life such as agriculture and industry, the advancement, to a large extent, is due to their efforts. I expect

from these scientists and engineers, as also from our universities and research institutions that with their research they should find solutions to
problems facing Pakistan. I have in mind particularly those research sectors on which depends our agricultural, industrial and mineral development.

Like Pakistan, other countries are also engaged in research in science and
technology to suit their requirements and meet their priorities. I feel very
strongly that it is necessary to benefit from the efforts of individual countries and to co-ordinate their work to mutual advantage. The Karakoram
Project 1980 is an admirable example of such international association and
co-operation. I hope that the Pakistani, Chinese and British scientists,
engineers and surveyors will consider all those aspects which are important
for the present and the future.

I have in view those research fields which relate to floods, mineral wealth,
underground resources of energy, hazards of earthquakes and their prevention. Surely, in your capacity as experts in Earth Sciences, when you undertake this expedition, new vistas of research will dawn on you.

I wish this tripartite association was extended to other countries and other
centres of research, so that more and more people should benefit from it.
The Pakistan Government will co-operate to the fullest possible extent.

The other thing I want to emphasise is the need for a constant liaison and
co-ordination between the sciences and their application. What I mean is
that the research should be put to test in practical life and the information
gathered in actual expeditions should be used in the advancement of research
itself. The Karakoram Project 1980 will, by the grace of God, prove extremely
useful in this respect. I hope that besides the civil, mechanical and electrical engineers, experts in Geology, Geophysics and Science of Survey who
are associated with this Project, will bring out important and useful conclusions. Moreover, it is a rare opportunity for Pakistani experts in scientific and research fields to work in co-operation with the delegates of the
People's Republic of China, on the one hand, and the representatives of
the developed countries, like Britain, on the other. We can learn a lot
from these countries in the field of science and technology.

I hope that Pakistani scientists and experts will not only benefit fully from
this Conference, but will also help make it a success.

For us, the Karakoram Range is important, as it links our country with
our great neighbour, the People's Republic of China, with whose co-operation
the great Karakoram Highway has been constructed. This mountainous region has been the centre of attraction to great scientists and travellers
of Pakistan in the past. The geographical and scientific information collected by those travellers and scientists, along with their experience, has made
a valuable contribution to the Earth Sciences.

I am sure that after the three-day Islamabad Conference the expedition
which will be sent to the Karakoram will be more fruitful and meaningful
than previous expeditions.

The inspiration behind this meeting, and the proposed expedition, is the
Royal Geographical Society, which has served, for the last 150 years, the
science of geography.

The Society has made a great contribution in collecting topographical information on land and sea and about new species of vegetation. I hope the
proposed expedition will be successful.

We have deep relations with the People's Republic of China resulting in useful co-operation and association in many fields. I am happy that our Chinese guests are present here. We appreciate their association with respect and expect that the friendship and co-operation between the two countries will be further promoted.

I also appreciate the efforts of the Quaid-i-Azam University, which has organised this important international Conference. Such a Conference is in accordance with its fundamental objectives. I hope that this University will continue to perform the same role in the field of research and knowledge.

I, once again. thank you for inviting me to inaugurate this Conference. I wish to pray that God Almighty may grant success to your efforts which should lead to the welfare of humanity.

Amen.

Address by
Lord Shackleton K.G.

Mr President, Your Excellencies, Ladies and Gentlemen:

We, the members of the British part of this international project, consider it a very great honour that you, Mr. President, should have consented to inaugurate the conference that initiates it. It is a notable fact that the Royal Geographical Society has always been able, because it is by its very nature, working for the increase of science and for the good of mankind, to have the support of responsible national leaders for its expeditions. I remember being told as a small child how the Queen at that time, that is Queen Alexandra, came on my father's ship before he went to the Antarctic, and only a fortnight ago we celebrated the 150th anniversary of the foundation of the Royal Geographical Society, and the present Queen of England and Prince Philip honoured us by their presence.

One hundred and fifty years ago, the scientists and adventurers of that day were concerned with some of the great geographical secrets; the discovery of the North West Passage, and then later on the discovery of the sources of the Nile and the Niger. The controversies then aroused (and that still go on) over the sources of these and other rivers such as the Oxus, have filled the minds of scientists and explorers for many years. Throughout this time, the Royal Geographical Society has sought to combine the requirements of high academic quality with a sense of responsibility for the scientific work that was being done; a combination of adventure and science. It is in keeping with these ideals that when Captain Scott's body was found in the Antarctic, there should have been found beside it the geological specimens that he brought back from his epic journey. We, who in a later generation were fortunate enough to take part in expeditions and explorations, particularly remember the work of some of the great pioneers in this part of the world also, in this wonderfully beautiful area that I was enabled to see today when I flew to Gilgit, men such as Colonel Mason, who was my tutor at Oxford and who taught me about the art of the surveyor. There will be some surveyors in the audience today, and they will be aware of the new techniques whereby they no longer have to carry a heavy theodolite to find their position, but can use other and more sophisticated devices. Indeed, some of the equipment that Professor Keith Miller is producing is so accurate that it would enable us to ascertain the exact position in latitude and longitude of where I am standing at this moment, to within a few metres.

So this expedition, this international project, has a new look about it; and it is significant that, in this 150th anniversary year of the Royal Geographical Society, we should seek both to carry on past traditions and to make new advances. The credit for the philosophy and for much of the thinking behind the project belongs to its leader, Keith Miller, who is trying not only to make progress in the quality of the sciences concerned, but to apply the results in technological terms; to apply them in a way that will be of practical value. For instance, the work of the surveyors will directly contribute to the research of other scientists in Pakistan. The work of the glaciologists will be related to the study of what is perhaps the greatest wealth that belongs to Pakistan, apart from her people, namely the water that is locked up in the glaciers. It is a resource that can be of great benefit, even though it can also, on occasion, do great harm. They will study these glaciers, working alongside the geomorphologists, in an area where the landscape changes its shape probably more frequently than anywhere else in the world. In this context of land movements, we may note that the seismologists, the Pakistani, the British and the Chinese (who have amassed so much valuable knowledge of earthquakes) will be looking at the fundamental problems resulting from the collision of the tectonic plates in the Karakoram.

Finally we hope to learn something about how the people who live in these earthquake-ridden areas are able to build houses that withstand earthquakes, and to produce information that will be of great application.

We are very grateful that we have been granted the privilege, in this 150th year of our Society, of co-operating in a great international project; to have the support of the Chinese Academy of Sciences, of the University Grants Commission of Pakistan, of the Quaid-i-Azam University, of the British Council and of our own Ministry of Overseas Development. Moreover, many other bodies in Britain, such as the National Environment Research Council and the Royal Society, have recognised the importance of this work and have made contributions, along with the world of business and industry, towards the finance and resources that are required. I believe that this project is an example of the way in which mankind can co-operate peacefully for the benefit of us all. In the Antarctic there is co-operation, as I so well know, and although this is a troubled world, it is my belief that here in Pakistan also this project will show that, by working together on research involving the inter-relationships of several disciplines, the scientists of three great nations can contribute to the welfare, not only of Pakistan, but also to the welfare, and to the peace, of mankind.

Thank you.

Address of welcome
by
Mr Muhammad Ali Khan
Minister for Education, Culture and Tourism

Honourable President, delegates, ladies and gentlemen.

It gives me profound pleasure to welcome you to this "International Conference on Recent Technological Advances in Earth Sciences" which is being held for the first time in Pakistan with the co-operation of experts from the United Kingdom, the People's Republic of China and our own scientists. The assembly of so many distinguished experts today is of special significance as they will be working jointly on a project which is expected to augment human knowledge.

Mr. President, we are grateful to you for sparing time out of your busy schedule to be with us at the inauguration. This should certainly inspire scientists in their efforts to develop new avenues of research. Your very presence here indicates the paramount importance you attach to the development of sciences and research in Pakistan.

This Conference is a prelude to the International Karakoram Project 1980, which is being sponsored jointly by the Royal Geographical Society, London, Academia Sinica, Beijing, and the Government of Pakistan. While this Conference will review the work done so far on Himalayan Studies and associated subjects, the Karakoram Project is expected to advance our knowledge of the Himalaya. The experts understandably will deliberate upon the store of knowledge pertaining to this great tertiary uplift and the treasure of precious material lying buried there. Their findings during discussions and later work in the field will certainly have a bearing on the development of resource material to be utilised for the promotion of scientific studies and the good of mankind.

Mr. President, this Conference, it is hoped, drawing upon the expertise of the three participating nations, will stimulate research in other countries. Besides the material benefit that might accrue to Pakistan and the neighbouring countries, the expedition will provide scientific data for study in the research institutions of the world. Local organisations, directly involved in this project in particular, will gain salient experience to improve their techniques and enter upon further research work.

Some fifty years ago, a Yale-Cambridge expedition for the first time explored the Himalayan glaciation under the leadership of H.D. Terra in co-operation with teams from the U.S.A., U.K., France, China and the Indo-Pakistan Sub-Continent. The team for the first time catalogued the flora and fauna of the glacial region in a chronological order. Their important work entitled "Study in the Ice Age in India and Associated Human Cultures" is the only standard work so far available. Since the early thirties of the present century when the expedition was organised, we have advanced progressively in our scientific and technological knowledge and also in developing new equipment which will be used in the present project. Research will un-doubtedly contribute to enriching our knowledge of the Great Himalayas lying between Pakistan and China. We have co-operated successfully with our great friendly neighbour China in different fields of mutual interest in the past. In this particular expedition too we are grateful to our Chinese friends for their collaboration. We are indebted too to the Royal Geographical Society, London, for sending its team of eminent scientists and technologists to take part in the project. I welcome the guest experts and wish them success in their deliberations.

The Government of Pakistan has been making efforts to support institutions within the country which are seriously engaged in this type of research. Among them the Survey of Pakistan and the Geological Survey of Pakistan were founded during the British days and have carried on their work in the light of new developments in science and technology. There are a num-ber of other institutions which Pakistan has established after independence. These have opened new fields of research for the benefit of the country. On the teaching and research side, we have established Departments and Institutions of Geology in most of our Universities. A Centre of Excellence for Minerology and Petrology has been established in the Baluchistan Univer-sity. Other organisations such as the Water and Power Development Authority, the Pakistan Mineral Resources Development Corporation, the Industrial Development Corporation of Pakistan and the Mineral Development Authorities in the Punjab and the N.W.F.P. have contributed as well towards our advance in this field. Of significance too, is the establishment of the Department of Geophysics under the Institute of Earth Sciences at the Quaid-i-Azam University. A proposal is also under consideration by the University to create a "Karakoram Research Cell".

We are obliged to the Royal Geographical Society, which is celebrating its 150th anniversary, for choosing our country as a base for research that is likely to benefit mankind in the long run.

I thank you once again Mr. President, for acceding to our request for inaugurating this Conference. I also thank the delegates and other guests who have made it convenient to be present here this evening.

Some recent technological advances applied to problems in earth sciences

K.J. Miller

University of Sheffield, England

Leader, International Karakoram Project 1980

OPENING REMARKS

This lecture inaugurates the International Karakoram Project, 1980. It is therefore opportune to emphasise the aims of the Project which were agreed two years ago by the Royal Geographical Society, to commemorate its 150th birthday. The Project was to facilitate international co-operation between scientists, promote inter-disciplinary research work, and seek inter-Governmental support, all of which have now been achieved. It is my hope that this venture will now set a pattern for others to build upon in future collaborative investigations.

I recognise that I am most fortunate in being both an engineer/materials scientist and a geographer/mountaineer. It is this dual existence that permits me to address you today on a topic that I consider to be closely related to the stability, peace and prosperity of mankind. In the International Karakoram Project, we have scientists from China, Pakistan, Britain and Switzerland, assisted by advisers from Italy and the U.S.A., plus many other countries. The papers read at this conference are concerned with geographical, geological, geomorphological, glaciological, engineering and human problems and they invoke theoretical and experimental studies in engineering, physics and chemistry which examine materials and structures ranging from atomic dimensions of ice crystals (10^{-10}m) through to the global dimensions of tectonic plates (10^{6}m) i.e. 16 orders of magnitude.

The fact that such a wide ranging Project and conference could be held is due to the efforts of countless people, many of whom had no hope of direct participation either at the conference or in the Project itself. Those who laid the foundations for our studies in the late 19th and early 20th centuries have presented us with a unique opportunity which we gladly accept. To all those past and present helpers, including those present here today, I, on behalf of the Project and The Royal Geographical Society, give thanks and hope that we may repay the trust given to us by producing scientific results of benefit to the people of Pakistan.

Einstein once said that Pure Science makes history, whilst Applied Science makes progress. I am tempted to say "Theoreticians tell us what is feasible, whilst technologists are responsible for providing society with an economic, practical solution". Thus engineers have to be scientific, practical and cost effective. They need mathematical skills, a knowledge of physics and chemistry and the ability to design, construct and commission safe and efficient engineering plant to provide the needs of society. Today, without the engineer, civilized society as we know it would retrace its path back towards the primeval forest.

This lecture discusses only a few areas in which technological innovation has had direct application to earth sciences. From these examples, and from the lectures to be given in the next few days, the importance of technological achievements and the ability of scientists of any nation to conduct inter-disciplinary studies will be emphasised by members selected to participate in this memorable occasion to celebrate the 150th birthday of The Royal Geographical Society.

INTRODUCTION

During the Cretaceous dispersal of Gondwanaland, the Indian/Pakistan tectonic plate moved across what is now the Indian Ocean from its position between Antarctica and Arabia, see Fig. 1, and collided with the Asiatic plate.

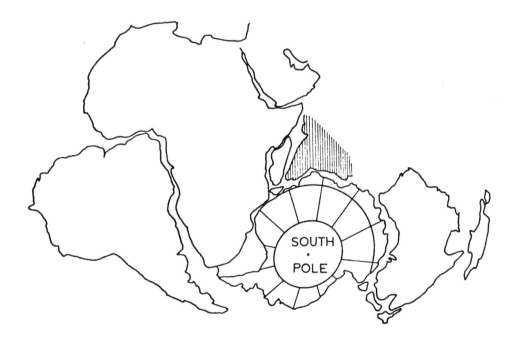

Fig. 1.   Reconstruction of Gondwanaland
before the Cretaceous dispersal

A consequence of that collision was the creation of the Karakoram, the world's most extensive and beautiful range of high mountains that includes K2 (8,611m), Nanga Parbat (8,126m), Rakaposhi (7,788m) and many other 7,000m plus peaks.

Along with this natural beauty however, come unstable glaciers and fast flowing rivers, earthquakes and landslides, extremes of temperature and climate, all militating against the construction of permanent settlements. That the peoples of the Hindu Kush and the Karakoram have resisted these powerful natural phenomena, have withstood catastrophic events and have frequently rebuilt their homes, is a testimony to the ability of mankind to challenge and sometimes win against the most persistent and intensive forms of natural power and destruction.

GEOTHERMAL POWER

Earthquake prone zones are also the sites at which nature exhibits one of its great natural resources; geothermal energy, see Fig. 2. Iceland, Japan, New Zealand, the U.S.A. and many other countries are now committed to large

Fig. 2 Geothermal power exhibited by "Old Faithful" in the U.S.A.

scale geothermal energy programmes. For example, the geysers of California now have enough generating capacity to satisfy the electrical power needs of San Francisco, and this supply capability could be tripled.

Reykjavik, the capital city of Iceland, population 120,000, is entirely heated by geothermal power. All its hot water supply comes from this natural resource. Agricultural, plus heavy industrial demands can be satisfied. Situated close to the arctic circle, farmers can grow tropical fruit inside high temperature, high humidity greenhouses. Heating for schools, hospitals, housing estates, churches, swimming baths and other public buildings, is also commonplace.

The total world geothermal energy utilization is indicated in the table below:

TABLE 1   The World (Past, Present and Future) Geothermal Energy
          Utilization in MW-year Units;   reference (1).

	1970	1975	1978	1982	2000	2020
Thermal	–	5,915	7,045	–	–	1,400,000
Electric	724	1,118	1,370	3,287	80,000	?

Table 2 gives details of the electric power capacity of several countries which is derived from geothermal activity.

TABLE 2   Geothermal electric-power capacity of several countries;
          reference (2).

COUNTRY	Units in Operation	Installed Capacity, MW	Immediate Additional Potential MW
El Salvador	2	60	35
Iceland	3	62	–
Italy	37	420.6	–
Japan	6	165	55
Mexico	4	150	30
New Zealand	14	202.6	–
People's Republic of China	1	1	–
Philippines	3	59.2	710
Soviet Union	1	5	–
Turkey	1	0.5	–
United States	13	663	1019
TOTALS	85	1788.9	1849

The potential of this energy form is obvious but the economic and techno-logical risks are high. One is never certain, before drilling, if a well, however shallow, can supply a user's needs. Figure 3 illustrates some of

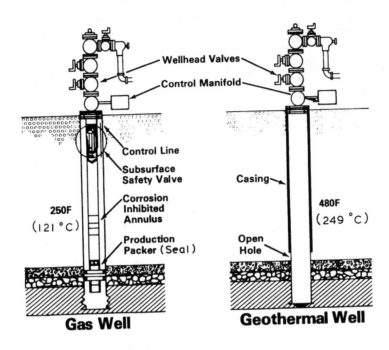

Gas Well   **Geothermal Well**

Fig. 3 Differences between gas and geothermal wells (1)

the differences between a gas well and a geothermal well. Gas well tech-nology is now well advanced in the U.K., alongside drilling techniques for oil under the North Sea. However, because less energy is available in a kilogramme of hot water than in the same quantity of gas, the geothermal wells are considerably bigger. Also, temperatures are higher, and in con-sequence studies on corrosion and fluid seals are at the frontiers of research. It can be seen from Fig. 3 that in order to avoid bore-hole pipe problems, no extraneous seals, safety valves or other control systems are inserted into the bore hole.

Probably the most difficult geological problem confronts the engineer who has to seal off the geothermal fluid from other geological formations, see Figs. 4 and 5. Normally an engineer, when designing seals, would avoid high temperatures, high pressures, reactive fluids, excessive clearances, linear motion and inaccessible locations. In geothermal plant none of these features can be avoided. A typical 2,000m deep bore will have a bottom temperature of around 300°C and pressure of 7.0 MPa, a casing variation of 5mm (due to taper and eccentricity) and changes in length of the tube (expansion/contraction) of about 6m. No single elastomer seal is suitable for all these service conditions at present. Unfortunately mechanical engineering types of failures of structures, as well as distortion of com-ponents, are still too frequent. Civil engineering problems are equally widespread as can be appreciated if one realises the geothermal plant at Krafla in Northern Iceland bestrides two tectonic plates that are moving apart!

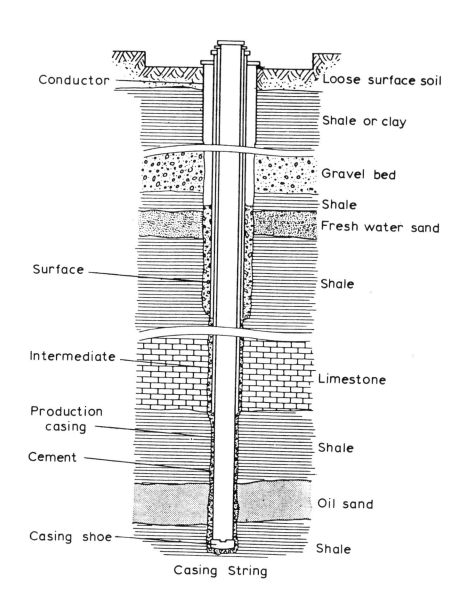

Conductor

Loose surface soil

Shale or clay

Gravel bed

Shale

Fresh water sand

Surface

Shale

Intermediate

Limestone

Production casing

Shale

Cement

Oil sand

Casing shoe

Shale

Casing String

Fig. 4   Various geological formations encountered in
developing geothermal wells (3)

18

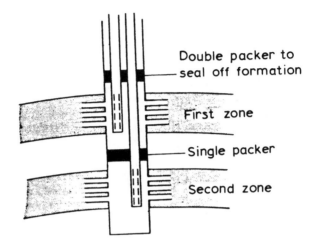

Double packer to
seal off formation

First zone

Single packer

Second zone

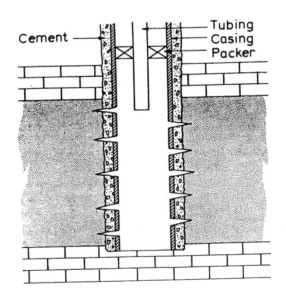

Cement

Tubing
Casing
Packer

Fig. 5 Seal systems for geothermal wells (3)

In Mexico, to overcome lack of corrosion protection from inhibitors at high temperatures, the outer case was completely covered by cement. Unfortunately nobody knows the corrosion resistance behaviour of cement under operating conditions. The corrosive nature of geothermal fluids can be gauged by examining the important chemical data given in Table 3.

TABLE 3   Important Chemical Data for Geothermal Fluids (1)

pH value	2-10	
H_2S	0-600 p.p.m.	(usually < 30 p.p.m.)
CO_2	0-1500 p.p.m.	(usually < 500 p.p.m.)
NH_3	0-300 p.p.m.	(usually < 20 p.p.m.)
CL^-	10-280,000 p.p.m.	(usually < 8,000 p.p.m.)

A schematic of a geothermal steam plant is given in Fig. 6 along with typical problems and their location, whilst some of the engineering materials problems and their solutions are presented in Table 4.

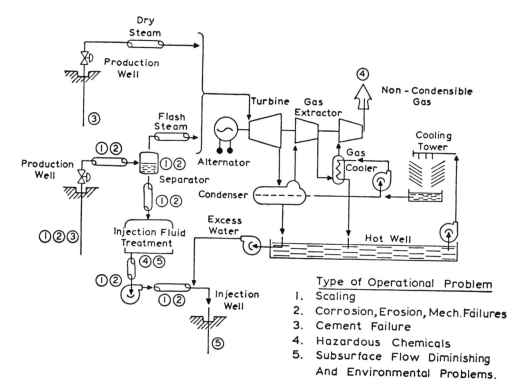

Type of Operational Problem

1. Scaling
2. Corrosion, Erosion, Mech. Failures
3. Cement Failure
4. Hazardous Chemicals
5. Subsurface Flow Diminishing And Environmental Problems.

Fig. 6   Schematic of a geothermal steam plant indicating problem areas (1)

Place	Problem	Solution
Italy	Erosion of valves	Tungsten carbide facing
	Pipe erosion	Thickness allowance and redesign
	High-temperature blowout preventers	High-temperature gaskets-Viton, lead rubberized canvas gaskets for $260^{\circ}C$
	Drill pipe fatigue, loss of circulation, well cementation	
	Casing collapse	Heavier well casing
	Thread parting	Longer coupling
	Erosion corrosion, last stages of turbine	Stellite shields
	Corrosion condensate; CO_2, H_2S, NH_3, BO_3	Austentitic, ferritic, and lead-clad stainless steels
	Turbine blade life	13% Cr, low C < 0.1% materials
	Erosion and cavitation of pump impellers	
	Construction cement and concrete deterioration	Coatings and materials R&D
	Stress, intergranular and localized corrosion of AISI 316	Low-carbon steel thick walls
	Wellhead equipment corrosion stainless steel 2 to 6-month life	Carbon steel, 9-year life
	Chloride attack - turbine blades	Higher Cr alloys passivate
	Erosion of lead on iron in condensers	Design
New Zealand	Sulphide stress cracking in medium and high-strength carbon and alloy steels used up to $190^{\circ}C$	Materials selection and caution
	Stress corrosion of austentitic stainless steels (aerated, chlorides, stress, geothermal media)	Keep bores hot, redesign, deaerate
	Surface tarnishing and erosion corrosion of commutators made of Cu, Ag, and alloys used in geothermal steam containing H_2S	Platinum contacts, cleaning continued protective maintenance
	Severe surface corrosion of carbon and galvanised steel exposed to salt spray	Painting, cladding, remote discharge
	Tarnishing and corrosion of overhead power cables, telephone cables, and controls exposed to atmosphere + H_2S	Remote discharge, special paints, aluminium conductors

(continued over)

Place	Problem	Solution
Mexico (electric systems)	Casing collapse and corrosion	Heavier casing and second-production casing cemented into outer casing; modified high-temperature cement
	Internal casing corrosion External wellhead and casing corrosion Crevice corrosion	Keep hot during standby High-temperature paints and lubricants, scraping Avoid metal-to-metal contact with cement and grease
	Turbine cooling system (Al, 304 stainless steel pitting)	Standby procedures plan, Ti substitution.
North California U.S.A.	Turbine erosion corrosion Corrosion condensate, low-pH sulphate-containing fluids Protective coating deterioration	13% Cr steel Materials selection coatings
	Electrical equipment (H_2S)	Aluminium transmission Clean rooms Pt and Au inserts on contacts
	Corrosion fatigue in 12% Cr stainless steel	Heat treatment changes
United States (nonelectric systems)	Downhole pump bearing failure Downhole heat exchanger corrosion (5 to 20-year life, black iron pipe)	Alumina prototype bearing Additives (oil, paraffin) Materials selection

TABLE 4: IMPORTANCE OF ENGINEERING MATERIALS SELECTION.

ICE POWER

To anyone who has traversed the fronts of glaciers and ice-caps and witnessed the movement and chaos of terminal moraines, there is no need to
describe the power of moving ice which in total represents the greatest of
all natural forces.

There are several papers in this Conference concerning glaciology. To understand the flow and fracture characteristics of ice, it is essential to appreciate the atomic, crystalline and continuum behaviour of ice. There are
nine known crystalline forms of ice, most of these being stable only at high
pressures. Our most complete understanding is of the open hexagonal atomic
structure, see Fig. 7, where every oxygen atom (large circle) is surrounded
by a tetrahedron of oxygen atoms. In most ice crystals the two hydrogen
atoms (small circles) are directed, at random, along two of the four possible
oxygen-oxygen vectors.

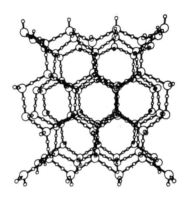

Fig. 7   A perspective view of the structure of ice
viewed down the hexagonal axis   (4)

The deformation of ice from the microstructural (atomic) viewpoint has been
studied extensively by Ashby who constructs diagrams of the form shown
in Fig. 8. These so-called deformation maps have been extensively used
in the study of high temperature materials for the nuclear and aerospace
industries (e.g. 304, 316 stainless steels, MARM 200 etc.). The difference
between the two diagrams presented here is the size of the crystals (grains)
which can vary substantially; see Fig. 9. The maps plot normalized shear
stress against normalized temperature and by considering the micromechanism
of deformation, including recrystallization effects, it is possible to determine
the dominant process responsible for flow.

For example, Lead is known to creep (extend) under its own weight (see
Fig. 10), a process that can be accelerated by heating. Analysis of a
60-year old hot water pipe indicates diffusional flow of atoms to be the
dominant process whilst an external drain of the same grain size (~50μm)
but lower temperature and higher stress creeps by a low temperature creep
process.

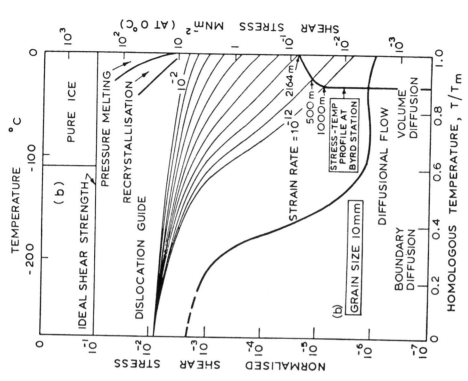

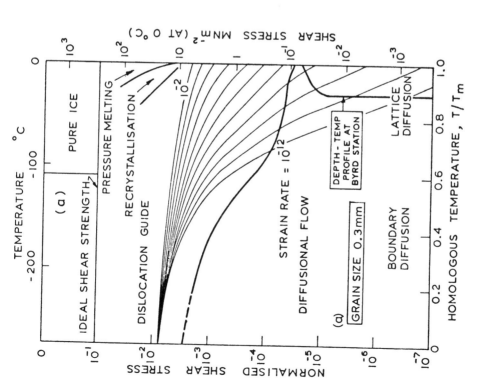

FIG 8 A deformation map for Ice (5)
(a) Grain size 0.3mm
(b) Grain size 10mm

24

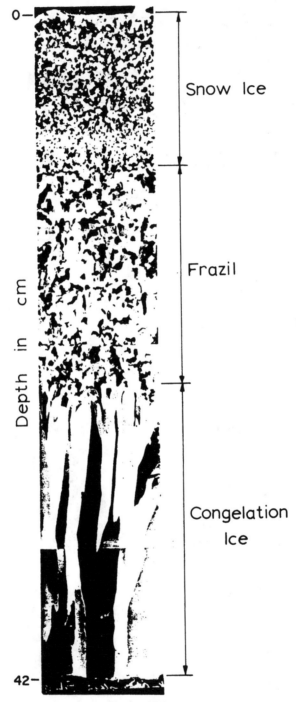

Fig. 9 Grain Size Variations in Lake Ice (6)

Fig. 10   Lead Pipes on a 75-year-old building in Southern England   (5)

The South Polar ice cap has been examined from such a viewpoint. A few tens of metres below the surface, temperatures do not fluctuate and the temperature gradually rises to 0°C at bedrock where geothermal heat may maintain a film of water. As with lead, the flow of ice is gravity driven and the shear stresses increase linearly with depth which may be determined from radar echo sounding (see later papers). Grain sizes change from ∿1 mm (surface) to 10mm (depths > 200m). From this information, deformation maps tell us that power law creep is the dominant mechanism and the creep strain rate at bedrock (highest temperature and stress) is approximately 10^{-9} sec^{-1}.

Whilst much of the earth's surface is covered by land ice (glaciers and ice-caps), floating ice on seas, lakes and rivers can present many formidable problems. Transportation across frozen lakes and along rivers depends upon the thickness of ice and the rate of accumulation of ice. The most important parameters to consider are the mean daily temperature, the water temperature and the latitude. Since 1959, records have been maintained at several Canadian arctic stations (see Fig. 11) which produce data typified by Figs. 12 and 13. When the break-up of ice occurs (see Fig. 14), thereby permitting ease of access to ships, other hazardous problems arise as the ice drifts towards shipping lanes and fishing areas. Of great concern is the possible collision of icebergs with offshore oil rigs. In this respect, the use of radar for locating and measuring the thickness of floating sea ice is most important. Recently some dangerous experiments have been conducted in towing away large icebergs from shipping lanes.

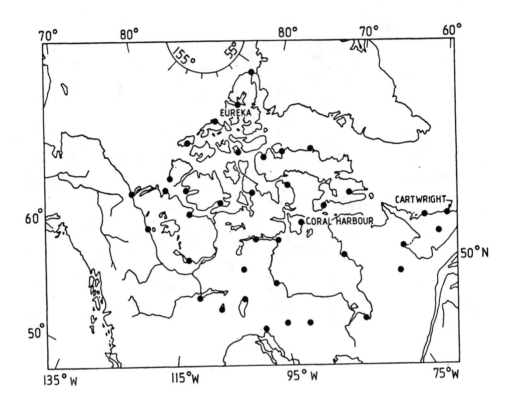

Fig. 11.   Location of Canadian research stations (7)

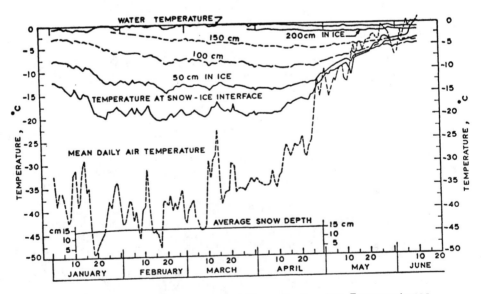

Fig. 12   Water, Sea Ice and Mean Daily Air Temperatures
at Eureka 1950-51   (7)

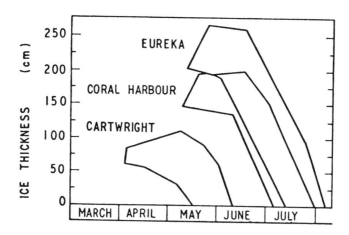

Fig. 13   Envelope Curves for Ice Decay at Three Research Stations   (7)

Conversely, iceberg towing could be positively helpful, since several countries (noteably Australia, Chile and Saudi Arabia) are now considering the towing of icebergs from Antarctica to their own shores to provide a fresh water source for the irrigation of their deserts. Statistical investigations are also being carried out on the size, shape, angularity, perimeter length, major chord orientation (relative to wind and waves) of ice-floes as well as the proportion of area below and above the water line. These research studies will permit environmentalists to study the effects of ice-floes on climate and marine biological productivity.

Ice can fail by a number of processes depending upon temperature and stress, see Fig. 15. At high stresses, dynamic fracture is a consequence of the propagation of an elastic stress wave. At high temperatures and low stresses, inter-crystalline creep fracture can occur due to cavitation, but most fracture processes in this brittle material are due to cleavage. Cleavage type 1 at low stress is due to pre-existing defects and fracture strength can be determined from fracture mechanics considerations. Cleavage type 2 requires cracks to be nucleated by deformation on slip systems whilst cleavage type 3 is due to slip generating cracks that propagate in an intergranular mode. Since the fracture behaviour of ice is the subject of two separate papers in this conference, it will not be commented on further, but it should be noted that this has important consequences on the size of icebergs and the depths of crevasses on glaciers and ice-caps.

The study of stress, strains and flow of ice in ice-caps and glaciers has also been boosted by the advances made by finite element methods of analysis. In engineering components of complex geometries (e.g. bearings, see Fig. 16) it is necessary to have a three-dimensional appreciation of stresses, particularly concentrations of stress, so as to eliminate initiation sites for fatigue cracks, fretting and wear of surfaces. Computer programs are exceedingly time consuming and very expensive, especially those that include elastic and plastic deformation behaviour, and at this moment, only two-dimensional and coarse grid elements are used to study ice problems, but I suspect that a three-dimensional program will have to be constructed to

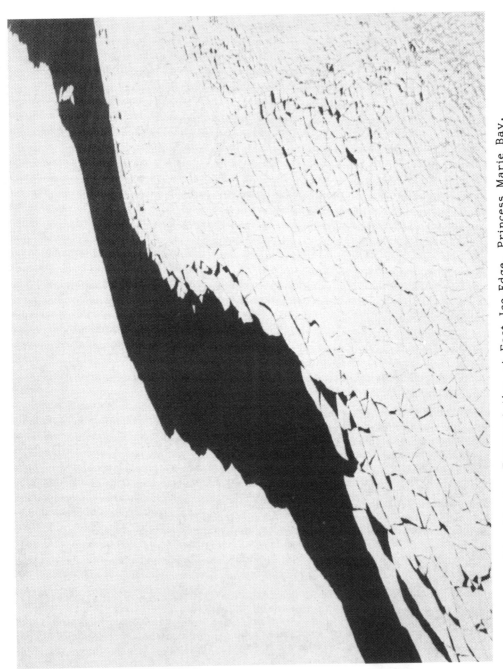

Fig. 14 Ice Fragmentation at Fast Ice-Edge, Princess Marie Bay, Eastern Ellesmere Island (8)

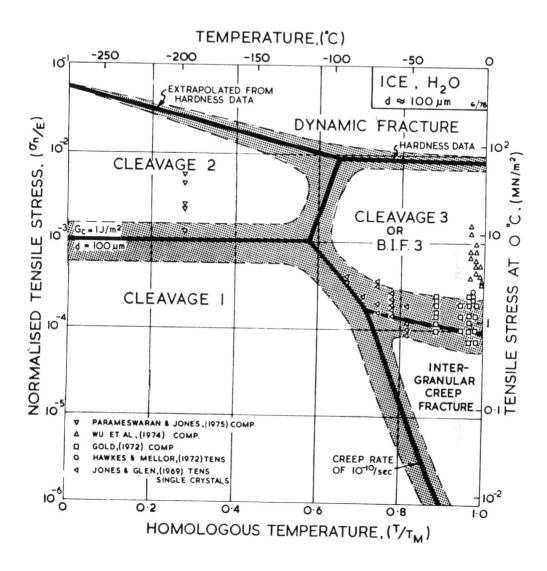

Fig. 15   A Fracture Map for Ice (grain size 100 μm)   (9)

study difficult glaciers, such as the Batura glacier, which can cause serious and expensive disruptions to communications. The Chinese investigations over the period of 1974–5 are an excellent base to start such a study, and the application of impulse radar systems, for accurately determining glacier ice depths, should be invaluable in these circumstances. Figure 17 shows a two-dimensional finite element idealization for a model glacier with basal sliding and Fig. 18 shows the Chinese map of the Batura glacier.

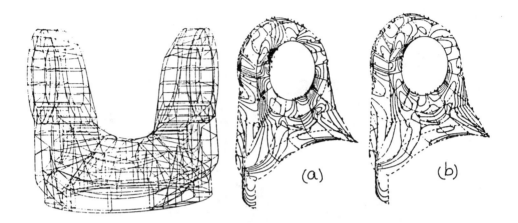

Fig. 16   Three dimensional finite elements used to study
(a) stresses, and (b) strains in a bearing yoke   (10)

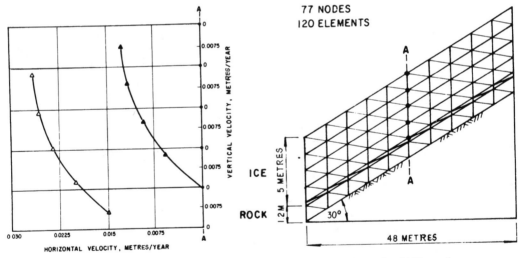

Fig. 17   Velocity distributions with and without basal sliding from a
two-dimensional finite element idealization of a glacier (11)

It is worth noting that deformation and fracture maps are also of interest
to the geophysicist who wishes to study the material that constitutes
the upper mantle of the earth.   In this case an added variable to shear
stress, temperature and time is pressure which, at depths of a few kilometres below the earth's surface, affects the mechanisms of fracture.
However, depths of a few hundred kilometres are required before pressure
affects the deformation behaviour of olivine.

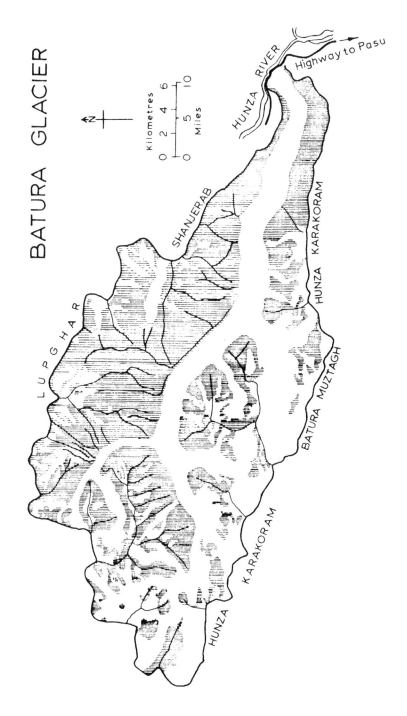

Fig. 18 A Map of the Batura Glacier prepared by the Chinese Investigation Team (12)
(see also the paper written by Shi Yafeng and Zhang Xiangsong presented in Volume 1 page 51).

Figure 19 illustrates the behaviour of olivine with a grain size of 0.1 mm and at a pressure of 81 kbar. Since both temperature and pressure increase with depth, a single diagram can describe the behaviour of the upper mantle e.g. see Fig. 20 which plots the upper and lower bound solutions. At the surface, and to a depth of at least 20 km, cleavage is dominant rather than plastic deformation. Below this depth plasticity replaces cataclasis as the dominant mechanism, whilst between 100 km and 400 km power law creep is the operative mode of deformation.

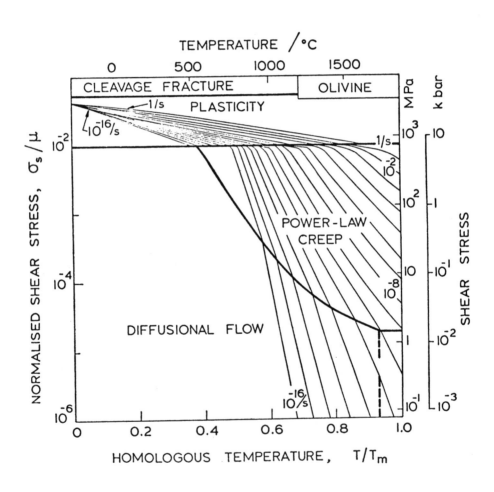

Fig. 19   A deformation map for olivine at a pressure of 81 kbar   (13)

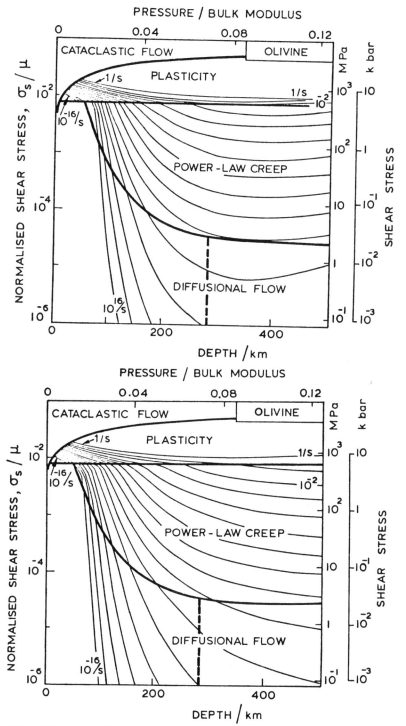

Fig. 20 Olivine Behaviour; Upper and Lower Bound Solutions (13)

WATER POWER

The life support system of Pakistan is its network of irrigation canals, fed by the Indus, one of the most important rivers in the world. The expertise of Pakistan engineers who design, construct and maintain these canals is internationally acclaimed.

In the villages of the Karakoram one may find many miles of aqueducts, built across vertical cliff faces, which supply water to different parts of the village, or even different villages. The yield of village crops is maximised by a judicious and intricate system of canals that starts at the top of the village, invariably sited on an alluvial fan formed at the base of a gorge. The water system descends to valley level in an ever expanding series of subsidiary canals that feed the multi-level garden plots, each of which has its own sluice gates that have to be opened and shut at specific times each day.

On towards the plains of Pakistan it is possible to build dams such as Mangla and Tarbela, the latter being the world's largest earth dam which supplies 2100 MW of electrical power to Pakistan. However, due to the steepness and ferocity of the Karakoram rivers, and the unstable slopes above the river beds, the Indus carries the greatest sediment load in the world. Figure 21 shows a rating curve for the sediment capacity of this river and the figures are frightening. For a flow rate of 300,000 cu. ft./sec. (8,500 m^3/sec) the suspended sediment in the river weighs 7,000,000 tons per day (70 GN). It is the deposition of this sediment at Tarbela (estimated at 320 million tons/year) that creates a major problem for the Water and Power Development Authority of Pakistan (WAPDA). The storage capacity of the 57 mile long Tarbela Reservoir is gradually being depleted by this sediment deposit which is accumulating at a rate sometimes approaching a depth of one metre a year. It is therefore important for Engineers and Geomorphologists to determine, not only the sediment capacity of the Indus and its tributaries, but also the movement of the bed load, which for Tarbela is about 5% of suspended sediment load. The movement depends on the size, density and shape of the bed load as well as the granular size, distribution of the sediment, the liquid-to-solid ratio of the flow and the three-dimensional geometry of the river. Thus, part of the sediment load is constantly being moved downstream towards the dam structures, a process which is accelerated by rapid fluctuation in reservoir height through the months of June to September, due to snow melt, then glacier melt, and finally monsoons, with the annual final shape of the deltas within Tarbela being formed in late September.

By using high frequency transducers with echo sounders, it is possible to determine the depth of water above the silt. With low frequency transducers fitted to modern equipment developed for deep ocean research, it is possible to obtain the depth of silt above the original river bed. It is therefore possible to monitor the elevation, depth and movement of sediment beds.

Our own project is to carry out investigations in the Hunza river, which carries one of the largest sediment loads feeding the Indus. This concentration of sediment is due to the huge catchment area of 5,080 sq. miles which contains some of the world's longest and deepest glaciers outside the polar regions. Other contributory factors are the lack of plant cover on the steep hillsides, the frequent landslips due to rain, earthquakes and the young, easily erodable rocks of the Karakoram. As a consequence of these features, the Hunza Valley is rapidly losing its top soil, but Pakistan is not alone in having problems of this magnitude. The River Thames of

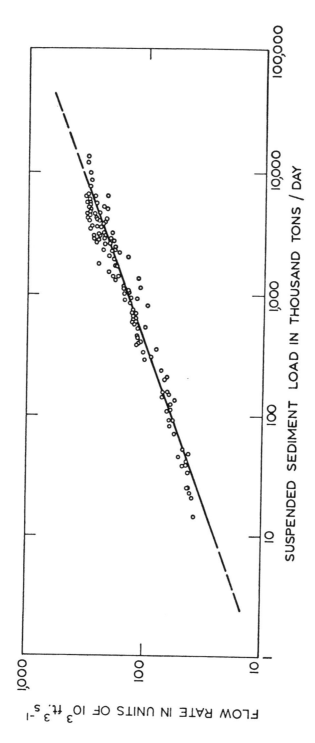

Fig. 21 Sediment Rating Curve for the Indus

Fig. 22   A Montage to Depict the Possible Problems of Flooding in London   (14)

England can cause flooding of some 50 sq. miles of Greater London's low-lying areas. For over 300 years, tidal flooding of the Thames has been recorded and is due to a combination of geological and meteorological factors. The maximum flood height has been increasing about 8.5mm per year since 1780. In consequence it has been decided to build the world's largest moveable barrier, which can be brought into operation within 15 minutes. Sited at Woolwich, not only will it prevent flooding, but it will also preserve navigation on the river. It is with some pleasure that I can report that Davy-Loewy of Sheffield is responsible for the manufacture, supply and installation of this barrier. However, climatic pressures, wind directions and moon gravity may combine to cause a flood surge, and bring a major catastrophe to London before the barrage is completed. Notices are on display in all London underground stations warning the population of this possible disaster.

CONCLUSIONS

Technology and earth sciences are inextricably bound together. In the Karakoram it is essential that scientists of all disciplines unite to help overcome some of the worst hazards to man on earth. The recent developments in engineering, quoted in this paper and in other papers in the conference, i.e. finite element analyses, impulse radar, satellite survey systems, new materials, deformation and fracture mechanisms etc. are but a few aspects of a rapidly developing engineering technology that can be of immense help to mankind.

REFERENCES

(1) R.R. Reeber. Geothermal Energy and Consensus Standards.
     A.S.T.M. Standardization News pp 6-10. October, 1979.

(2) R. DiPippo. International Developments in Geothermal Power.
     ibid pp 19-28 1979.

(3) D.L. Hertz Jr. Developing Standards for Geothermal Seals.
     ibid pp 15-18 1979.

(4) W.C. Hamilton and J.A. Ibers. Hydrogen Bonding in Solids.
     W.A. Benjamain Inc. New York, p. 191 1968.

(5) M.F. Ashby and H.J. Frost. Seven case studies in the use of deformation maps and the construction of transient maps and structure maps.
     Cambridge University Engineering Department Report, CUED/C-MATS/TR26 1976.

(6) R.E. Bates. Winter thermal structure, ice conditions and climate of Lake Champlain.
     Cold Regions Research and Engineering Laboratory. CRREL Report 80-2 1980.

(7) M.A. Bilello. Maximum thickness and subsequent decay of lake, river and fast sea ice in Canada and Alaska.
     ibid, CRREL Report 80-6. 1980.

(8) A.M. Cowan. A sea ice photo-reconnaissance flight to the North Pole with the Royal Air Force: a preliminary report.
     Scott Polar Research Institute (Sea Ice Group) Report, Cambridge. 1980

(9)   C. Ghandi and M.F. Ashby.   Development of Fracture Mechanism Maps
      for Ceramics.
      Cambridge University Engineering Department Technical Report
      CUED/C/MATS/TR 48   1978.

(10) L.G. Fisher.   Lightweight propeller shafts.
      Automative Engineer, 4 (3) 56–57   I.Mech.E., London.   1979.

(11) J.J. Emery and F.A. Mirza.   Finite Element Method Simulation of Large
      Ice Mass Flow Behaviour
      IUTAM Symposium "Physics and Mechanics of Ice" Copenhagen,   1978.

(12) Members of the Chinese Study Group.   Investigation Report on the
      Batura Glacier (1974–75) in the Karakoram Mountains, the Islamic
      Republic of Pakistan.

(13) M.F. Ashby and R.A. Verrall.   Micromechanisms of flow and fracture
      and their relevance to the rheology of the upper mantle.   Phil. Trans.
      R. Soc.1 Lond. A288, 59–95 1977.

(14) F.R. Higgins.   The Thames Barrier.   The Chartered Mechanical
      Engineer.   May 1980, pp 49–54.

ACKNOWLEDGEMENTS

I have freely used data from many sources, most, if not all, of which are
quoted in the Reference List above.

I record my most grateful thanks to these authors, who so assisted me in
preparing this paper.

Recent variations of some glaciers in the Karakoram mountains

Zhang Xiangsong

Co-leader of the Chinese team and Professor of Glaciology
Lanzhou Institute of Glaciology & Cryopedology

ABSTRACT

In 1974 – 1975 and in 1978, the author was a member of the Batura Glacier Investigation Group of the Karakoram Highway Engineering Headquarters of the People's Republic of China. Studies of contemporary glaciation along the Karakoram Highway have been actively carried out and some valuable data on glacial variations gathered. During our investigations we were warmly supported by our Pakistan friends. This article is based on the geomorphological investigation together with documentary records with especial comparison to the landsat images in the 1970's and topographic maps in various years, and aims at explaining recent variations.

RECENT VARIATIONS OF GLACIERS IN THE NEIGHBOURHOOD OF KHUNJERAB PASS

Friendship No. 1 Glacier is located on the right side of the Khunjerab Pass, looking towards Pakistan. It is a small valley glacier. By estimation, the height of the snow line is 5200 m. On both sides of the glacier in the middle and lower reaches, the large lateral moraine ridges have a relative height of some 70 m, and continue towards the upper reaches. Inside the lateral moraine, there are four steps at 4620 m, 4650 m, 4700 m and 4775 m respectively, reflecting three previous stages of ice surface after the maximum of the last advance. The terminus height of the ice tongue is 4720 m.

In 1978 we carried out an investigation of the terminus of this glacier and compared it with the 1:10000 topographical map drawn by Pakistan in 1966. We discovered that the position of the glacial terminus had advanced about 320 m (40 m/yr) and descended 118 m in height from 1966 to 1978 (Fig. 1).

Friendship No. 2 Glacier is located on the left side of the Khunjerab Pass. It is a small hanging glacier with a short ice tongue. By estimation, the height of the snow line is 5100 m. There are three dark brownish terminal moraine loops visible at the end of the glacier. Heights of the terminal moraine are 4650 m, 4680 m and 4700 m respectively. From

40

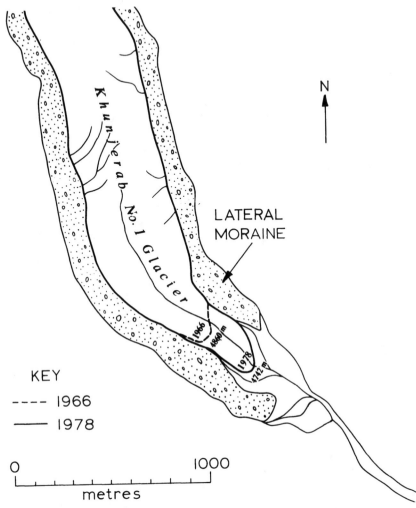

FIG. 1. VARIATIONS IN THE POSITIONS OF THE
KHUNJERAB NO. 1 GLACIER FRONT.

there upwards, there is a greyish terminal moraine at 4780 m. At present,
the terminus of the ice tongue is at 4795 m. The relative height of the
ice cliff at the terminus is 60 m. For this glacier there is no complete
subglacial drainage system, but drainage by surface and lateral channels
are dominant. The terminus of the ice tongue has risen 90 m and retreated
210 m according to the topographical map (1:10000) compiled by Pakistan
in 1966 (Fig. 2).

Friendship No. 3 Glacier is a larger hanging glacier. The glacial
terminus was at a place 200 m away from the Hunza River in 1966. On
July 30, 1978 we found that the terminus on the right side of the ice tongue
had retreated 600 m and risen 204 m. At the same time, the terminus
on the left side had retreated 290 m and risen 120 m (Fig. 3). Because
the ice tongue had retreated quickly a vast stretch of moraines was left
on the slope. We suppose that the glacier had advanced before 1966 and

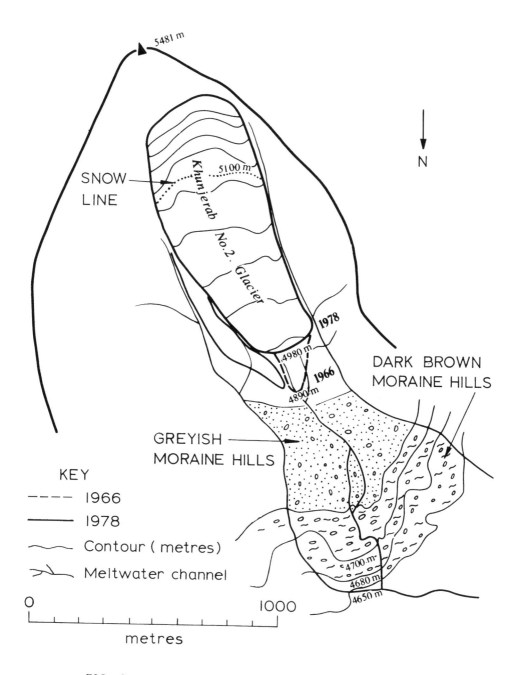

FIG. 2.   VARIATIONS IN THE POSITIONS OF THE
KHUNJERAB NO. 2 GLACIER FRONT.

42

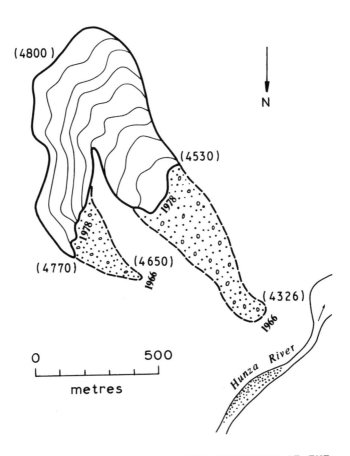

FIG. 3. VARIATIONS IN THE POSITIONS OF THE
KHUNJERAB NO. 3 GLACIER FRONT.

reached its maximum by 1966. Afterwards, the glacier retreated.

RECENT VARIATIONS OF THE BATURA GLACIER

The Batura Glacier is one of eight large glaciers with a length more than 50 km at middle and low latitudes on earth. There is a special report on its recent variations; reference (1). For comparative purposes the present author briefly reports as follows.

During 1885 - 1925, according to documentary records by Woodthorp (1885), E. F. Knight (1891), Etherton (1909), K. Mason (1913) and Visser (1925) (see reference (1) and Bibliography), the Batura Glacier's terminus either advanced or was relatively stationary. The positions of the ice tongue wandered close to the Hunza River and had slight variations. Aged local residents in the villages of Pasu and Khaibar had witnessed the advance of this glacier. They said, "50 - 60 years ago, in one autumn between 1910 - 1930, the glacier advanced and blocked the Hunza River. The river water, however, could still pass through under the ice. In the spring of next year, the glacier retreated again and the place blocked was left on the left side of the mouth of the present-day Batura stream, leaving very large boulders at two places."

After the 1930's, the glacier began to decline. Thirty-years ago, the large ice cliff at the terminus was on the lower side of the damaged 30 m bridge. A German-Austrian Himalaya-Karakoram Expedition report in 1954, reference (2), said "the terminus of the glacier was on the right side of the Hunza River in 1944 and retreated to a place 300 m away from the river in 1954." The large ice cliff retreated to a position about 800 m away from the Hunza River according to the topographical map (1:10000) compiled in Pakistan in 1966 (Fig. 4). The key to Figure 4 is as follows:- 1. Displacement of contours (ft) and date; 2. Displacement and date of terminal ice cliff; 3. Bore hole; 4. Highway, temporary roadway, bridge; 5. Planned alternative line; 6. Drainage cavern and its date; 7. Marked advance and thickening area of the Batura Glacier at present-day; 8. Thick greyish moraine hills with buried ice; 9. Without buried ice; 10. Yellowish moraine hills formed two centuries ago.

In 1974, after surveying the topography of the glacier terminus, we discovered that the position of the large ice cliff had advanced 90 m, compared with that of 1966. In the same period, within the radius of 1 km of the upper position of the ice cliff, the ice surface rose by an average of 15 m, while both sides of the glacier were continuously declining. We measured the area again in 1975 and found that the large ice cliff had advanced further by 9.6 m compared with that of 1974. We measured again on May 29, 1978 and found that the large ice cliff had advanced by 32.55 m as compared with that of 1974, while the small ice cliff at the upper reaches of the main drainage channel on the south side of the glacier had retreated at least 230 m between the end of 1975 and early June 1978.

RECENT VARIATIONS OF THE PASU GLACIER

According to measurements from the topographical map compiled in Pakistan in 1966, the terminus of the Pasu Glacier was at 2550 m a.s.l. and 1150 m from the jeep road. Based on our investigation and survey of July 3, 1978, we found that:-

44

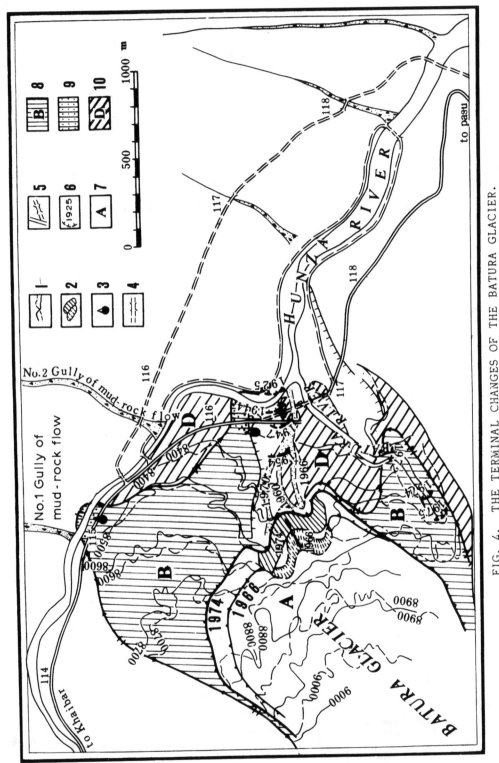

FIG. 4. THE TERMINAL CHANGES OF THE BATURA GLACIER.

Note:- The key to this diagram is given in the text on the previous page.

(1)   The drainage ice cavern of the Pasu Glacier had retreated to 1400 m from the position of the old bridge, showing that the draining ice cavern receded 140 m in the period 1966 to 1978.

(2)   The ice tongue on the left side of the Pasu Glacier had extensively receded.   Ice hills had disappeared completely, while the ice tongue on the right side had only declined slightly and retreated 30 m.

(3)   The large ice cliff had retreated to the upper reaches of the ice threshold at 2640 m.   The height of the draining ice cavern was at 2615 m measured by a barometer.   If the draining ice cavern acts as the mark of the glacial terminus, the height of the Pasu Glacier's terminus had therefore risen by 65 m and the glacial length had shortened by 150 m (12.5 m/yr.).

RECENT VARIATIONS OF THE GULKIN GLACIER

The terminus of the Gulkin Glacier had been separated into two parts as shown on the Pakistan topographical map in 1966.   The terminus height of the ice tongue on the northern side was 2520 m a.s.l., some 500 m away from the left bank of the Hunza River.   On the other hand, the tongue on the southern side was 2520 m and 660 m respectively.   This indicated that the Gulkin Glacier had advanced in the period 1925 – 1966.   But the range of this advance was less than that in early 20th century.

We determined by theodolite on November 5, 1974, that the height of the ice cavern on the southern side was 2534 m a.s.l. and 800 m from the Hunza River.   This indicated that the ice tongue on the southern side receded 180 m (22.5 m/yr.) between 1966 and 1974.   On July 3, 1978, we made repeated measurements at the fixed station and discovered that the distance of the southern ice cavern from the left bank of the Hunza River was now 860 m.   The draining ice cavern had receded again by 20 m as compared with that of 1974.   At the same time, the height of the northern ice tongue was 2600 m.   The terminus height had risen 140 m and the position of the terminus had receded 40 m when compared with 1966.   On July 13, 1978, Professor Shi and I made aerial observations on the middle and upper reaches of the Gulkin Glacier and found that the ogives were well developed below the threshold of the upper section of the ice tongue. The high lateral moraines on both sides of the glacier extended upwards to higher altitudes.   The large lateral moraines are far higher than the ice surface.   The glacial surfaces in the middle and upper reaches of the ice tongue are clean with only a few superglacial moraines.   Surface slope is relatively gentle with few crevasses.

RECENT VARIATIONS OF THE HASANABAD GLACIER

W. Pillewizer and others; reference (2), used terrestrial stereophoto-grammetric methods for surveying and drew a topographical map of the large area of drainage of the Hunza River.   The sketch map published in 1959 showed clearly that the Hasanabad Glacier had separated into two glaciers.   The measurements on the map indicated that Shispar Glacier (east branch) and Mutschual Glacier (west branch) had receded 4.5 km and 7 km respectively as compared with the positions of 1929.

We made aerial observations by helicopter on July 13, 1979 and discovered that the long separated east and west branches had met together again and extended towards the lower reaches well below the confluence. According to the different colours of the superglacial moraines, the author supposed that the west branch advanced before the east branch and first extended below the confluence, while the east branch advanced only to the confluence at that time and that its ice tongue had overlapped the west branch. On August 25 – 30, the author and others investigated the Hasanabad Glacier along the Nala River. Based on the data of field geomorphological investigations and documentary records, we determined that the retreating period of the Hasanabad Glacier was in 1953 – 1954. In the retreating process, there were two longer periods of relative stationary and surface decay. This is evidenced by the moraine hills between 2350 – 2400 m and 2400 – 2500 m a.s.l. in the valley. According to the geomorphological character, we determined the corresponding position of the ice tongue terminus at that time, and decided that the ice tongues of the east and the west branches advanced 1.5 km and 4.8 km respectively in the period 1954 – 1978. But, we do not know when the advance began. The Hasanabad Glacier is still advancing at present.

RECENT VARIATIONS OF THE MINAPIN GLACIER

The Minapin Glacier is a surging glacier like the previously mentioned Hasanabad Glacier. In August 1978 the author investigated the Minapin Glacier and found that the glacier was advancing. The shape of the ice tongue terminus had undergone an obvious change, especially the left flank with the draining ice cavern distinctly displaced forwards. The ice tongue had advanced at least 150 m. The ice cliff of the glacial terminus is very steep and its surface height had risen almost to the top height of the lateral moraine formed about two hundred years ago. The terminus position of the ice tongue (at 2350 m a.s.l.) had advanced to almost the same place as in 1954.

RECENT VARIATIONS OF THE PISAN GLACIER

The Pisan Glacier is a small valley glacier and lies on the west side of the Minapin Glacier. There are no documented data describing the variation of its terminus. The glacial termini showed on the 1:10000 topographical map by Pakistan in 1966, were at heights of 2070 m and 2220 m a.s.l. respectively on the right and left side. In August 1978, the author investigated in more detail the lower reaches of the Pisan Glacier. It was discovered that the ice tongue had distinctly retreated since 1966. The considerable length of the lower section of the ice tongue had been divided into two parts by a channel of melt water. The ice hills with relative heights less than 15 m wholely consist of black dirty ice. There are large lateral moraines on both sides of the glacier, which seemed to be formed approximately two hundred years ago and made a deep impression on us. Both sides of the lateral moraine are not symmetrical; steep inside but gentle outside. This indicates that the lateral moraine had been shorn during the last glacial advance. The terminal moraine and lateral moraine meet together in front of the ice tongue, but are cut by melt water. The terminal heights of the tongue on both left and right sides are 2450 m and 2420 m a.s.l. respectively. This indicates that the terminus had been evidently rising. According to the relative positions of a ground object point, the sides of the ice tongue of the Pisan Glacier receded 300 m and 370 m respectively in the period 1966 – 1978.

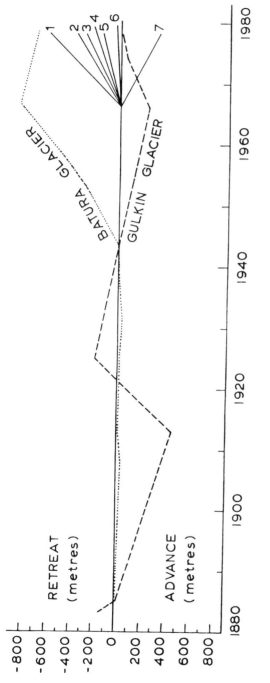

FIG. 5. RECENT VARIATIONS IN THE ADVANCE AND RETREAT OF THE ORDINARY GLACIERS
         IN THE WESTERN PART OF THE KARAKORAM MOUNTAINS.

KEY: (1) Khunjerab Pass No. 1 Glacier (right)
      (2) Pisan Glacier (left)
      (3) Pisan Glacier (right)
      (4) Khunjerab Pass No. 1 Glacier (left)
      (5) Pasu Glacier (left cavern)
      (6) Pasu Glacier (right side)
      (7) Khunjerab Pass No. 2 Glacier

48

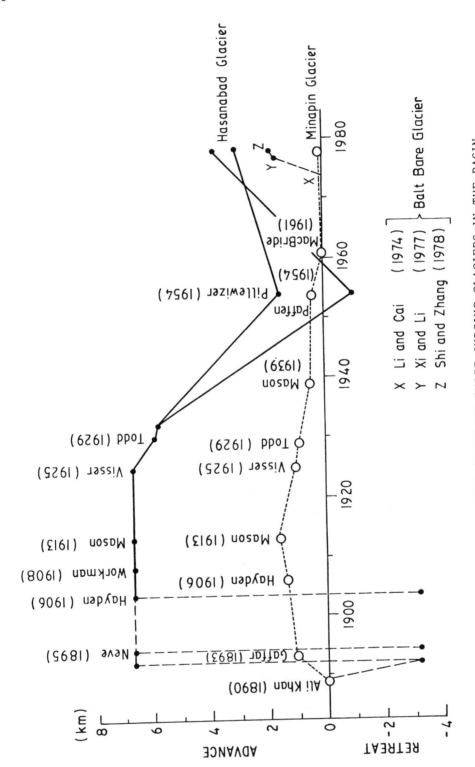

FIG. 6.   RECENT FLUCTUATIONS OF THREE SURGING GLACIERS IN THE BASIN
OF THE HUNZA RIVER, KARAKORAM MOUNTAINS.

RECENT VARIATIONS OF THE GLACIERS ON THE
NORTHWEST OF QOGIR FENG

From documentary records, and comparisons with landsat images in the 1970's and two maps of the 1930's and 1960's, recent variations of the Insukati Glacier and adjacent glaciers can be discussed. Three out of five glaciers on the northwest of Qogir Feng are in recession and two glaciers are advancing. The Insukati Glacier with a length of 41.5 km is the longest glacier in China. Though its terminus was covered by thick superglacial moraines and it was stable in the period 1937 – 1968, the lowest position of bare ice now stretches downward about 5 km. The Braldu Glacier located in Pakistan is 35 km long with thick debris cover at the terminus also. In the perod 1937 – 1968, the position of its terminus was unchanged, yet the lower limit or bare ice now stretches 4 km downward and in 1968 – 1973, the terminus advanced a distance of 900 m. In contrast, the Skyang Glacier, K2 Glacier and Westm-i-Yaz Glacier had retreated 4210 m (137 m/yr.), 1700 m (50 m/yr.) and 550 m (18 m/yr.) respectively. In addition, the Skyang Glacier had retreated again by 1000 m (200 m/yr.) in the period 1968 – 1973. However, the recession rate of the K2 Glacier obviously reduced during this period.

CONCLUSION

In summary recent variations of some of the glaciers of the Karakoram Mountains are shown in Figures 5 and 6, which indicate that the glacier termini in the Karakoram Mountains have undergone many changes during recent times. From the 1880's to the 1920's – 1930's, the glaciers either advanced in general or were relatively stationary. From the 1930's to 1960's, the glaciers usually retreated. From the 1960's on, the changes in the positions of glacier fronts have presented a very heterogenous picture, as for instance, a lot of glaciers in the western section of the Karakoram Mountains have been advancing and others retreating. This reflects the inherent difference of the time interval between change input and the achievement of new equilibrium by the various glaciers.

REFERENCES

1) SHI YAFENG and ZHANG XIANGSONG, (1978). 'Historical variations in the advance and retreat of the Batura Glacier in the Karakoram Mountains', Acta Geographica Sinica, Vol. 33, No. 1.

2) PAFFEN, K. H., PILLEWIZER, W. and SCHNEIDER, H. J., (1954). 'Investigations in the Hunza-Karakorum: preliminary report on the scientific work of the Austro-German Karakorum Expedition 1954', Erdkunde, Vol. X, No. 1, pp 1 – 33.

BIBLIOGRAPHY

Permanent Service on the Fluctuations of Glaciers of the IUGG-FAGS/ICSU 'Fluctuations of the Glaciers 1959 – 1964; 1965 – 1970; 1970 – 1975', IAHS (ICSI) – UNESCO.

MASON, K., (1930). 'The glaciers of the Karakoram and neighbourhood', Records of the Geological Survey of India. Vol. LXIII.

SCHNEIDER, H. J., (1969). 'The Minapin Glacier and men in the N W Karakorum', Die Erde 2 - 4, pp 265 - 286.

SHIPTON, E. E., (1938). 'The Shaksgam Expedition, 1937', Geographical Journal, Vol. 91, pp 313 - 339.

Some studies of the Batura glacier in the Karakoram mountains

Shi Yafeng and Zhang Xiangsong

Institute of Glaciology and Cryopedology, Lanzhou

People's Republic of China

ABSTRACT

In 1974 and 1975, using terrestrial stereophotogrammetry, the Glacier Investigation Group of China made a survey of the Batura Glacier drainage area and drew a 1:60,000 map. A glacial inventory has been accomplished and much data on morphological measurements obtained. The Batura Glacier has some continental-type characteristics but there are a number of maritime features; we call the Batura "a complex type glacier". According to our forecast, the glacier will advance another 180-240 m but in the 1990's it will once again be on the decline.

INTRODUCTION

Both the ancient Silk Road and the present Karakoram Highway linking China and Pakistan, pass by the terminus of the Batura Glacier. The international overland route through this region has long been influenced by the advance and recession of this glacier and the migration of its meltwater channels. Since 1885, many travellers and explorers have given their accounts and impression of the Batura Glacier. The chief ones among them were: Ph. C. Visser, who led a Dutch scientific expedition in 1925 and reported the first arrival at the upper reaches of the Batura Glacier (1); K. Mason, who made a comprehensive review of the glacier in 1930 (2); and W. Pillewizer and H. J. Schneider, who organized respectively the 1954 and 1959 German-Austrian Glacial and Geological Expeditions and who made brief reports of their findings (3)(4). All these provided us with preliminary and historical knowledge of the Batura Glacier.

From 1974 - 1975 and 1978, a glacier investigation group was dispatched by the Chinese government to carry out a detailed study of the Batura Glacier. The group received much support from Pakistani friends. The present article is a preliminary report on the major results of our three years work.

CONTROL SURVEY AND A TOPOGRAPHIC MAP OF THE DRAINAGE AREA OF THE BATURA GLACIER

Within the distance of 20 km between the terminus and cross section XI, see enclosed maps, the extended side of a rhomboid baseline network

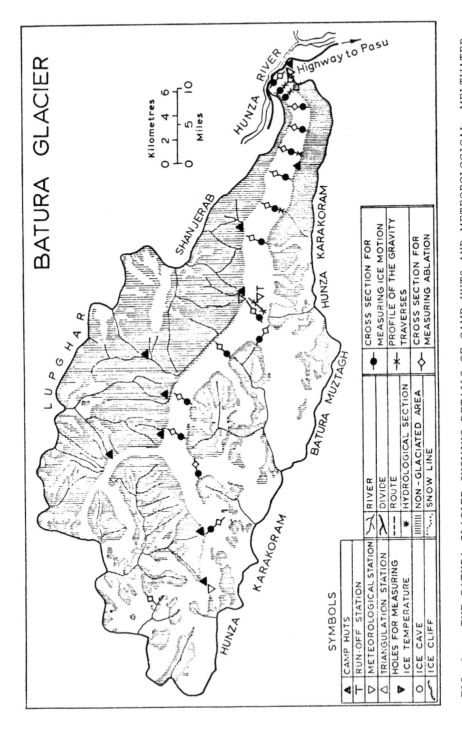

FIG. 1a. THE BATURA GLACIER SHOWING DETAILS OF CAMP HUTS AND METEOROLOGICAL, MELTWATER GAUGE, AND ICE-VELOCITY STATIONS, TOGETHER WITH CROSS-SECTIONS FOR GRAVITY AND ABLATION MEASUREMENTS.

53

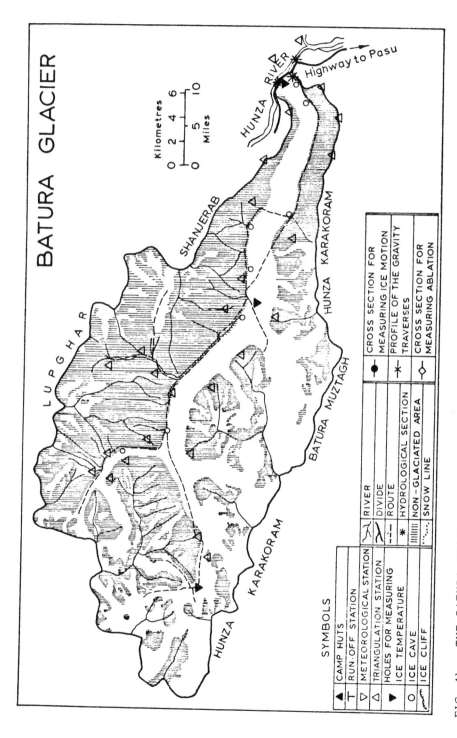

FIG. 1b. THE BATURA GLACIER SHOWING DETAILS OF ROUTES, ICE CAVES AND STATIONS FOR ICE-HOLE TEMPERATURE MEASUREMENTS AND SURVEY.

54

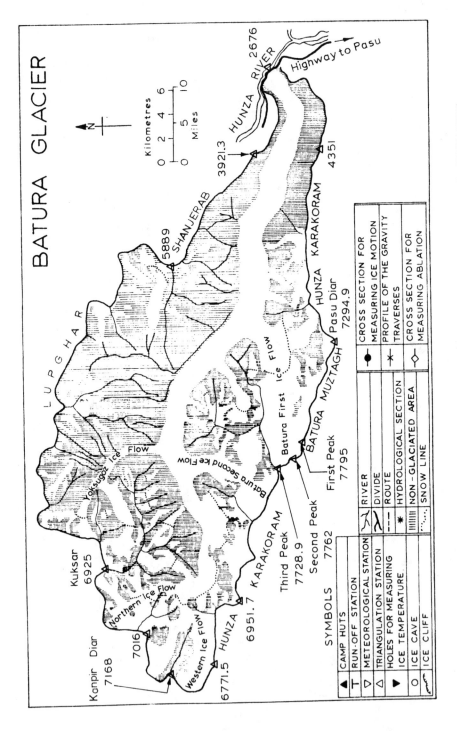

FIG. 1c. THE BATURA GLACIER SHOWING DETAILS OF FIVE LARGE TRIBUTARIES,
THE SNOW LINE AND THE HEIGHTS OF PEAKS.

has been taken as base control. Between the terminus and a point 40 km upstream, a control network has been established, amounting to 25 stations. The accuracy of the weakest side is 1/21,700. From control points of the network, we measured the main peaks in the drainage area of the Batura Glacier and so the control network covered the whole Batura Glacier.

In 1974 – 1975, Chen Jianming and Zhang Huaiyi et al (5), worked on terrestrial stereophotographic mapping from 65 photographic base lines. The average height of the control points and photographic points was 4000 m above sealevel. The average length of the photographic base lines was 320 m. The maximum photographic distance was 14 km, but in general it averaged 8 km. The angular error of the photographic base line was 1.1" and photographic dead space about 7%. They utilized landsat images and camera pictures for subsidiary detail. On the basis of this, the topographic map of the drainage area of the Batura Glacier was compiled to a scale of 1:50000 and a colour map published to a scale of 1:60000.

The map of Batura Glacier employed four methods of morphological illustration, namely, contour lines, conventional signs, hatching and shading. Efforts have also been made to coordinate harmoniously all points, lines, and planes in appropriate colours and shades, with a view to show, as precisely as possible, the glacial type and pattern, and other peculiarities of the Batura region (6).

According to the topographic map, the existing glacial inventory of the drainage area of the Batura Glacier has been accomplished and much new data on morphological measurements of the Batura Glacier obtained.

BASIC FEATURES OF THE BATURA GLACIER

Size and topography of the glacier

The Batura Glacier lies on the northern slope of the main ridge of the Karakoram, i.e., the Hunza Karakoram that runs from E to W and consists mainly of granodiorite. The highest peak, the Batura Muztagh, reaches 7795 m above sealevel. The drainage area of the Batura Glacier totals 687 km^2, while the glaciated area is 332 km^2, occupying 48% of the drainage area. On its southern side, 12 large and small glaciers flow into the main ice stream, whereas on its northern side, only 4 glaciers flow into it. In addition, there are 55 small glaciers with a total area of 47 km^2, lying apart from the Batura Glacier, yet their meltwater enters beneath the ice into the Batura Glacier's channel. With a length of 59.2 km, the area of the Batura Glacier is 285 km^2. The firn line reaches 4700 – 5300 m a.s.l. The accumulation area above it totals 144 km^2, not including the area (70 km^2) of exposed rock. The ablation area is 141 km . Hence, the glacial ratio (K) of the Batura Glacier is approximately unity. The variation of the widths of the glaciers is within the limits of 1.2 to 2.5 km. The Batura Glacier descends from the 7795 m high Batura Muztagh to the bank of the Hunza River, 2540 m a.s.l. This is the greatest fall in elevation of a glacier between its snow field and terminus in the world. The space extension below the snowline is as high as 2460 m. Of the eight largest glaciers in medium and low latitudes, the terminus of the Batura Glacier is the lowest in elevation.

According to our observations, two-thirds of the Batura Glacier's ice tongue is covered with surface moraines. The last 3 km of the glacier are almost entirely capped by superglacial moraines. An ordinary glacial surface is often uplifted by medial moraines, but on the lower reaches of the Batura Glacier, the position of medial moraines is often depressed. Though the ablation in the middle and lower portions of the ice tongue is very intense, there is no surface drainage system so the meltwater penetrates the ice body and forms complete subglacial drainage channels. At the head of the glacier where various ice flows converge, the ice layer is subjected to great pressure. It cracks and folds and as a result, a lot of ridges, pinnacles and seracs, forerunners of ice pyramids, appear. The First and Second Ice Flows which originate from the Batura Muztagh, form attractive wave ogives below the great icefalls. Judging from the character of the glacial surface, the Batura Glacier is not of a periodic surging type.

The developmental conditions of the glaciers

The Karakoram Mountains are situated in the interior of Central Asia. The climate in the deep valleys is dry. The annual precipitation recorded at Bunji, Gilgit, Batura and Misgar in the Hunza valley is merely 100 mm on average. It is obvious that such a meagre precipitation cannot create these big glaciers. Some researchers such as D. N. Wadia (1953) and others (7)(8) thought that "they are without doubt the leftovers of the Quaternary Ice age", and that "under the present-day climatic conditions, it was impossible for them to grow to such great proportions". In August 1974, in order to find out the secret of the alimentation of these glaciers, we followed them to their upper reaches and measured the thickness of the five annual layers of ice in two crevasses at a height of 5000 m a.s.l. on the north-eastern tributary of the Northern Ice Flow, and translated it into annual net accumulation. The ice thickness was equivalent to a water layer of 1030 - 1250 mm. In addition, we measured the 1973 net accumulation, which was equivalent to a water layer of 1034 mm, at the altitude of 4840 m in the western small cirque of the Western Ice Flow. As the places mentioned above are all near the snowline, a considerable amount of the accumulated snow and ice melts and evaporates in summer every year. Therefore, the actual annual precipitation of this region must be much greater than the 1000 - 1300 mm we measured. Even so, though less than the actual amount, it is still ten times the annual precipitation recorded in the Hunza valley. This shows that there is an exceptionally abundant precipitation in the upper reaches of the Batura Glacier. In addition, the distribution of precipitation on the Batura Glacier differs from that of the Tian Shan in Asia and that of the Alps in Europe, where the zone of maximum precipitation occurs at a medium height (2000 - 3000 m a.s.l.), whereas precipitation in the Batura Glacier valley increases from 3000 m upward and reaches its maximum over 5000 m.

The high, steep mountains provide an ideal cold environment and abundant precipitation for the development of the Batura Glacier. According to the annual temperature (about $10^{\circ}C$) recorded at the glacier's terminus (2563 m a.s.l.) and the lapse-rate of temperature relative to elevation, the annual temperature near the snowline (5000 m) should be $-5^{\circ}C$ or so. The annual $0^{\circ}C$ isotherm is at an altitude of 4200 m.

Thus there is not only an abundant snow fall, but also very cold conditions in the upper reaches of the Batura Glacier. Coupled with its favourable terrain, the region is capable of developing large glaciers. The lower reaches of the glaciers descend into dry valleys, quite incongruous with and contradictory to the surrounding desert outlook, and creates a false relationship between the glaciers and present-day climatic conditions.

Ice formation and temperature

A study of the snow layers was made in the firn-basins of the north-eastern small ice flow, the western small cirque and the North Ice Flow. In the firn-basin, of the first zone, there was a layer of deep hoar on the dirty ice, formed by infiltration and congelation. The layer ran to a thickness of scores of centimetres. The crystals were mainly cup-like and with a framework as a core. The biggest ones reached a diameter of 5 mm. Column-shaped crystals were comparatively few. In two sections of the second zone and four sections of the third zone, we also saw remaining crystals of deep hoar deformed by infiltration of meltwater. The wide distribution and good development of cup-like deep hoar showed that the evolutionary process of snow layers at the source of the Batura Glacier belongs to the cold metamorphic type. Sharp temperature gradients exist between these layers.

According to our observations in the firn-basin, seasonal superimposed ice layers develop widely below the equilibrium line. The thickness of such ice layers is 50 cm at an elevation of 5200 m on the Northern Ice Flow and 30 cm at 4890 m on the Western Small Ice Cirque. This super-imposed ice belongs to the ice zone of infiltration and congelation. The altitude of its distribution is 5200 to 5300 m on the Northern Ice Flow and 4850 m to 4900 m on the Western Small Cirque. The vertical height of the superimposed ice zone is 50 – 100 m. Above this zone is the infiltration zone. The thickness of these types of firn layers increases with increase in elevation.

Ice temperature is one of the important indices reflecting the physical features of the glacier. In the summer of 1974, Zhang Jinhua et al bored holes at three different altitudes and measured their temperatures (9). The results are shown in the table below.

Altitude (m)	Depth of Hole (m)	Date	Ice Temperature (°C)
2560	8	June 1 – July 31	0.0
3300	13	July 18 – Aug. 31	−0.5 to −0.9
4500	4.7	Aug. 11 – Aug. 20	−1.5 to −1.7

ICE TEMPERATURE OF THE BATURA GLACIER
AT DIFFERENT ALTITUDES (1974).

From this table, we know that the middle part of the Batura Glacier is still a cold glacier and it gradually turns into a temperate one as it approaches its lower reaches. This is due to there being a very great difference in elevation between the glacier's head and its terminus, producing very different climatic conditions.

The Glacier's velocity

Wang Wenying [see (10)] measured, at various times, the glacial motions at 129 velocity stations on 18 cross sections of the Batura Glacier. The maximum surface velocity was 517.5 m/yr, which was just below the icefall of the First Ice Flow. In general the maximum longitudinal surface velocity is near the firn-line, the velocity decreases in speed towards the lower reaches. But in the lower and middle portion there appear two anomalous areas. One of them is at cross section IX with its maximum velocity of 179 m/yr, 21 m/yr more than that at cross section X, 4 km away. The other is at cross section III; its maximum velocity reaches 85.4 m/yr, also 21 m/yr more than that at cross section IV, which is 1.3 km upstream from the former.

The surface velocity varies with the change of seasons. In most cases, the velocity is higher in summer. For instance, at station IX (3), from June to September, 1975, it was 16% higher than the annual velocity during 1974 - 1975. But the velocity in summer at the terminus was a little lower in speed than the annual one. In 1974 and 1975, the change of speed observed at most stations on the Batura Glacier was, as a rule, less than 20%.

Superglacial Ablation of the Batura Glacier

At various times in the summers of 1974 and 1975, Zhang Jinhua et al (11) set up stakes at 66 ablation stations on 16 cross sections on the surface of the glacier to measure the superglacial ablation and to obtain comprehensive data in a period of intensive ablation. They took a whole year record at four stations at the glacial terminus. It appears that the mean daily temperature reflects the most suitable index for surface ablation of the Batura Glacier. From this index, and taking the value of actual measurements as a basis, we calculated the annual amount of ablation at various stations in the lower portion of the Batura Glacier. The features of superglacial ablation of the Batura Glacier are as follows:

(1)   The surface ablation is very intensive. For example at station I(29), some 2644 m a.s.l., the annual ablation is 18.94 m. The ablation of bare ice decreases in proportion to the increase in height.

(2)   The distribution of superglacial moraine has a very strong influence upon ablation but the moraine thickness of various ice flows shows a greater difference. Thus, the ablation isoline in the lower portion of the Batura Glacier appears a special phenomenon since to some extent it is parallel to the direction of ice flow.

Solar radiation is the main source of heat for the ablation of the Batura Glacier. According to observations carried out during a short period of time on the ice surface in July–August 1974, Bai Zhongyuan found that 89.2% of the total heat was derived directly from solar radiation and 10.8% from heat conduction of the air and condensation heat of vapour (12). The maximum balance value of the radiation observed was 666.4 cal/cm per day (August 18). Of the total heat output, 83.3% was spent on ice surface ablation while the rest penetrated into the ice layer to raise the ice temperature. On the ice surface that was covered with debris 2 - 3 cm thick, the rise of temperature of the debris layer accounted for 9.1% of the heat due to conduction, while the spending of heat on ice surface ablation decreases to 78.5%. The thicker the surface moraine, the more the amount of heat spent therein, and the weaker the ablation becomes.

In conclusion the Batura Glacier has some characteristics of a continental type of glaciation but there are quite a number of maritime features. We, therefore, call the Batura "a complex type of glacier".

GRAVIMETRIC DETERMINATION OF ICE THICKNESS OF THE BATURA GLACIER

We used a Chinese made ZS -67 quartz spring gravimeter for measuring the ice thickness of the Batura Glacier. Five gravity traverses were established along Profiles II, III, VI, VIII and XI. In general, the distance between two profiles was 100 - 200 m. After having made a terrain correction, Bouguer correction and latitude correction to the measured gravity value, Su Zhen et al calculated the ice thickness at all five gravity traverses in the lower reaches of the Batura Glacier (13).

At Profile XI, the average ice thickness is 310 m while the maximum is 432 m. With an increasing ablation towards the lower reaches, the glacier gradually becomes thinner, and at Profile II near the glacial terminus the average ice thickness drops to 85 m while the maximum one is not more than 115 m. From the given rate of motion and ice thickness, we have calculated the annual ice discharge of the various profiles. In an area of 39.8 km^2 between Profiles II and XI in the lower portion of the Batura Glacier, the average thickness is 229 m and the total storage of ice amounts to 9134 million m^3.

The distribution of ice thickness in the lower reaches of the Batura Glacier shows two features :

(1)   The distributions of ice thickness in the south and north are asymmetric. The maximum thickness points of various profiles are not in conformity with the centre line of the glacier. There is an obvious difference between the two sides. From Profile XI downward, the average ice thickness on the northern side of the centre line is 180.5 m while that on the southern side reaches 215.7 m. The reason for the asymmetry between the southern and northern sides in the lower portion of the Batura Glacier is, in the main, the uneven supply and waste of ice between them.

(2)   The bend of the glacial bed has no direct influence on the distribution of the thickness of ice.

VARIATIONS IN TERMINAL POSITION OF THE BATURA GLACIER

According to repeated measurements at fixed stations, geomorphological investigations, documentary records and recalls of the local residents, the Batura Glacier's terminus has undergone many changes during the Quaternary and recent time.

Glacio-geomorphological evidence indicates that at least three glaciations occurred in the drainage area of the Batura Glacier during the Pleistocene (Shanoz, Yunz and Hunza). Since the last glacial epoch of late Pleistocene, the glacier oscillation amplitude has been smaller. We can see clearly from geological evidence that the ancient Batura Glacier of the late Pleistocene was 1.52 times longer than today, with the snow line lower by about 700 - 800 m.

During the Neoglaciation, the Batura Glacier advanced to a position 2 km lower than the present entrance of the Batura discharge channel. It left several dark brown moraine hills with a relative relief of some 60 - 70 m on either side of the Hunza River. Later, the glacier retreated by a big margin. The mudrock flow on the left bank developed fan-like deposits.

About two hundred years ago, the glacier advanced again, and the lateral moraines transcended five canals which were built by the local residents of Pasu village on the above-mentioned dark-brown moraine hills some 200 - 350 years ago. At the glacial terminus, there formed on the right bank of the Hunza River, yellowish moraine hills, and below the moraines, there occurred occasionally buried ice. The obvious polished surface and the clay mud on the bedrock wall showed that the ice surface at that time was 50 - 60 m higher than today, and that the lateral moraines formed at the same period in the middle reach of the glacier were 20 - 40 m higher than the present day. Afterwards, the glacier retreated again.

From the end of last century to the beginning of the present one, the glacier, however, advanced again. It was recorded in many documents. For example, in 1925, Ph. C. Visser wrote, "The glacier shows strong advance"; "The position of the end of the glacier is now such that the Batura stream unites with the water of the Hunza River almost without ever having seen the sight of day" (1).

The aged local residents in the Pasu and Khaibar villages witnessed this glacier advance. It left a vast stretch of grey moraine hills with uninterrupted buried ice.

From the 1930's on, the glacier began to decline. Thirty years ago, the large ice cliff of the glacier terminus was on the lower side of the damaged 30 m bridge. A German-Austrian Himalaya-Karakoram Expedition reported in 1954, "The terminus of the glacier was on the right side of the Hunza River in 1944 and retreated to a place 300 m away from the Hunza River in 1954". The large ice cliff retreated to a position about 800 m away from the Hunza River according to the topographical map (1:10000) compiled in Pakistan in 1966.

In 1974, after surveying the topography of the glacier terminus, we discovered that the position of the large ice cliff had advanced 90 m, compared with that of 1966. In the same period, within the radius of 1 km from the upper position of the ice cliff, its ice surface rose by an average of 15 m, while both sides of the glacier were continuously declining. We took measurements again in 1975 and found that the large ice cliff had advanced again by 9.6 m as compared with that of 1974. Similarly, on 29 May, 1978, we found that the large ice cliff had advanced again by 32.55 m as compared with that of 1975, while the small ice cliff on the upper reaches of the main drainage channel on the south side of the glacier had retreated at least 230 m between the end of 1975 and early June 1978.

FORECASTS OF GLACIER CHANGES

So far, only a few publications have dealt with the theory of variations in mountain glaciers (14) – (18). N. Untersteiner and J. F. Nye have, with this theory, forecast the movement of the Berendon Glacier in Canada (19); see also references (20) and (21).

In accordance with our calculations on glacial motion, glacial ice thickness, superglacial ablation, variations in the positions of the glacial terminus plus tree ring data, together with the calculation of the glacial advance made by knowledge of the decay rate in velocity at the terminus, and the fluctuating ice discharge balance, we predict that the Batura Glacier will continue to advance by 180 – 240 m, i.e., about 300 m from the highway in 16 – 22 years, so that by 1991 – 1997, the maximum advance will have occurred. It is judged that the advancing glacier will not endanger the highway. From the 1990's, the glacier will once again decline. Up to a point between Profiles III and XI, the ice balance presents a deficit state. During our investigations in 1974, we actually surveyed the glacial surface velocity on some profiles, 20 km upstream from Profile XI, but we found no high speed area whose speed exceeded that of Profile IX. In other words, in the middle and lower portion of the Batura Glacier, we could find no kinematics wave conducive to make it advance once again. So we come to the basic conclusion that the decline of the glacier, beginning from the 1990's, will last for a considerable time, i.e., more than 20 – 30 years.

The annual rings of trees are an important source of paleoclimatic data. A description is made of a tree-ring record of 484 years from sabina on the northern side of the Batura Glacier at an elevation of 3800 m a.s.l. We found three cycles of 145 – 150 years. From past climate variations by tree-ring records advance and retreat of the Batura Glacier usually lags behind the change of climate by about 60 years. Many investigators point out a worldwide air temperature decline during the forties and fifties of this century which will continue to the 2000's. If so, we may suppose that the next glacial advance maximum will appear about the 2060's.

After a further three years, the Chinese Glacier Investigation Group observed, in 1978, that the ice cliff at the terminus had advanced continuously a further 32.55 m. This approaches our forecast, but the increment value of ice flow velocity at Profile II is lower than that forecast. This means the advance of the glacier may finish before the time previously estimated and that the magnitude of advance is smaller than the above-mentioned 180 m.

REFERENCES

1) VISSER, P. C. and VISSER-HOOFT, J., (1938). 'Karakoram; Scientific Results from the Netherlands Expedition to the Karakoram and adjacent areas in 1922 – 25, 1929 – 30 and 1935 (3 vols.) (in German).

2) MASON, K., (1930). 'The glaciers of the Karakoram and neighbourhood', Records of the Geological Survey of India, Vol. LXIII.

3) PAFFEN, K. H., PILLEWIZER, W. and SCHNEIDER, H. J., (1956). 'Research in the Hunza Karakoram: Preliminary Report on Scientific work of the German-Austrian Himalaya-Karakoram Expedition 1954', Erdkunde 10 (1), p 1 – 33.

4) SCHNEIDER, H. J. 'The Mountain World, 1960/61'.

5) CHEN JIANMING and ZHANG HUAIYI, (1980). 'The terrestrial stereo-photographic mapping of the drainage area of the Batura Glacier'. In Professional Papers of the Batura Glacier in the Karakoram Mountains. Science Press, Beijing. (In Chinese with English abstract).

6) WANG YIMOU, PENG QILONG and FENG YUSUN, (1980). 'The cartographic methods of the map of the Batura Glacier'. ibid.

7) WADIA, D. N., (1953). Geology of India, London.

8) BURRARD, S. G., (1933). 'A sketch of the geography and geology of the Himalaya Mountains and Tibet', Delhi.

9) ZHANG XIANGSONG, CHEN JIANMING, XIE ZICHU and ZHANG JINHUA, (1980). 'General features of the Batura Glacier'. In Professional Papers of the Batura Glacier in the Karakoram Mountains. Science Press (in Chinese with English abstract).

10) WANG WENYING, (1980). 'The ice movement of the Batura Glacier'. ibid.

11) ZHANG JINHUA and BAI ZHONGYUAN, (1980). 'The surface ablation and its variation of the Batura Glacier'. ibid.

12) BAI ZHONGYUAN and ZHANG JINHUA, (1980). 'Some features of radiation and heat balance of the Batura Glacier'. ibid.

13) SU ZHEN, ZHANG XIANGSONG and GU ZHONGWEI, (1980). 'The thickness and the quantity of ice of the Batura Glacier'. ibid.

14) NYE, J. F., (1960). 'The response of glaciers and ice-sheets to seasonal and climatic changes', Proc. Roy. Soc. A 256, pp 559 - 584.

15) NYE, J. F., (1963). 'Theory of glacier variations'. In Ice and Snow: Properties, Processes and Applications, pp 151 - 161. M.I.T. Press.

16) NYE, J. F., (1963). 'On the theory of the advance and retreat of glaciers', Geophysical Journal of the Royal Astronomical Society, Vol. 7, pp 431 - 456.

17) NYE, J. F., (1963). 'The response of a glacier to changes in the rate of nourishment and wastage', Proc. Roy. Soc., A 275, pp 87 - 112.

18) NYE, J. F., (1965). 'The frequency response of glaciers', Jour. Glaciol. 41, pp 567 - 587.

19) NYE, J. F., (1965). 'A numerical method of inferring the budget history of a glacier from its advance and retreat', ibid, pp 589 - 607.

20) NYE, J. F., (1965). 'The flow of a glacier in a channel of rect-angular, elliptical or parabolic cross-section', ibid, pp 661 - 690.

21) UNTERSTEINER, N. and NYE, J. F., (1968). 'Computations of the possible future behaviour of Berendon Glacier, Canada', Jour. Glaciol. Vol. 7, No. 50, pp 206 – 213.

Some observations on glacier surges, with notes on the Roslin glacier, East Greenland

A.J. Colvill

Institute of Arctic and Alpine Research and
Department of Geography, University of Colorado, Boulder,
Colorado 80309, U.S.A.

ABSTRACT

The occurrence of ice surges in both ice caps and valley glaciers is noted, and their characteristics with particular reference to valley glaciers are described. Possible causes and mechanisms of surging are discussed briefly. Some observations which have been made on a surge-type glacier in east Greenland are detailed, but further studies of both surging and non-surging types of glacier are required before a complete understanding of the dynamics of surging can be obtained.

INTRODUCTION

Glacier surges occur in many parts of the world, and may affect ice bodies as diverse as ice caps and narrow valley glaciers. Some 204 surging glaciers in western North America alone have been inventoried (1), and their distribution appears not to be random; they are common in some areas and absent in others. Both long and short glaciers may surge, and usually the glaciers appear to be of subpolar type. The existence of a temperate surging glacier has not yet been thoroughly verified by temperature measurements (2), though other investigators refer to such glaciers; there may be a problem of definition. Tributary glaciers often surge after the main glacier has done so. This paper suggests that a tributary glacier may surge alongside the main glacier without the latter surging. Such an implied dislocation of adjacent ice bodies is not inconsistent with the surging of parts of ice caps. There is evidence of surging of parts of the Barnes Ice Cap (Baffin Island) in the past, and it has been suggested that climatic changes are the cause (3). Brúarjökull (part of Vatnajökull) has also been observed to surge (4), while it has been suggested (5, 6) that surges of the Antarctic ice sheet may have been the cause of the Pleistocene ice ages. Surges may have been a factor in the rapid decay of the Laurentide ice sheet (3).

While the importance of ice sheet surges, with their consequential effects on sea level and global climate, is clear, smaller scale glacier surging can be of more immediate concern. In inhabited areas cultural features such as roads, dams, irrigation channels and pipelines may be threatened. In more remote regions mineral extraction or transportation may be at risk.

The causes of surges are not fully understood, although some explanations and models have been proposed. A number of surge-type ice bodies have been studied, many in the purely descriptive sense. In other cases attempts have been made to deduce possible causes and mechanisms of surging.

CHARACTERISTICS OF SURGES

The observed characteristics of surging glaciers include the following features: chaotically crevassed surfaces; rapidly opening crevasses; sheared margins and sheared-off tributaries; bulging, over-riding, advancing fronts; large vertical and horizontal displacements of the ice; and very often a higher discharge of silty meltwater. Distinctive surface features exhibited during the quiescent phase include repeated loops, folds, or irregularities in the medial moraines, curious pits in the surface, distinctively contorted ice foliation, and the virtual stagnation of exceptionally large portions of glaciers. Longitudinal profiles locally steeper than the profiles of adjacent lateral moraines or trimlines may be noticeable just before a surge (1, 7).

In their studies of surging glaciers in western North America, Meier and Post (1) and Post (8) identified a number of typical characteristics. It seems likely that these characteristics apply to surging glaciers in general, and they are now stated here.

All surging glaciers surge repeatedly. Most surges are uniformly periodic, and the period appears to be rather constant for a particular glacier. All active surges take place in a relatively short period of time (< 1 to ~ 6 yr), after which the glacier lapses into a quiescent phase lasting for a much longer time (~ 15 to > 100 yr). The time for a complete cycle, or for the active phase, has no simple relation to the length, area or velocity of the glacier. Probably the period is, in general, a complex function of the total ice displacement during the surge and the net balance rate during quiescence.

Velocities of ice flow during the active phase are always much faster, perhaps at least an order of magnitude faster, than flow rates at similar locations on the same glacier during the quiescent phase. Flow rates during surges can be as high as 5 m hr^{-1} for short intervals of time, and can average more than 6 km yr^{-1} at a fixed location for a year or more.

An ice reservoir and an ice receiving area can be defined for all surges. The reservoir may be entirely within the ablation zone. During the quiescent phase the reservoir thickens, and the longitudinal profile in the lower part of the reservoir area continuously steepens. The active phase usually appears to begin with rapid movement where this steepening has occurred. The rapid movement propagates quickly up glacier to include all the ice reservoir area and down glacier into the receiving area. The rapid flow removes ice from the reservoir area causing vertical lowering of the surface in the order of tens or hundreds of metres, and adds ice to the receiving area causing a similar or greater amount of thickening there. As the surge proceeds the zones of lowering and thickening may move down glacier, so that in some areas the ice surface first rises and later falls.

No abrupt bedrock sills or depressions are suggested by the profiles of most surging glaciers, nor are 'ice dams', which might weaken to trigger a surge, usually evident. Practically all kinds and sizes of glaciers surge, and surges occur in almost all climatic environments. No particular shape of glacier is necessary. No special mean air temperature, precipitation, altitude or activity index is favoured. No specific rock type is indicated. Distinctive differences in permeability or surface roughness of bedrock in areas where surging glaciers are located do not appear likely. Geological faults are sometimes present and sometimes not.

The periodic activity indicates the instability of the longitudinal profile of surging glaciers. As the ice reservoir thickens the basal shear stress apparently tends to a critical value, at which the surge begins. As the surge propagates the basal shear stress in the reservoir area decreases, although the total shear stress on the bed may still be increasing owing to the very large longitudinal stresses. This rapid flow condition indicates a decoupling of the ice from the bed. The active phase of the surge appears to end when the basal shear stress, or the total bed shear stress, reaches a certain low value. This low value may be due to a low thickness in the reservoir area and a low slope in the receiving area.

POSSIBLE CAUSES AND MECHANISMS

Three main causes have been proposed for the occurrence of glacier surges: (a) stress instabilities; (b) temperature instabilities; and (c) water-lubrication instabilities. Causes (b) and (c) also involve stress instabilities, but these are secondary effects that arise from other causes (9).

The advance of a glacier snout during each surge reaches approximately the same limits as the earlier advances. This suggests that approximately the same volume of ice is involved in each surge, and this point may be of critical importance in deciding the merits of different theories. Both causes (b) and (c) depend on climatic factors; in (b) the factor is the basal temperature of the ice which results from a combination of mean annual temperatures over some time and the role of meltwater percolation in firn zones. The thickness of a water film under a temperate glacier also depends on surface ablation processes. In both cases, Robin states, it is difficult to suggest a climatic control which would be sufficiently regular to lead to glacier surges involving the same volume of ice on each occasion, although one could postulate a regular period of surges from temperature instabilities in thick, cold glaciers. On the other hand, build-up of a stress instability behind some obstruction to glacier flow would seem likely to lead to a history of surging involving a similar volume on each occasion, but not necessarily a regular period between surges.

A number of papers have discussed various aspects of the initiation and propagation of surges. An early study (10) showed how a rise in basal ice temperature, initially below melting point, might lead by a sort of chain reaction to the basal temperature reaching melting point, followed by relatively fast ice flow. This could occur in ice sheets or glaciers where the accumulation area is sufficiently cold to ensure the supply of firn being appreciably below freezing point.

Later papers, rather than concentrating on the thermal properties of ice masses, investigated the possibilities of lubrication at the glacier bed. Weertman (11) sought to show how a water film sufficiently thick to overcome obstacles could form at the ice/rock interfaces, while Lliboutry (7) studied the effects of cavitation behind obstacles. Both studies were heavily dependent on the nature of the glacier bed, a characteristic not easily determined in practice.

Another study dependent on a lubrication factor has been proposed (12), in which the sliding velocity of a glacier is assumed to be a function of the basal shear stress, which is given approximately by the formula

$$\tau = \rho \, g \, h \, \sin \, \alpha$$

where τ is the basal shear stress, ρ the density of ice, g the gravitational acceleration, h the ice thickness, and α the slope of the upper ice surface. The basal shear stress increases with time above and immediately below the firn line, after a surge. It decreases in the snout region. This leads to stress instability, but the model requires the introduction of a water lubrication factor in order to trigger the surge. An assumption that subglacial drainage is limited is required in order for sufficient water to accumulate.

Two further studies (9, 13) discussed the initiation and propagation of surges in terms of longitudinal stresses. The comment is made (13) that the adhesion component of friction at temperatures close to melting point appears to be so low that the earlier proposed lubrication models (7, 11) would apply even in the absence of an appreciable quantity of water. It is noted (13) that the model is based on stress instabilities which are closely related to bedrock irregularities, whereas other models discuss rapid sliding due to stress instabilities caused by thickening water films or changing basal temperatures.

A model for sliding based on the properties of the glacier as a whole has been developed (14). This model is applicable to all temperate glaciers, and it is the bed profile and balance profile which determine whether the glacier is of the cyclically surging type or not. Characteristics at a point are not directly relevant. The model does require that frictional lubrication be present to allow the local lowering of basal stress. Evidence is cited to support the frictional lubrication basis for this surging model, and the model can be made to match real surges by choosing appropriate values for average viscosity and the frictional lubrication factor.

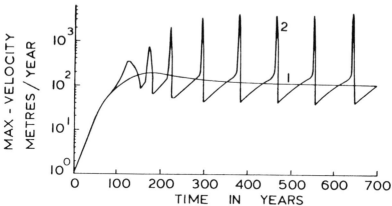

Fig. 1   The growth towards steady state for an ordinary glacier (1) and a surge-type glacier (2), in terms of maximum velocity and time; after (14).

For this surging model, all the following principles are adopted:

(a) With constant input consisting of the bedrock configuration, the accumulation/ablation balance distribution, and the properties of ice, a steady-state glacier does not result, but instead, the glacier builds up slowly then surges rapidly, stagnates, then builds up, and repeats the process periodically. Figure 1 illustrates this cyclic behaviour.

(b) The same laws and ice properties apply to the first order to all temperate glaciers. The final state of a glacier, be it surging or non-surging, would then depend on the glacier's bed profile and balance profile.

(c) Surging does not require as a necessary criterion any special features other than sufficient accumulation for the given bedrock profile. Other properties such as special bed smoothness, high geothermal flux, etc., are considered as secondary contributors to surging conditions.

(d) Surging is a large-scale phenomenon which can be adequately represented at least to the first order by a numerical model using discrete values at spacings comparable to the ice thickness. It follows that small-scale bed features are not relevant except in so far as they contribute to bulk averages over larger scales.

A SURGE-TYPE GLACIER IN EAST GREENLAND

The Roslin Glacier is a subpolar glacier draining southeastwards from the Staunings Alps, at $71°44'N$ in east Greenland. The mountains at the head of the glacier rise to about 2500 m. The glacier is some 38 km long, and below the firn line has a somewhat constant width of 1.7 km. The elevations at firn line and snout are about 1300 m and 130 m respectively. Few meteorological data are available for the glacier itself, but data from a weather station 60 km to the north show a mean annual temperature of $-9.7°C$ between 1952 and 1961. The mean temperature for the coldest month was $-24.3°C$, and for the warmest month $+5.9°C$. Mean annual precipitation was 372.5 mm of water equivalent, the maximum usually occurring in autumn and winter. Since the end of the 18th century there appears to have been a significant climatic warming which has been reflected in a general retreat. In recent decades, however, a general cooling has been apparent in high latitudes. Recent data for northeast Greenland suggest a continuing downward trend in mean temperatures (15).

Oblique aerial photography taken on 19 August 1950 (Fig. 2) shows the Roslin snout area with the chaotically crevassed appearance typical of a surging glacier, with ice cliffs also visible at the terminus. It is suggested that the glacier last surged not long before this time. Vertical photographs dated 14 July 1968 (Fig. 3) show the crevasses and cliffs to have disappeared, although the location of the ice margin is unchanged. An ablation rate of 0.3 m in 15 days has been obtained near the firn line, but no data are available for the snout area. The aerial photography reveals three elongate lobes on the north side of the glacier at somewhat regular intervals, and two tear-shaped lobes on the south side. A number of less prominent lobate or curvilinear structures are also visible in Fig. 3. The lobes are part of the ice structure, but their visibility is enhanced by the moraines which are caught up in them, outlining their form. The elongate lobes in each case appear to be derived from the medial moraine whose origin is at the junction of the Roslin and Dalmore Glaciers, while the smaller lobes on the south side appear to originate from two small tributary glaciers below the firn line. The features described have not moved any significant distance downstream between 1950 and 1968, confirming the expected stagnant regime of the glacier following a surge. As would also be expected, the 1950 photography appears to show a surface lowering above the firn line only, and a surface raising below it.

The Dalmore Glacier is a large tributary entering the Roslin from the north just below the firn line. Its velocity just above the confluence was measured at 33.2 m yr^{-1} between 1970 and 1973, while that of the Roslin just above the confluence was 10.0 m yr^{-1} (16). Just below the junction the ice velocity varies from 4.0 m yr^{-1} on the Dalmore side to 9.6 m yr^{-1} in the

Fig. 2   Roslin Glacier Snout Area (1950) Oblique

Reproduced by permission (A301/80) of the Geodætisk Institut, Denmark

Fig. 3   Roslin Glacier Snout Area (1968) Vertical

Fig. 4   Transverse profile of the Roslin Glacier, taken
         just below the point where the Dalmore Glacier
         joins it from the left; after (17).

main Roslin ice stream.   The widely differing Dalmore velocities occur only
2.2 km apart.   This would suggest that if the Dalmore ice is constrained to
flow into that part of the Roslin north of the medial moraine which separates
the glaciers, then the Dalmore ice must be damming up at the junction.   This
northern stream of the Roslin is in fact narrower than the Dalmore but the
question arises whether the surface boundary indicated by the medial moraine is any guide to englacial or subglacial conditions.   Figure 4 shows a
transverse profile of the Roslin Glacier, obtained about 0.5 km below the
line where the velocity of 4.0 m yr^{-1} was observed.   The profile indicates
that there is a well-defined subglacial ridge which separates the two ice
streams, and suggests that they maintain their own identity farther down
the valley.   The profile was obtained by radio echo sounding using a frequency of 440 MHz (17).

Ice depth contours in Fig. 5, show a bedrock ridge extending in from the
edge of the Dalmore just at the junction, and ice surface contours in Fig.
6 show that the slope is steeper at this point (16). A low sill is also indicated
at the lower end of the Dalmore.   The bedrock configuration and velocity differences suggest that a pronounced damming effect occurs where the Dalmore
enters the Roslin.   The conditions seem to be consistent with a probable
surge regime in the Dalmore.   A longitudinal surface profile of the Roslin
from near the firn line to the snout, obtained in 1976, does not reveal any
prominent 'step', a feature sometimes found prior to a surge (and hinted at
on the Dalmore).

The ablation area of the Roslin Glacier lies within a long narrow valley
while the accumulation area is fairly extensive.   This general configuration
restricts the glacier's capacity to handle its discharge, and is probably the
main cause of periodic surges.   The Dalmore Glacier also surges, and may
do so with a greater frequency than the Roslin.   Neither glacier can surge
without affecting the other, but the relationship between the two is uncertain.
The velocity of the Dalmore in comparison with that of the Roslin, and the
bedrock configuration, seem to indicate sufficient cause for a surge-type
regime to exist in the Dalmore Glacier on its own account.   The Dalmore may
be much closer than the Roslin to its next surge.

The pattern of lakes and outer moraines of the Roslin (Figure 3) suggests
a stagnation topography which is a result of surging at a period of greater
mass balance than the present, most likely during the 19th century glacial
maximum.

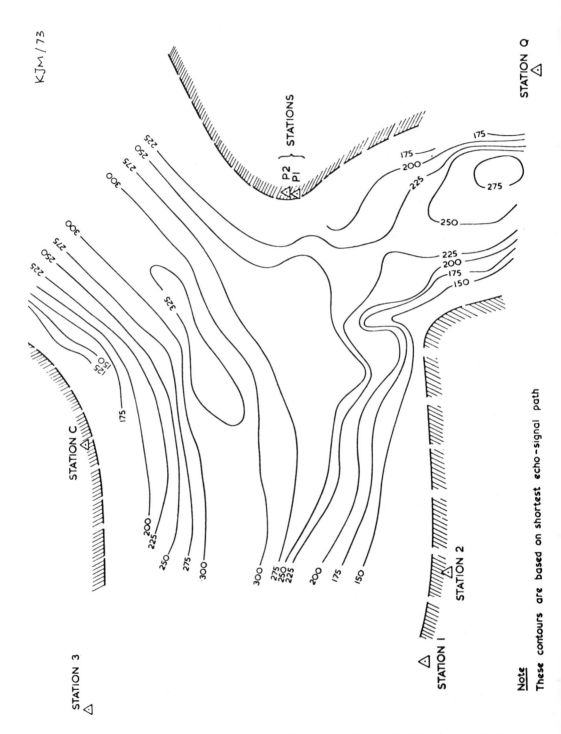

Fig. 5   25 m ice depth contours at the junction of the Roslin (top left to bottom)   and Dalmore (lower right to left) Glaciers.

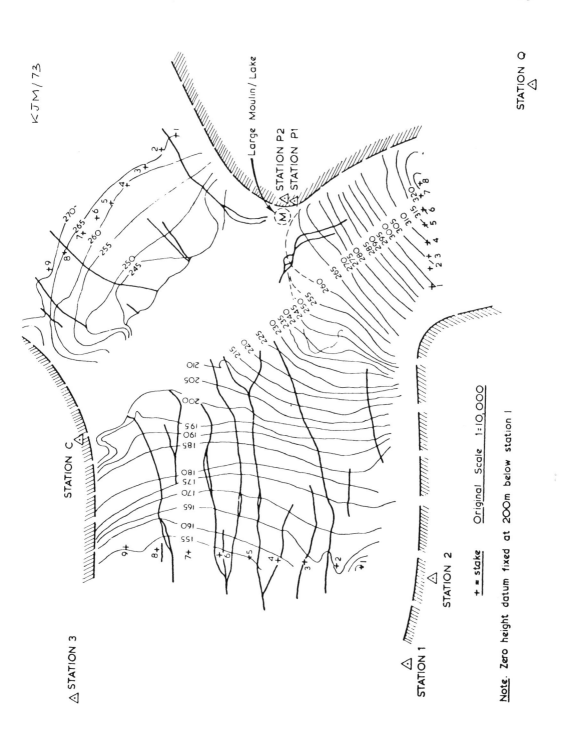

Fig. 6 Ice surface contours and melt streams, Roslin and Dalmore Glaciers

73

CONCLUSIONS

Glacier surges appear to be due to instabilities related to temperature, water lubrication or mechanical stresses, or a combination of these. These instabilities occur because, for reasons probably mainly dependent on the bedrock profile and net mass balance, the glacier cannot regularly discharge the volume of ice required to maintain it in a steady-state mode. The accumulation zone thickens and the ablation zone thins until at some critical point the resulting stress instabilities cause a catastrophic failure. Some dislocation of the glacier ice from its bed must occur to explain the fast velocities achieved, and some lubrication mechanism is therefore implied. Most of the theories outlined above seek to explain the mechanisms involved in the building up of the stress instabilities, and in the provision of lubrication.

Some studies have concentrated on the bedrock detail as a factor in the lubrication mechanism, while others have shown a progression towards the broader view of the glacier as a whole. By dealing with the properties of the whole glacier, a general solution has been obtained, and comparison with real surges are encouraging.

In the case of the Dalmore Glacier, the bedrock configuration and ice velocities measured at different locations seem to provide ample evidence for the build-up of mechanical stresses sufficient to cause a cyclical surge regime to exist. Whether water or temperature instabilities are also a factor is not known. A similar situation may occur in the case of the Roslin Glacier, but the indications are less obvious.

The flow laws of ice, whether applied to surging or non-surging glaciers are not yet fully understood. Detailed fieldwork on all types of ice bodies is a necessary adjunct to theoretical studies and the formulation of realistic models. It is in this light that past and present studies of such glacier as the Roslin, summarised in this paper, and Batura must be seen. The latter, although not of the surging type, terminates very close to the Karakoram Highway, and has been the subject of detailed studies (18) According to Hewitt (19), the Karakoram and the Alaska-Yukon region between them account for perhaps 90% of known surging glacier events, though it should be added that some of the Karakoram examples are not scientifically documented.

REFERENCES

(1) Meier, M.F., and Post, A. 1969. What are glacier surges? Canadian Journal of Earth Sciences, 6, No 4, Part 2, 807-816.

(2) Bindschadler, R., Harrison, W.D., Raymond, C.F. and Gantet, C. 1976. Thermal regime of a surge-type glacier. Journal of Glaciology, 16, No 74, 251-259.

(3) Holdsworth, G. 1977. Surge activity on the Barnes Ice Cap. Nature, 269, No 5629, 588-590.

(4) Thorarinsson, S. 1969. Glacier surges in Iceland, with special reference to the surges of Brúarjökull. Canadian Journal of Earth Sciences, 6, No. 4, Part 2, 875-882.

(5) Wilson, A.T., 1964. Origin of ice ages: an ice shelf theory for Pleistocene glaciation. Nature, 201, 147-149.

(6) Hollin, J.T., 1964. Origin of ice ages: an ice shelf theory for Pleistocene glaciation. Nature, 202, 1099-1100.

(7) Lliboutry, L.A., 1971. The glacier theory, In Advances in hydro-
science. 7. Edited by Ven Te Chow. Academic Press, New York.
81-167.

(8) Post, A., 1969. Distribution of surging glaciers in western North
America. Journal of Glaciology, 8, No. 53, 229-240.

(9) Robin, G.de Q., 1969. Initiation of glacier surges. Canadian Journal
of Earth Sciences, 6, No. 4, Part 2, 919-926.

(10) Robin, G.de Q., 1955. Ice movement and temperature distribution in
glaciers and ice sheets. Journal of Glaciology, 2, No. 18, 523-532.

(11) Weertman, J., 1969. Water lubrication mechanism of glacier surges.
Canadian Journal of Earth Sciences, 6, No. 4, Part 2, 929-939.

(12) Robin, G.de Q. and Weertman, J., 1973. Cyclic surging of glaciers.
Journal of Glaciology, 12, No. 64, 3-18.

(13) Robin, G.de Q. and Barnes, P., 1969. Propagation of glacier surges.
Canadian Journal of Earth Sciences, 6, No. 4, Part 2, 969-976.

(14) Budd, W.F., 1975. A first simple model for periodically self-surging
glaciers. Journal of Glaciology, 14, No. 70, 3-21.

(15) Allen, C. R., O'Brien, R.M.G. and Sheppard, S.M.F., 1976. The
chemical and isotopic characteristics of some northeast Greenland
surface and pingo waters. Arctic and Alpine Research, 8, No. 3,
297-317.

(16) Miller, K.J. (Ed.), 1973. Reports of Cambridge University East
Greenland Expeditions 1970, 1972, 1973. Cambridge University
Engineering Department Library.

(17) Davis, J.L., Halliday, J.S. and Miller, K.J., 1973. Radio echo
sounding on a valley glacier in east Greenland. Journal of Glaciology
12, No. 64, 87-91.

(18) Batura Glacier Investigation Group, 1979. The Batura Glacier in
the Karakoram Mountains and its variations. Scientia Sinica, 22,
No. 8, 958-974.

(19) Hewitt, K., 1969. Glacier surges in the Karakoram Himalaya (Central
Asia). Canadian Journal of Earth Sciences, 6, No. 4, Part 2,
1009-1018.

A surging advance of Balt Bare glacier, Karakoram mountains

Wang Wenying

Huang Maohuan

Chen Jianming

Institute of Glaciology and Cryopedology, Lanzhou

People's Republic of China

ABSTRACT

In November 1977, it was discovered that the Balt Bare Glacier's terminus had moved downwards 1.6 km since 1974. This movement took place mainly in 1976. During the investigation in October 1978, it was found that the terminus had further advanced a distance of 0.4 km but had begun to thin. The glacier terminus, 2.5 km in length, was surveyed and mapped at a scale of 1:5000 and a short-term measurement of surface velocity and temperature on the glacier was taken. We believe that the temperature in the surveyed section is high, and at melting point on its bottom. When the thickness and the basal sliding velocity of the glacier are estimated, one would consider that basal sliding was the dominant mechanism of the surge. It was predicted that the glacier would be thinner in the near future and that this surging advance would arrest.

INTRODUCTION

On April 12, 1974, an extensive mud-rock flow burst out from the Balt Bare Valley above Shishkat Village through which the Karakoram Highway passes. The mud-rock flow, with tremendous force, swept away a household and a pedestrian. With a maximum discharge of about 6300 m^3/s, it poured out 5 million m^3 or more of earth, debris and rock after running out of the mouth of the valley. The Hunza River was blocked and a debris-rock dam was formed which caused a lake to form 8 km in length which flooded over a 120 m long bridge and a 4 km section of highway so disrupting communications in the Hunza Valley. This disaster forced both Chinese and Pakistan highway builders to observe repeatedly, from helicopters and on-the-ground investigations, the stability of the area. In December 1974, November 1977 and October 1978, the Karakoram Highway Engineering Headquarters of the People's Republic of China repeatedly sent out engineering and technical staff to investigate the valley, in co-operation with Pakistani friends. They found that the Balt Bare Glacier, the source of the problem, was of a surging type. During the last study they particularly observed the dynamic state of the glacier terminus. They also collected data on the mud-rock flow, surveyed and mapped the glacier terminus some 2.5 km in length to a scale of 1:5000, and took short-term measurements of surface velocity and temperature of the glacier.

THE DRAINAGE AREA

The Balt Bare Valley terminates in several high peaks, the highest of which is estimated to be 7000 m. The valley starts near to Shishkat Village (lat. 36°20'N., long. 74°52'E.). From satellite photographs we found the length of the valley to be 20 km and the drainage area about 50 km^2, including an area of 18 km^2 covered by snow and ice; see Fig. 1. The mean gradient of the channel is about 23%. Most of this

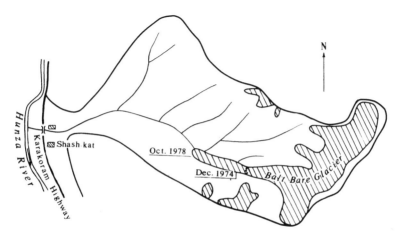

FIG. 1.   A SCHEMATIC MAP OF BALT BARE VALLEY.

snow and ice resource converges into the Balt Bare Glacier, which is about 8 km in length and which was considered to have initiated the mud-rock flow of 1974. The equilibrium line of the glacier is estimated to be at 4800 - 5000 m. On both sides of the glacier there are several cirque glaciers. There is an obvious firn basin and ice tongue, and a few hanging glaciers of small size. The channel below the glacier is very narrow and steep with many drops. In the channel there are four canyons, the narrowest opening of which is no more than 10 m wide. Nobody can reach the glacier from the Village during the melt period that lasts from April to August every year.

THE RECENT ADVANCE

During the investigation in December 1974, the glacier terminus lay at the bottom of the canyon, 3800 m a.s.l. But the glacier, during the investigation in November 1977, was noted to have advanced 1.6 km when compared with 1974, and this was witnessed to be a continuous advance. Mr. Mahar Ban, a herdsman of livestock for over 30 years said that the glacier terminus always advanced and retreated about the canyon bottom (where the terminus was located in 1974) ever since his father was able to remember, but that they had never seen an extra advance like the late one. He added that the recent advance took place mainly in 1976;

in the summer of that year, the glacier was raised and came close to his cattle-shed which was sited on the lateral moraine. The result was that the ice surface was higher than his shed by more than 10 m. He was so frightened that he dared not go to sleep inside his shed. During the investigation in October 1978, the terminus was found to be advancing continuously over a distance of 0.4 km, but by reading the photograph taken in 1977, we found that the ice surface was lowered in the middle and the upper parts of the surveyed section and that Mr. Mahar Ban's shed was now higher than the ice surface. According to traces on both sides of the glacier, the surface was judged to be 20 - 30 m lower than its highest level. From these facts, we consider that the Balt Bare Glacier is of a surging type.

THE BEHAVIOUR IN 1978

Figure 2 shows a map of the glacier terminus, 2.5 km in length, surveyed to a scale of 1:5000 during October 1978. As stated above, the middle and the upper part of the surveyed section began to thin but the lower part thickened and advanced forwards. There was an ice fall above 3800 m with a slope of approximately 25°.

Debris was spread all over the 2.5 km length, but the debris cover was not very thick and some exposed surfaces existed. Figure 3 shows the surface velocity and also the distribution of the main crevasses. In general, the velocity is rather high with a maximum of 110 cm/day; a velocity which has never been measured in China. Since this speed was towards the end of the surge, one can imagine the speed attained during the period of rapid advance. There are two comparatively high velocity areas in Fig. 3, one is the ice fall, another between the cross section II and III of Fig. 3 (points 1, 2, 12 and 14 respectively), where the valley is neither steep nor narrow. In the ordinary course of events, the velocity should not increase toward the lowest zone and so perhaps there is a kinematic wave here. Be that as it may, the second comparatively high velocity area possessed sufficient kinetic energy to push the lowest part forward continuously. Above this area there was a region of extending flow with a series of tranverse extension crevasses whilst below this area there was a region of compressed flow with many divergent crevasses in the lowest part.

Of course, the area of the ice fall was very broken and full of crevasses. At the base of the ice fall several longitudinal crevasses appeared.

At cross-sections A-A and B-B of Fig. 3 several additional points were studied to examine tranverse variations in velocity but the variation was not remarkable and this implied the existence of marginal sliding. Along both sides of the glacier one could clearly see the shear faults, i.e., the interface between the main rapid sliding glacier and the unmoved bank.

Ice temperature was taken at 3540 m a.s.l. At a depth of 3 m the temperature was -2.4 °C on October 29 when the ice surface had not yet been covered by snow and melting was going on weakly but had not yet disappeared. On the No. 1 Glacier of Urumqi He, Tian Shan, China, at a depth of 3 m a temperature of -2.8 °C was measured at the end of September 1962, when snow covered the glacier and melting was absent: reference (1). It is evident, therefore, that the Balt Bare Glacier's

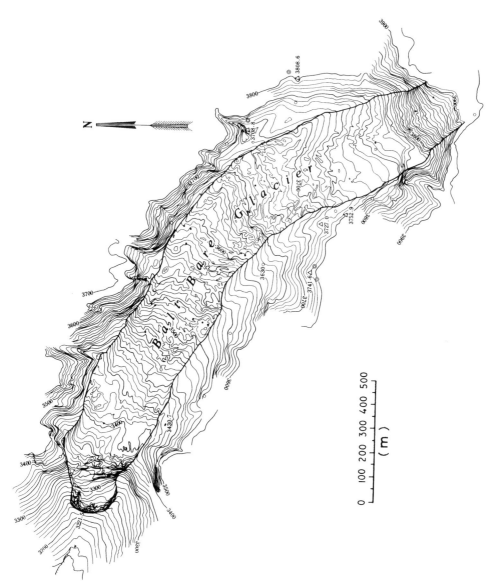

FIG. 2.   A MAP OF BALT BARE GLACIER'S TERMINUS.

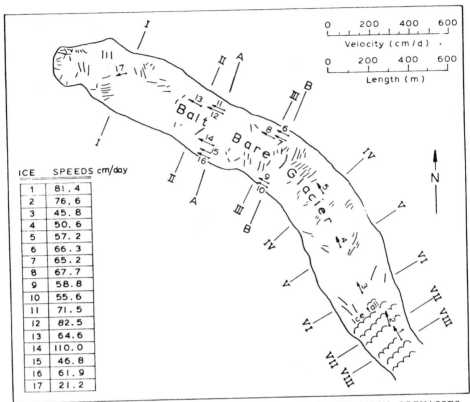

FIG. 3. THE VELOCITY AND THE DISTRIBUTION OF MAIN CREVASSES
ON BALT BARE GLACIER'S TERMINUS.

terminus had a long melt period and a high temperature as compared with
No. 1 Glacier of Urumqi He. Furthermore, at an altitude of 3300 m on
the neighbouring Batura Glacier the ice temperature, taken from July 18
to August 31 at a depth of 13 m, was -0.5 to -0.9°C; reference (2).
One can perceive that at this altitude the glacier had already transformed
into a temperate one. From these facts we consider that the base of the
glacier was at melting point, and that basal sliding took place there.

Now we can estimate the glacier thickness on the centreline, H, by
means of a perfect plasticity model. The formula is

$$H = \frac{\tau_b}{F \, g \, \sin\alpha} \qquad (1)$$

where τ_b is the shear stress on the glacier bed, which should be equal
to the plastic yield stress of ice (usually being taken as 1 bar) when
basal sliding exists; g is acceleration due to gravity; α is the slope
of the glacier surface; F is a shape factor relating to the sides of the
valley (3). Assuming the cross-section of the glacier to be a semi-ellipse
and applying equation (1) to the lower part of the Batura Glacier where
a gravity survey was taken (2), the calculated value was roughly coincident
with the measured value of H the error being not more than 20%.

We can calculate the velocity due to ice deformation, V_d, with the simplified formula of Nye (4),

$$V_d = \frac{2A}{n + 1} \tau_b^n H \quad (2)$$

where H and τ_b are same as in equation (1), and A and n are the parameters in the flow law of ice. For calculation we adopt the parameters from Nye (5), i.e., A = 0.173 bar$^{-3.07}$ yr.$^{-1}$ and n = 3.07, and assume τ_b = 1 bar (6).

Thus the basal sliding velocity, V_b, was found by means of subtracting V_d from measured surface velocity, V_s, i.e.,

$$V_b = V_s - V_d \quad (3)$$

Drawing eight cross-sections through the surveyed points in Fig. 2, we can measure the width and slope of each cross-section and obtain H on the centreline from equation (1), V_d on the centreline from equation (2), V_s on the centreline by means of extrapolation of measurements to the centreline and V_b from equation (3). These data are listed in Table 1, which gives an extraordinarily high rate V_b/V_s of up to 91 - 99%.

Cross Section	I	II	III	IV	V	VI	VII	VIII
H (m)	85	52	70	51	60	66	32	36
V_s (cm/d)	23	72	72	58	50.6	45.8	78	83
V_d (cm/d)	2.0	1.2	1.6	1.2	1.4	1.5	0.7	0.8
V_b (cm/d)	21.0	70.8	70.4	56.8	49.2	44.3	77.3	82.2
V_b/V_s (%)	91	98	98	98	97	97	99	99

TABLE 1: ESTIMATED GLACIER THICKNESS AND COMPONENT
VELOCITY ON CENTRELINE.

Generally speaking, this rate should be 80 - 90% (possibly 95%) in the upper part of the ablation area to indicate that basal sliding is the major mechanism of glacier movement (6). It follows that on the Balt Bare Glacier basal sliding was even more dominant.

DEVELOPMENT OF ANALYSIS

Applying the measured and the above calculated data, we can make a rough estimation about the variation in ice discharge and glacier thickness on the surveyed section. Although the Balt Bare Glacier was in a turbulent mood, for simplification we can suppose it to be in a steady state. The result is therefore approximate and only significant for the glacier state at the end of October 1978.

Assuming that the eight cross-sections in Table 1 have the shape of a semi-ellipse, we can calculate the cross-section area from measured widths and values at H in Table 1. Referring to Raymond (7) the average velocity in the cross-section is V_a = 0.415(V_s + V_b). Thus V_a A is the ice discharge through the cross-section. The discharge difference between the front and the back section divided by the horizontal projection of the surface area between the two sections is the variation rate of glacier thickness, $\partial h/\partial t$ (cm/d; positive shows raising), i.e.,

$$(\partial h/\partial t)_{i,\ i+1} \simeq (V_{a,i+1}A_{i+1} - V_{a,i}A_i)\ t^{-1}_{i,\ i+1} \qquad (4)$$

The estimated result is given in Table 2.

Section	Below I	I - II	II - III	III - IV	IV - V	V - VI	VI - VII	VII - VIII
$\partial h/\partial t$ cm/day	4.37	3.07	1.84	-2.70	0.35	-0.52	-3.54	0.71

TABLE 2: THE ESTIMATED VARIATION RATE OF GLACIER THICKNESS.

If the glacier was in a steady state, the annual raising value should be counterbalanced by annual ablation, b, i.e.,

$$365\ \partial h/\partial t = b \qquad (5)$$

We have not been able to obtain the value of b of Balt Bare Glacier, but systematic ablation measurements were taken on the Batura Glacier in 1974 (2). The annual ablation of the Batura Glacier was 800 - 810 cm on the surface without debris at an altitude of 3500 m, and about 750 cm at 3650 m a.s.l. The ablation condition of glaciers is not necessarily identical but assuming similarity conditions at the same altitude and taking b on the surveyed section of Balt Bare Glacier as 800 cm, this means 2.19 cm per day on average. By subtracting 2.19 cm/d from $\partial h/\partial t$ in Table 2, one can see that the middle and the upper part of the surveyed section would begin to thin whilst the lower zone would thicken. This agrees with the present study conclusions namely that when the lowest part of Balt Bare Glacier's terminus was advancing continuously with a rather high velocity it was diminishing in strength; the remainder had already begun to thin and would get thinner.

ACKNOWLEDGEMENTS

The research on Balt Bare Glacier was organized by the Karakoram Highway Engineering Headquarters, the People's Republic of China. Many Pakistan friends co-operated with us in the field work. Comrades Sun Zuozhe, You Genxiang and Cai Xiangxing worked as field associates during the investigation. The authors gratefully acknowledge all the above for their unfailing support.

REFERENCES

1) HUANG MAOHUAN and YUAN JIANMO, (1965). 'The temperature regime of the surface snow and ice layer on No. 1 Glacier, Urumqi He', The Study on Glaciology land Hydrology in the Drainage Area of Urumqi He, Tian Shan, Science Press, Beijing, p 25 (in Chinese).

2) THE BATURA GLACIER INVESTIGATION GROUP, LANZHOU INSTITUTE OF GLACIOLOGY AND CRYOPEDOLOGY. Academia Sinica, (1979). 'The Batura Glacier in the Karakoram Mountains and its variations', Scientia Sinica, Vol. XXII, No. 8, p 958.

3) PATERSON, W. S. B., (1969). 'The Physics of Glaciology', Pergamon Press, Oxford, England, pp 89 - 113.

4) NYE, J. F., (1952). 'The mechanics of a glacier flow', Journal of Glaciology, Vol. 2, p 82.

5) NYE, J. F., (1953). 'The flow of ice from measurements in glacier tunnels, laboratory measurements and the Jungfarufirn borehole experiment', Proceedings of the Royal Society of London, ser. A., Vol. 219, No. 1139, p 477.

6) HODGE, S. M., (1974). 'Variation in the sliding of a temperate glacier', Journal of Glaciology, Vol. 13, p 349.

7) RAYMOND, C. F., (1971). 'Flow in a transverse section of Athabasca Glacier, Alberta, Canada', Journal of Glaciology, Vol. 10, p 55.

The distribution of glaciers on the Qinghai-Xizang plateau and its relationship to atmospheric circulation

Li Jijun, Xu Shuying
Laboratory of Glaciology and Cryopedology
Lanzhou University, P.R.C.

INTRODUCTION

The glaciers in the marginal mountainous regions of the Qinghai-Xizang (Tibetan) plateau have been studied for about a century or so and increasing attention has been given to the glaciers of the Himalayas and the Karakoram where P.C. Visser, K. Mason, G. Dainelli and A. Desio, et al have made considerable contributions. However, for many years the glaciers in the interior of the plateau were ignored; not only their properties and trends but even the details of their distribution were unknown.

In the past twenty years or so Chinese glaciologists have done much fieldwork on the glaciers on the Qinghai-Xizang Plateau. In this paper we describe the distribution of the glaciers on this plateau and examine the relationship of that distribution to atmospheric circulation.

THE DISTRIBUTION OF GLACIERS ON THE QINGHAI-XIZANG PLATEAU

Reliable data about the distribution of glaciers on the Qinghai-Xizang Plateau can be obtained from fieldwork studies, topographic maps published in China, aerial photographs, satellite pictures, etc. Statistics show that the total area of the glaciers on the plateau within Chinese territory is 47,113 km^2, amounting to 50% of the total area (94,554 km^2) of all the glaciers on the whole plateau and the peripheral mountainous regions. The data about the glaciers in the peripheral mountainous regions lying outside China's border are cited mainly from H. von Wissmann (1) and S.V. Kalesnic (2). The area of all the glaciers on the Asian Continent is 117,200 km^2 so that the glaciers on the Qinghai-Xizang plateau occupy 81% of the total area of Asia's glaciers. Most of the great rivers of Asia have their sources in the glaciers on this plateau. The water equivalent value of ablation of the glaciers within Chinese territory is at least about 45,000,000,000 m^3 per year amounting to 15% of the total annual run off of the rivers on the plateau. On the northern side of the plateau, glacial meltwater makes up the bulk of the fluvial run-off and plays a vitally important role in the economic life of the arid regions. On the southern side, glacial meltwater, in addition to its importance in irrigation, often causes glacier block sliding and glacier surging, which

in turn give rise to the bursting of glacial lakes and debris   flows, resulting in severe natural calamities.

TABLE 1   Glaciers on the Qinghai-Xizang Plateau
and its Marginal Mountain Ranges

Mountain Range	Glacier Area		Glaciers within China's boundary
	Area (km^2)	%	Area (km^2)
Himalayas	29685	31	11055
Karakoram	17835	19	3265
Kunlun Shan	11639	12	11639
Pamirs	10304	11	2263
Nyenchin Tangla Shan	7536	8	7536
Hindu Kush	6200	7	0
Chang Tang	3566	4	3566
Gangdise Shan	2188	2	2188
Tanggula Shan	2082	2	2082
Qilian Shan	2063	2	2063
Heng Tuan Shan	1456	2	1456
TOTAL	94554	100	47113

Note: Data about glaciers inside China were collected by Lanzhou Institute
of Glaciology and Cryopedology and Lanzhou University; those outside
China are based on references (1) and (2).

Table 1 provides statistics of the glaciers on the Qinghai-Xizang plateau,
whilst Fig. 1. is a map indicating the distribution of the glaciers on the
plateau and its marginal mountain ranges.   The glacier area of each zone
is indicated by the shaded circles which are two times the actual size. From
Table 1 and Fig. 1 it can be recognized that the distribution of the glaciers
on the plateau is extremely uneven.   The main features are:-

(i)   The largest glacial regions lie mainly on the northwestern part
of the Plateau.   Here are concentrated many of the biggest glaciers
of the middle and low latitude regions, e.g. Siachen (75 km),
Fedchenko (77 km), Baltoro (66 km) and Batura (59.2 km).   The
longest glacier in China, Insukaiti glacier, having a length of
41.5 km, is also found here.

(ii)   The southeastern part of Xizang, close to the big bend of the
Yarlung Zangbo River, is another zone in which glaciers are concentrated.   Qiaqing glacier (35 km) near Yigong lake is the longest
in the Xizang region.

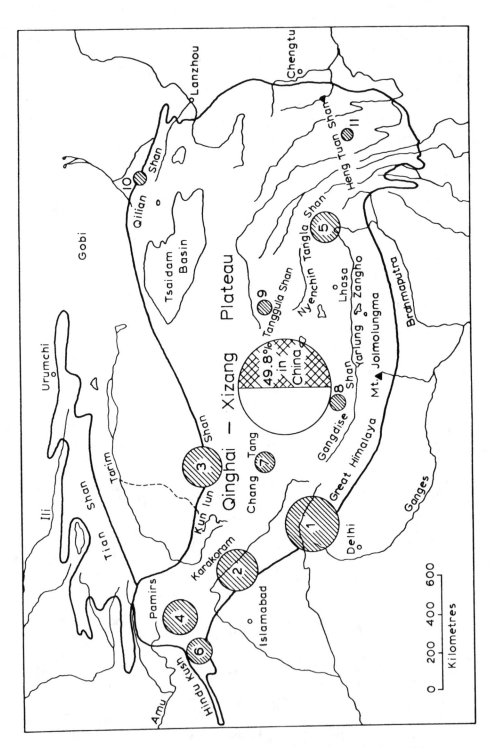

Fig. 1 Distribution of Glaciers on the Qinghai-Xizang Plateau

(iii)   The Himalayas, the highest and greatest mountain range on Earth, has the largest ice-covered area on the whole plateau. But the glaciers here are scattered, not concentrated, the lack of large or long ones appearing anachronistic.

(iv)   In the Kunlun Shan, the longest mountain range in Asia, an impressive concentration of large glaciers occurs in its western part near the upper reaches of Kalakash and Yulungkash rivers. However especially in the central and eastern Kunlun, an outstanding feature is that the glaciers are scattered. This is also the case with the glaciers of the Qilian Shan and the Aerjing Shan.

(v)   In the interior of the plateau glaciers are sparse, the total number only amounting to 10% of all the glaciers on the whole Plateau. Nevertheless, the glaciers in the interior of the plateau have their own distinguishing features. For example there often occurs tens or hundreds of square kilometres of ice caps or flat-top glaciers, being quite different from the marginal mountains where valley glaciers are predominant.

AN INTERPRETATION OF THE DISTRIBUTION OF THE GLACIERS ON THE QINGHAI-XIZANG PLATEAU

For many years, a number of scientists have explained the distribution of the glaciers of the Qinghai-Xizang plateau on the basis of the "masserhebung" effect. Evidently, this effect is very important. The spreading of the snow-lines on the Qinghai -Xizang plateau takes the form of irregular concentric circles, clearly demonstrated by the rise in the snowline altitudes from all sides towards the interior. That the glaciers in the interior make up only 10% of the plateau's ice cover is a result of this effect. However, the glaciers in the marginal mountains are also distributed very unevenly. Other causes have to be sought for this, and an obvious one concerns atmospheric circulation. In recent years a lot of research work had been done by meteorologists, both at home and abroad, into the climate and weather of the Qinghai-Xizang plateau. Examination of these results provides a basis for evaluating glacier distribution on the Qinghai-Xizang plateau.

Geological data show that it is only in the Quaternary period, about the last 2,000,000 years, that the Qinghai-Xizang plateau was sharply elevated. During the Pliocene tropical forests grew on the plateau, and hipparion, giraffe and antelope roamed across it, rhinoceros dwelling on the banks of marshes and lakes. The height of the plateau at that time, excepting the mountain ranges, did not go beyond 1000m.(3) The elevation of the plateau from modest altitudes to what is now called the "roof of the world" at an average height of 4500-5000 m, has created acute changes in the natural environment of the plateau itself and its adjacent regions. The elevation of the plateau and its mountains has resulted, not only in glacier growth, but also in changing the atmospheric circulation and climate of the region.

The major effect on atmospheric circulation due to the rise of the plateau is its powerful heating effect. Especially in summer there forms over the plateau a special atmospheric condition, being characteristic of a low near the surface of the plateau and a warm high above it, which punctuates the subtropical planetary, high-pressure zone. This induces, or at least enhances, the Indian Ocean Monsoon circulation of Asia, which is completely opposite to the famous Hadley's circulation in its direction. The southern slope of the Qinghai-Xizang plateau, and its southeastern part in particular, is thus an area of heavy monsoon rain, producing the monsoon maritime glaciers which are very active. When the vigorous westerly circulation

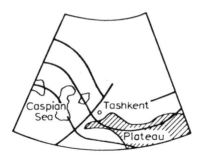

Fig. 2   The westerly trough induced and enchanced by the plateau's
blocking action and so increasing snow accumulation on the
glaciers in Hindu Kush, Karakoram and the Pamirs.

moves southward to dominate the whole of the plateau in winter an abundant
precipitation of rain and snow occurs in the western mountains of the plateau,
the Pamirs, Karakoram, Hindu Kush and Western Himalayas, owing to the
disturbance caused by the blocking action of the plateau and the natural
thermodynamic effect.   These give rise to the development of the most powerful glacial region in the middle and low latitudes of the earth.   But the
vast interior of the plateau, relatively speaking, lies in a rainshadow area
where the glaciers are very much underdeveloped because of low rates of
nourishment.   In general terms the distribution of humidity on the very high
Qinghai–Xizang plateau is a miniature of that of Eurasia.   If the heavy
snowfall brought about by the frequent depressions of the west–European
winter are the mainstay of the glaciers from Scandinavia to the Alps similarly
in the northwest of the Qinghai–Xizang plateau, because of the sharp rise
in altitude, a low trough forms in the west–wind zone which provides the
mainstay of the glaciers in the Pamirs, Karakoram, Hindu Kush and the
Western Himalayas.   The simulated experiment conducted by Ye Duzheng and
Zhang Jieqian proved that in the "dead–water area" in the west a low trough
forms in the circulation zone even without the heating effect of the plateau,
and this trough appears more distinct when heated (4).   In fact when the
big long–wave trough in the westerlies approaches the plateau, the velocity
initially decreases and then the vortex is cut off and deepens.   As a result
the Pakistani trough forms in the west and south of the plateau in winter.
In summer when the westerlies migrate northward, the Tashkent low also
forms.   These are the two main sources of depression rainfall in the west
of the plateau.   One should not be misled by the very low annual precipitation on record in the western mountain valleys or plains near the plateau
(for example Gilgit and Leh have an annual precipitation no more than 100mm)
because in the alpine zone abundant snowfall occurs.   This is caused, on
the one hand, by water vapour carried by the upper westerlies and, more
significantly, by the fact that as the deepened westerly trough stands to the
west of the plateau, it extends southward, and not only brings warm and
humid air flow from the subtropics to the western mountains along the front
ahead of the trough but it also generates the cloud vortex originating in the
Arabian Sea.   Both of these give rise to abundant snowfall.   For example,
on 27–29 November, 1972, the Arabian vortex moved along the west–wind
trough and climbed on to the plateau at 40°N, resulting in a snowstorm over
a large area, the snow cover remaining unmelted for months.   Again, on
26–29 December, 1977, a particularly strong high altitude west–wind trough
climbed the plateau and caused a snowstorm the like of which had never been
seen for a hundred years, the snow cover being 15 cm deep over the whole
landscape, setting a new record for the longest–lasting snow cover.   This
was associated only with the action of the low trough.   It is because of this

Fig. 3. Azha glacier within dense   Fig. 4. The snout of Azha glacier
         forest, Zayii Country, Xizang   extending to 2400 m a.s.l.

that in the high glaciated mountain ranges an unusual annual precipitation
occurs.   For instance, the firn basin of the Fedchenko glacier (altitude
5,000 m) has an annual snowfall of 1,500 mm (water equivalent).   In 1974
Chinese glaciologists discovered that near the firn limit of the Batura glacier
the net accumulation was 1030-1250 mm (water equivalent), and considering
the hydrological records, they inferred that the annual precipitation above
the firn limit should be more than 2,000 mm.   In 1930 R. Finsterwalter
studied the glaciers on Nanga Parbat and inferred that the annual precipitation above the snow line could reach 6400mm.   This seems quite possible.
According to recent studies, a wet zone runs from Afganistan and Pakistan
to the plateau all the year round, and the humid centre and humidity vary
with the seasons. In January, it is situated on the western bank of the Indus,
the specific humidity being 4g/1000g.   In April, it moves eastward to the
upper reaches of the Indus, with an increase in the specific humidity to
its highest level of 11g/1000g, after which is decreases.   In addition, in
January, April and October, a northward-trending stream line is seen.   This
agrees exactly with the circulation situation previously mentioned.   D. N.
Wadia once said "These giant ice-streams of the Karakoram are doubtless
survivors of the last Ice Age of the Himalayas, since the present-day precipitation of snow in this region is not sufficient to feed these great rivers of
ice".   Obviously, this is incorrect.

The monsoon from the Indian Ocean is what nurtures the glaciers on the
southern slope of the Qinghai-Xizang plateau.   In particular, the glaciers
in the southeast of the plateau are not only large in scale, but also have
the lowest snow line altitudes and are the warmest glaciers on earth.   The
end of the Azha glacier lies at 2400 m, the annual mean air temperature
being 11.5 °C.   Rice and tea can be grown at this altitude, and dense forests
grow on all sides; see Figs 3 and 4.   The concentration of glaciers in this
region and the special features which they display can be explained by
atmospheric circulation patterns.   Situated in the southeast of the Qinghai-
Xizang plateau, the Yarlung Zangbo-Brahmaputra river valley forms a reversed V-shaped gap which drives water vapour northward onto the plateau;
see Fig. 5.   In winter and spring, the vigorous southern branch of the
westerlies, after rounding the Himalayas, often maintains a trough here, and
this again supplies snowfall to southeastern Xizang.   The proportion of

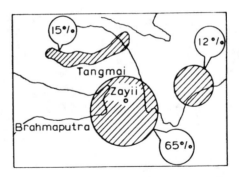

Fig. 5   Centres and frequency of cloud vortexes in south-west China
          supplying abundant snow to the glaciers of south-eastern
          Xizang and also adding to the precipitation of western
          Sichuan and north-western Yunnan.

precipitation during the winter half-year is 56% and 35% for Chayu and Pome
respectively.   In summer, the reversed V-shaped gap acts as a funnel
through which the monsoon from the Bay of Bengal travels northward.   From
the satellite cloud chart it can be seen that the Yarlung Zangbo-Brahmaputra
river valley is an area where the monsoon cloud concentration is at its greatest.   And it is mainly from here that the water vapour in summer is carried
onto the plateau.   Moreover, the Yarlung Zangbo-Brahmaputra river valley is
the chief source of the southwestern low vortex which is one of the main
sources of summer precipitation in China, its influence reaching not only the
southwest, but the eastern areas of China.   As a result, the southeastern
part of the Qinghai-Xizang plateau is the wettest part of the plateau, thus
lowering the snow line and maintaining glaciers.

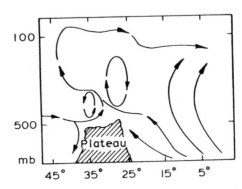

Fig. 6   Vertical meridional circulation of Qinghai-Xizang plateau
          in summer.   Note that the descending current on the
          northern flank results in very low accumulation on the
          glaciers there.

But the circulation pattern is not always favourable.   There are few glaciers
in the northern part of the plateau, and the snow line rises higher there.
This is because, on the one hand, it is situated on the lee side of the Asian
monsoon in summer, and on the windward side of the monsoon in winter,
whilst on the other hand, the high altitude circulation pattern also has an
effect.   It has been stated in the recent study by Ye Duzheng et al (5) that
the air flow that converges on the plateau from the north in summer occurs

only on the thinner and shallower 500mb surface, and, in addition, it develops a branching flow which slides downward along the northern slope. Such descending flow is frequently seen on the section along 80°-90°E. Moreover the Qinghai-Xizang high pressure zone high above the plateau in summer splits into two sections, and there are sinking air currents which flow down on the northern and southern margins of the plateau. With reference to Fig. 6 if the southern subsiding air current is checked because of the deep and thick southwestern monsoon which climbs up the southern slope of the plateau, then the northern branch on the northern plateau slope converges with and is enchanced by the low altitude downward sliding wind below the 500mb surface. Thus the northern part of the plateau is the driest area in the interior of the Asian continent due to this control of the high pressure zone. As a result, the glaciers here are of the extreme continental type because of very low alimentation. In Laohogou glacier in the western sector of Qilian Shan, the ice temperature close to the snow line has been measured as –12.8°C, which is, at present, the lowest recorded ice temperature in the mountain regions of the middle and low latitudes. The annual mean temperature near the termini of the ice-tongues is –6°C to –8°C, and a vast perennial frost zone occurs here, making a striking contrast with ice temperatures at the pressure melting point in the southeast of the plateau; see Figs. 7 and 8.

Fig. 7 The snout of Laohogou glacier in   Fig. 8 Middle reaches of Laohogou
        Qilian Shan surrounded by   glacier.
        barren rocks and permafrost.

Besides the effect produced by the general circulation pattern on the development of glaciers, it should be pointed out that the local circulation also affects the development of glaciers. In summer, each and every mountain peak is a "warm island", and at the same time a centre for condensation of water vapour, so that it is a "moist island" as well. In studying the glaciers of the Himalayas, the scientists of our country once pointed out that there exists a "second large rain zone" at high levels. In recent years Japanese scholars have also noticed this effect on the southern slope. In the southeastern part of Xizang, the precipitation of this second wet zone is as high as that of the wet zone in the middle altitudinal range, the precipitation of the former generally reaching 3,000 mm; see Fig. 9. Each "warm island"

92

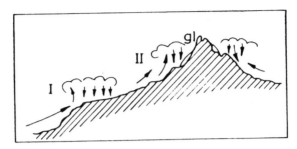

Fig. 9.   The two zones of high precipitation.   Zone II is the most
favourable to maintenance of glaciers on Qinghai-Xizang plateau.

through sublimation and coagulation, continuously releases large quantities
of latent heat into the atmosphere with the rising air columns maintaining
the powerful Qinghai-Xizang high pressure in summer in the upper stratum
of the troposphere.   This large scale transportation of energy provides
the glaciers with a plentiful snow precipitation.   Therefore, any single
massif supplies not only the appropriate topographical conditions for
glaciers to form but also generates the necessary precipitation.   Figs.
2, 5, 6 and 9 show the effects of the four kinds of circulation discussed
on the distribution of the glaciers on the Qinghai-Xizang plateau.

Finally, according to the source and properties of the precipitation,
the glaciers on the Qinghai-Xizang plateau may be divided into the
following three regional types;   see Fig. 10.

    I   Mediterranean Maritime glacier and polythermal glacier
    II   (a) Monsoon Maritime glacier
          (b) Monsoon Continental glacier
    III   Inland Continental glacier.

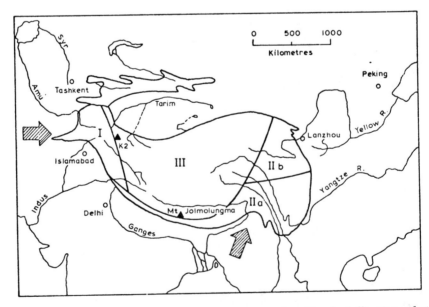

Fig. 10.   The division of glacial regions on the Qinghai-Xizang plateau
according to their properties and origin of precipitation.

REFERENCES

1) WISSMAN, H. Von., (1959). 'The present glaciers and snow line in High Asia', press of the Institute of Science and Literature in Mainz. (in German)

2) KALESNIC, S. V., (1963). 'An Introduction to Glaciology', National Press of Geographic Literature. Moscow (in Russian).

3) LI JIJUN, WEN SHIXUAN, ZHANG QINGSONG, WANG FUBAO, ZHENG BENXING and LI BINGYUAN, (1979). 'A Discussion on the Period, Amplitude and Type of the Uplift of the Qinghai-Xizang plateau', Scientia Sinica 22 (11), pp 1314 - 1328. (in English)

4) YE DUZHENG and ZHANG JIEGIAN, (1974). 'Simulation Test on the Heating Effect of the Qinghai-Xizang Plateau upon the Atmospheric Circulation in East Asia in Summer', Scientia Sinica 17 (3), pp 301 - 320. (in Chinese)

5) YE DUZHENG, YANG GUANGJI and WANG XINGDONG, (1979). 'The Average Vertical Circulation over East Asia and the Pacific Area (I) in Summer', Scientia Atmospherica Sinica 3, pp 1 - 11. (in Chinese)

Techniques for the study of glacial fluctuations

F.A. Perrott
A.S. Goudie
Dept. of Geography, University of Oxford

ABSTRACT

The Karakoram Glaciers have fluctuated over a variety of timescales from decades to millennia. This paper outlines the techniques that are available to study such fluctuations: mapping and stratigraphy; relative dating (moraine morphology, rock weathering and soil properties, loess deposition, lichenometry and extent of plant cover); absolute dating techniques (historical surveys, dendrochronology, radiocarbon dating); and the study of glacier equilibrium-line elevations.

INTRODUCTION

The glaciers of the Karakoram Mountains are broadly known to have fluctuated on various different time scales. Since the first explorers saw them in the mid-nineteenth century they have shown measurable advances and retreats lasting a few decades; see reference (1). There is a strong probability that since the end of the Pleistocene, 10,000 years ago, they have undergone more substantial phases of advance (Holocene neoglaciations) (2); while in the Pleistocene the glaciers were greatly expanded during cold phases and may have reached down to the Potwar Plateau and even to the site of the Tarbela Dam. The appropriate techniques for studying these fluctuations depend very much upon the time scale of interest, but the general aims of the present project can be summarised as follows:

(a)   To map the glacial landforms and deposits in selected valleys and to establish their lateral and vertical relationships.

(b)   To place the mapped features in a relative age sequence using time-dependent criteria such as surface weathering, soil development and vegetation cover.

(c)   To calibrate this sequence as far as possible using radiocarbon dating, and

(d)   To reconstruct the variations in glacial extent and equilibrium-line altitudes through time, as a guide to the magnitude of past climatic fluctuations in the Karakorams.

The most important objective is to obtain good dating control, for this is essentially lacking for this area, and indeed for Central Asia as a whole (3).

MAPPING AND STRATIGRAPHY

Morphological features of glacial origin, such as lateral and terminal moraines and outwash terraces, will be mapped on the available topographic base maps, together with any older patches of glacial sediments without surface expression. One of the prime tasks of the geomorphologist seeking to identify the extent of former glaciation is to be able to recognise ancient glacial sediments and to differentiate them from other diamictites such as mudflow deposits. Detailed mapping and sedimentological analyses will provide information on the basis of which such differentiation can be made.

The sequence of glacial advances through time will be established from the evidence of the lateral contacts or cross-cutting relationships between the various fill and outwash bodies, and from any exposure created by streams or along the new highway. In the Swiss Alps, a very detailed neoglacial chronology has been established by dating the buried soils and organic deposits which separate the numerous superimposed morainic units exposed in natural or artificial sections (3).

We will also be searching for localities where glacial deposits are overlain by lakes, bogs or former lake sediments, since these provide a source of datable material. Lakes are a common component of glaciated landscapes as a result of uneven glacial erosion and deposition. The palynological and chemical characteristics of the lake sediments are a valuable palaeoclimatic indicator in their own right, as they provide a record of conditions since the basin was formed.

RELATIVE DATING TECHNIQUES

A wide variety of age-dependent criteria can be used to differentiate glacial deposits. They are reviewed by Birkeland (4)(5) and Burke and Birkeland (6). It has been found that different techniques vary in their applicability, both from area to area and according to the age range of the deposits. Modern approaches involve the measurement of a variety of indices at each site, followed by statistical analysis to determine (a) the most useful techniques for the area being studied and (b) the relative age sequence of the deposits.

Moraine morphology

In many alpine areas, young moraines tend to have steeply sloping sides and sharp, continuous crests, whereas older moraines have been degraded by surface erosion (7)(8)(9). Maximum slope angles and crest widths may therefore be a good guide to age.

Rock weathering and soil properties

It was noted by the Chinese team which investigated the Batura Glacier (10) that moraines of different ages showed different colours resulting from different degrees of weathering. We shall be using a variety of indices to measure the degree of surface weathering in the field, including:

(a) the percent of fresh, weathered and pitted granitic boulders;

(b) the change in angularity of boulder corners with age;

(c) the development of weathering rinds, particularly their thickness; and

(d) the extent of soil profile development and weathering of subsurface clasts.

This will be followed up by laboratory analyses of soil properties.

Loess deposition

Both to the north and south of the Karakoram mountains, loess (wind blown aeolian silt) is known to have been deposited in the past. Several authors have reported that loessic material has accumulated to varying degrees on surfaces of different ages (7)(8). Rock glaciers and moraines appear to be good traps for loess, and thus the presence of absence of loess, as well as its thickness, may be an index of relative age.

Vegetation

(a) Lichenometry

Since the 1950's the measurment of lichens on surfaces of known age (e.g., monuments, grave-stones, cairns, etc.) has enabled the construction of lichen growth curves which can be extrapolated to date surfaces of unknown age (e.g., deglaciated rock surfaces, moraines, etc.). Such techniques are not without problems (11), and will not be applicable to the limestone lithologies of the area, but have been tried out successfully in Nepal (J. Benedict, personal communication). This technique may permit the establishment of a relative chronology for the past few centuries, even if no "absolute" dates are available.

(b) Plant Cover

The percentage area of rock surfaces covered by lichens, and the development of a cover of higher plants such as grasses and shrubs, are often useful indices of the age of neoglacial moraines and outwash (8)(9)(12)(13), and will be tested in the Karakorams.

Statistical Analyses

Two computer programs developed in the USA, known as CHARANAL and GRAPH, are particularly useful for the analysis of relative age data from small glacial deposits (12)(13). We hope to use these packages to divide the deposits into groups of different ages.

"ABSOLUTE" DATING TECHNIQUES

Historical surveys

The surveys made in the period since the mid-nineteenth century enable a comparison to be made between the glacier's state then and now. Detailed chronologies and maps are available for some of the glaciers in the area, e.g., Batura, Minapin, Pasu, Hasanabad, Figs. 1 – 3, and it is our

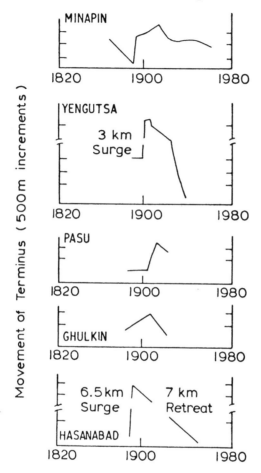

FIG. 1.   GLACIER FLUCTUATION IN THE KARAKORAM MOUNTAINS
(AFTER MAYEWSKI AND JESCHKE, (1)).

intention to bring these up-to-date. We shall also establish some appropriate cairns and markers to provide additional control for future work. Detailed geomorphological maps of glacier front positions will serve a similar purpose. The present behaviour of the glaciers is of great significance because of their effects on road construction, river flooding, etc. (see, for example, reference (14)).

Dendrochronology

In temperate latitudes tree-ring series, and in some instances, patterns of width variation, have provided a direct means of establishing time spans of direct relevance to moraine studies. In most environments trees do not live for more than several centuries and consequently direct application of this method is limited to that time scale. By counting tree rings, dates of geomorphic surfaces may be obtained. In addition, however, some sensitive species vary the nature of their ring growth according to environmental conditions, such that the nature of ring growth can be

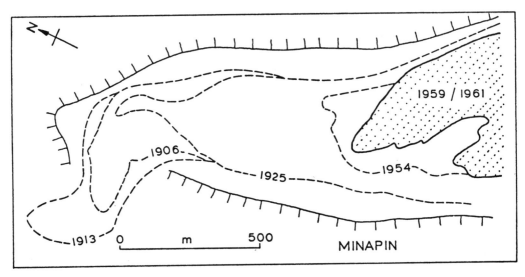

FIG. 2. MAP OF FLUCTUATIONS IN THE POSITION OF THE MINAPIN GLACIER
IN THE 20TH CENTURY.

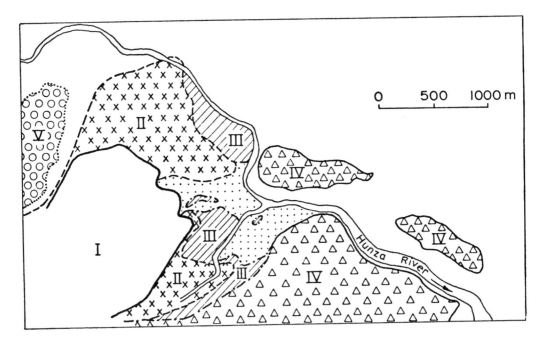

FIG. 3. MAP OF BATURA GLACIER DEPOSIT
(FROM CHINESE INVESTIGATION TEAM (10)).

KEY: I Present glacier snout
II Dead ice covered with grey till
(ca. 1885–1913, 1925 AD)
III Yellowish moraines formed two
centuries ago

IV Dark brownish moraines of
early neoglacial age
V Pleistocene moraines
Area scoured by meltwater

utilised to interpret past environmental fluctuations.

Radiocarbon dating

All the dating techniques discussed so far are only applicable to a very short time span, or yield only relative ages. For the study of the earlier Holocene and late Pleistocene events, geochronometric techniques are necessary. In particular, we shall be searching for datable materials, such as wood, charcoal, soil humus or lake sediments, which can be used to date the moraine or outwash deposits by C14. This may enable ages to be obtained back to about 40,000 years before the present.

LONG-TERM FLUCTUATIONS IN GLACIERS AND SNOW LINES

Using data on moraine ages and elevations, it is possible to estimate the extent of the glaciers and their equilibrium-line elevations during individual glacial maxima (2)(7)(15). These are two important measures of the magnitude of past climatic changes and may also yield useful information about the spatial variability of net accumulation in the past.

REFERENCES

1) MAYEWSKI, P. A. and JESCHKE, P. A., (1979). 'Himalayan and Trans-Himalayan glacier fluctuations since AD 1812', Arctic and Alpine Research 11 3, pp 167 - 187.

2) FUSHIMI, H., (1978). 'Glaciations in the Khumbu Himal (2)', Seppyo 40, pp 71 - 77.

3) GROVE, J. M., (1979). 'The glacial history of the Holocene', Progress in Physical Geography 3 1, pp 1 - 5.

4) BIRKELAND, P. W., (1973). 'Use of relative age-dating methods in a stratigraphic study of rock glacier deposits, Mt. Sopris, Colorado', Arctic and Alpine Research 5 4, pp 401 - 416.

5) BIRKELAND, P. W., (1973). 'Weathering and soil development with time '. In 'Pedology, Weathering and Geomorphological Research', Chapter 8, pp 153 - 180.

6) BURKE, R. M. and BIRKELAND, P. W., (1979). 'Re-evaluation of multiparameter relative dating techniques and their application to the glacial sequence along the eastern escarpment of the Sierra Nevada', Quaternary Research 11 1, pp 21 - 51.

7) PORTER, S. C., (1970). 'Quaternary glacial record in Swat Kohistan, West Pakistan', Geological Society of America Bulletin 81, pp 1421 - 1446.

8) IWATA, S., (1976). 'Late Pleistocene and Holocene moraines in the Sagarmartha (Everest) region, Khumbu Himal', Seppyo 38, pp 109 - 111.

9) FUSHIMI, H., (1977). 'Glaciations in the Khumbu Himal (1)', Seppyo 39, pp 60 - 67.

10) CHINESE INVESTIGATION TEAM, (1976). Investigation report on the

Batura Glacier, in the Karakoram Mountains, the Islamic Republic of Pakistan.

11) WORSLEY, P., (1981). 'Lichenometry'. In 'Geomorphological Techniques', (A.S. Goudie, Ed.) pp 302 – 305.

12) MILLER, C. D., (1971). 'Quaternary Glacial Events in the Northern Sawatch Range, Colorado'. Ph.D. Thesis, University of Colorado, 86 pp.

13) CARROLL, T., (1973). 'Relative age dating techniques and a late Quaternary chronology, Arikaree Cirque, Colorado', Geology $\underline{2}$ 7, pp 321 – 325.

14) MASON, K., (1935). 'The study of threatening glaciers', Geographical Journal $\underline{85}$, especially pp 29 – 32.

15) ANDREWS, J. T., (1975). 'Present and paleodistributions of ice-bodies and snowlines'. In 'Glacial Systems: An Approach to Glaciers and their Environments', pp 6 – 60.

Survey and analysis systems for the Vatnajökull ice-depth sounding expedition 1977

J.F. Bishop

Sir Alexander Gibb & Partners, Reading

K.J. Miller

University of Sheffield, U.K.

ABSTRACT

Traversing an ice-cap under predominantly poor visibility conditions requires innovation in both recording survey data and its subsequent analysis. This paper presents the method of integrating the results from pressure and odometer readings, theodolite surveys and satellite doppler recordings on magnetic tapes. The analysis includes adjustments of field data and implementation of conversion parameters from WGS 72 to Icelandic Geodetic and Icelandic Lambert co-ordinates. The paper concludes with a table of fixed points on the 295 km traverse on which ice-depth profiles were subsequently based.

INTRODUCTION

In 1977 an expedition from Cambridge University went to the Vatnajökull ice-cap in Iceland (Fig. 1) in order to test and develop equipment for measuring the considerable thickness of ice found there (1) (2). During the course of the expedition the equipment, carried in a Weasel snow-tractor, traversed some 295 km of ice-cap measuring the ice thickness (Fig. 2). Results of these measurements are presented in another paper at this conference (3).

METHOD OF SURVEY

The requirement of the survey was to fix the track of the Weasel and its attendant antenna array during traverses over the ice-cap, so that the position of every ice thickness measurement was known. In addition, vertical heighting would help to define the actual elevations of the snow surface and bedrock. Positions had to be related to local maps and also to the excellent LANDSAT satellite imagery of Vatnajökull. The radio-echo equipment, using a low frequency band, is limited in the size of the object that it can register and there is little to be gained in fixing a horizontal position of a measurement point to be better than 100 m. This also represents a reasonable plotting limit for available maps and imagery. A standard error in heighting of 5 m was attempted since electronic and recording errors are of the order of 0.1 µs (about 10 m).

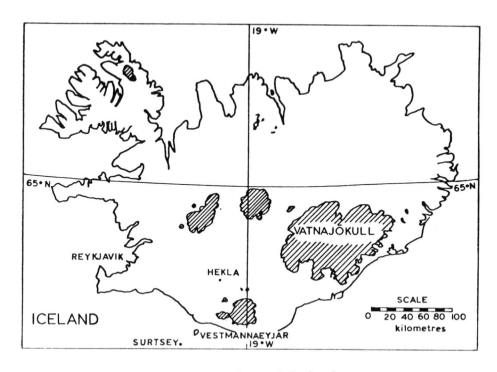

Fig. 1   Ice-Caps of Iceland

Fig. 2   Vatnajökull, showing track of 1977 expedition

An ice-cap differs from the more commonly known valley glacier in having a central accumulation area from which ice flows in all directions. Mountains tend to be visible only around the perimeter of an ice-cap where the ice is thin, but nearer the centre a few may protrude through forming nunataks. Thus it was expected that within the central region of Vatnajökull theodolite resections used to fix the track of the Weasel might be impossible. It would, of course, be ideal to fix the track by reference to some external reference frame. Sun sights, beside being time consuming, would not be sufficiently accurate. It was realised that fixes made off the Transit system of navigational satellites, which enable ground receivers to be fixed to a world-wide reference frame, would be ideal. Decca Survey Limited were approached and agreed to loan one JMR-1 doppler satellite receiver for the duration of the expedition and J.F. Bishop attended a two-week course covering the principles and operation of the instrument.

The principles of satellite survey are relatively simple. It is in the analysis and technology that complications arise. Five satellites circled the earth in polar orbits and the positions of these were known accurately relative to fixed tracking stations. As the satellite passed above the horizon, the receiver measured the doppler shift between itself and signals from the satellite, using a very stable frequency. The pattern of this doppler shift was used to fix the receiver's position. Two types of orbital data were available: one was predicted and was broadcast continuously from each satellite; the other was precise and was obtained directly from observations at the tracking stations. However, this latter information was difficult to obtain and was available only for two satellites.

The receiver, powered by 12 V batteries, recorded orbital and doppler data on cassette tape for later processing and was easy to operate. The antenna was of simple construction; a folding tripod which supported an extendable pole. Within five minutes, the antenna could be erected and connected to the receiver, the latter being ready for the next satellite pass having had temperature, pressure and site data entered through its keyboard. There was also a facility for waking the receiver at pre-selected times. This saved on battery power and allowed some discrimination , as potentially good passes could be selected from previously prepared data sheets of satellite passes.

The receiver was used in two modes: single pass and multiple pass. Single passes lasting 20 minutes were used whilst traversing the ice-cap. During traverses the receiver, lead-acid battery, theodolite, tripod, tent and emergency equipment travelled on a small sledge behind a skidoo. Multi-pass sites, where more than one pass was received from more than one satellite, were typical of where we stopped overnight. The adjustment of measurements from these sites is described later in the paper. The receiver could be left in the open through driving rain and drifting snow and proved excellent for our operations. And, of course, it was possible to fix our position during the thickest of fogs or the foulest of weathers at which times conventional survey was impossible.

Wherever sufficient mountains or survey stations were visible, theodolite resections were made and elevations computed from the verticle angles. A Wild T2 theodolite was used. Fixes were made every 3-5 km during a traverse, with the Weasel following a straight course in between stations. A series of points defining the track was thus obtained. Two further measurements enabled the position and elevation of intermediate ice thicknesses to be interpolated.

Trailing behind the Weasel was a bicycle wheel odometer which triggered a counter in the vehicle at every revolution. The readings from this were noted every two minutes and could be related directly to film records of ice depths. Simultaneously a precision aneroid (Baromec) was used to obtain pressure readings which could be reduced to elevations. An attempt was made to measure the course travelled using a reconditioned aircraft compass which was fixed on the Weasel roof and a repeater mounted in front of the driver. Unfortunately the magnetic dip was so large and the vibration of the Weasel so great that the compass drifted with gay abandon.

ADJUSTMENT OF FIELD DATA

Except for the recordings on magnetic tape from the satellite doppler receiver, all survey results came back to England in the form of notes. One of the first tasks was to establish methods and formats for putting this information onto a computer file which could then be accessed during later processing and adjustment. Fig. 3 shows a flow chart of the processes whereby three independent sets of data were adjusted and merged to produce a coherent record of the track of the antenna. A brief description of some of these steps is given below.

1) Satellite Doppler Measurements

Recordings made in the receiver on the ice-cap consisted of header information, measurement of doppler shift and satellite orbit predictions. The contents of many cassette tapes were first transferred by Decca Survey Limited onto 1600 bits per inch computer compatible magnetic tapes. Initial processing indicated that some corruption of the data had occurred and the tapes were sent to us in Cambridge where online editing of the large data files could readily be carried out. The program used in the computation was a commercial one which could not be released, and for this reason computations were carried out by Decca Survey Limited on their computer in Leatherhead. Positions are computed in the satellite ellipsoid, WGS 72.

Where only one or two passes of the navigational satellites had been recorded, elevation had to be treated as a fixed quantity to allow a sufficiently accurate determination of horizontal position. Where many passes had been recorded (a multiple pass site), elevation could be adjusted with horizontal position and the results gave each component to within 5 m.

2) Theodolite Measurements

Both horizontal and vertical angles were observed with the Wild T2 theodolite. Reduction involved the identification of targets, whose elevations and positions in the Icelandic Lambert grid were already known, and the computation of resections. About 50% of the sites could be treated in this way. The rest were adjusted in groups using a powerful computer program developed at the British Antarctic Survey in Cambridge and additional survey information. Typically this comprised of an additional azimuth or odometer distance between sites or the intersection of common points. Thus many sites were resolved from poor or ambiguous resections.

Once the horizontal positions had been established, adjustment of the vertical angles could proceed. As most sites had a surfeit of vertical angles, the necessary correction for refraction and curvature in terms of seconds

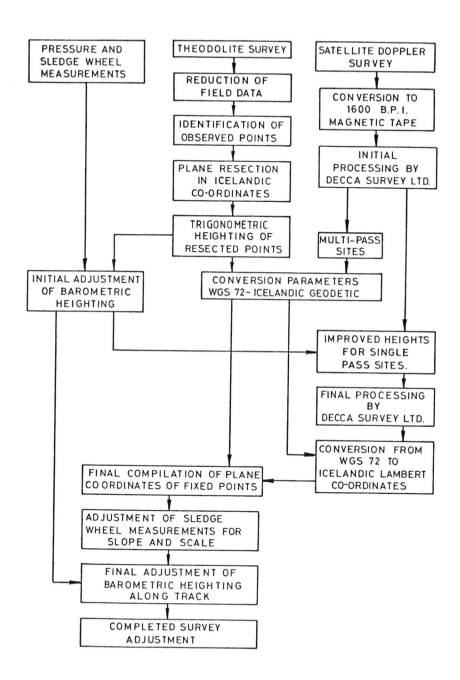

Fig. 3 Flow chart of survey analysis

of arc per kilometre was obtained by adjustment to give a least-squares fit. The corrections used range from 9.6 to 17.0 seconds/kilometre with a mean of 12.2 seconds/kilometre.

3) Conversion from WGS 72 to Icelandic Lambert co-ordinates

Fixes obtained from the doppler satellite measurements are found in terms of WGS 72, a global ellipsoid developed in the course of ranging to orbiting satellites. However, no conversion could be traced from WGS 72 to the Icelandic Ellipsoid, the International Ellipsoid, based on a datum at Hjorsey near Reykjavik (4). If common points to both ellipsoids could be found, the necessary translation parameters between the origins of the two ellipsoids could be computed. Three sites on Vatnajökull were used, one at Grimsfjall in the centre and two on the ice-cap to the west, Camps 2 and 3. The positions of Camps 2 and 3 were established in the Icelandic datum by very careful resections to a large number of stations on and off the edge of the ice-cap. Grimsfjall was part of the Icelandic triangulation network.

Thus positions in Lambert co-ordinates for the Icelandic datum were converted into geocentric co-ordinates in the International Ellipsoid (a = 6378388.0m, e = 0.00672667) and doppler derived latitudes and longitudes in WGS 72 were converted into geocentric co-ordinates in WGS 72 (a = 6378135.0m, e = 0.006694318). By subtraction, the necessary ΔX, ΔY and ΔZ translation co-ordinates were found with a standard error of 13.8 m, adequate for this survey. The co-ordinates were, for translation from WGS 72 to the International Ellipsoid, Hjorsey datum:

$$\Delta X = + 81 \text{ m}$$

$$\Delta Y = - 50 \text{ m}$$

$$\Delta Z = + 69 \text{ m}$$

These agreed well with translation co-ordinates found for a site occupied in Reykjavik, 240 km to the west of Vatnajökull.

Using the translation parameters, the remaining doppler sites then had their position established in terms of the Icelandic Lambert co-ordinates. The geoid to ellipsoid separations on Vatnajökull for the International Ellipsoid in the Icelandic datum was interpolated from various spot measurements around Vatnajökull and taken to be -8.0 m.

4) Barometric Height Adjustment

During the planning of the expedition, it was realised that deviations on the ice-cap could best be measured by barometric heighting between points of known elevation. To measure barometric variations during such measurements, a barograph was installed in the Icelandic Glaciological Society's hut at Grimsfjall and left running while the traverses of 15th - 19th June were carried out. Unfortunately due to a malfunction, these measurements could not be used. An attempt was therefore made to allow for pressure fluctuations by fitting measured pressures to measured altitudes whilst ambient pressure was allowed to vary with time. Diurnal variations were represented by the expression

Table 1

Lambert coordinates and heights of traverse fixes

("stakes") established on Vatnajökull

Reference Point	Northings	Eastings	Elevation
Camp 3	425761.0	497613.0	1180.0
A1	426345.0	500076.0	1111.0
A2	428435.0	497404.0	1192.0
Camp 2	430802.0	495204.0	1263.0
Camp 4	430741.0	492585.0	1303.0
B1	430746.0	491212.0	1334.0
B2	430674.0	489507.0	1355.0
B3	430456.0	486445.0	1391.0
B4	430164.0	483313.0	1439.0
B5	429799.0	479394.0	1496.0
B6	429690.0	474724.0	1571.0
B7	429600.0	472830.0	1627.0
B8	430448.0	471222.0	1693.0
Grimsfjall	434095.0	464646.0	1719.0
C5	434541.0	462166.0	1534.0
C4	437421.0	459421.0	1596.0
C3	438413.0	459041.0	1626.0
C2	438833.0	458614.0	1643.0
C1	439895.0	457406.0	1650.0
Camp 7	451635.0	446039.0	1612.0
11	454998.0	441659.0	1718.0
12	453229.0	444310.0	1628.0
20	448676.0	444851.0	1609.0
21	445619.0	443503.0	1597.0
22	442924.0	442280.0	1597.0
23	439676.0	440608.0	1541.0
24	435260.0	438779.0	1525.0
25	431071.0	436983.0	1524.0
26	426334.0	434764.0	1576.0
27	421457.0	433499.0	1612.0
28	425171.0	430313.0	1566.0
29	428970.0	427193.0	1515.0
30	432675.0	424055.0	1389.0
Camp 8	431318.0	409298.0	1446.0
34	432456.0	407586.0	1534.0
35	435952.0	410643.0	1387.0
36	435866.0	415082.0	1366.0
37	435823.0	419521.0	1345.0
41	435015.0	442909.0	1536.0
42	434802.0	447447.0	1546.0
43	434556.0	452435.0	1539.0
Camp 9	434403.0	457275.0	1559.0

$$\Delta p = a_1 \cos (\pi t/12) + a_2 \sin (\pi t/12) + a_3 \cos (\pi t/24)$$
$$+ a_4 \sin (\pi t/24)$$

and the relationship between pressure and altitude was taken to be

$$E = E_{ref} + 29.22 \ T \ \ln \ (P_{ref}/(P_{meas} - \Delta p))$$

where

T	=	temperature in K
t	=	time elapsed
E	=	required elevation
E_{ref}	=	reference altitude, usually the mean of the measured altitudes
P_{ref}	=	reference pressure accorded to the reference altitude (one of the variables)
P_{meas}	=	measured pressure
Δp	=	diurnal pressure corrections

Where the elevation of a fixed point had been derived from theodolite observations or from a multi-pass doppler site, this elevation was combined with a pressure reading at that point and a time of reading, to produce equations of the type shown above. These equations were weighted according to the quality of the observations and a least-squares fit applied to give a_1, a_2, a_3, a_4 and P_{ref} for each day's traverse. The residual was typically near 3 m. Intermediate points where pressure and time had been measured could then be heighted. The odometer and barometer observations, made every two minutes whilst the Weasel was travelling, thus enabled a more detailed surface profile of the various traverses to be computed and later used with measurements of ice depth to obtain bedrock levels. Improved heights for single pass doppler sites also resulted and co-ordinates of the sites were recomputed by Decca Survey Limited.

At this stage the adjustment of fixed points was complete and these final co-ordinates are presented in Table 1. We needed to plot these points not only onto existing maps, but also onto available satellite imagery. Techniques developed for transferring these points and a Lambert grid onto the satellite imagery are described in the final part of this paper.

MAPPING ONTO SATELLITE IMAGERY

Whilst the expedition was making its traverses over Vatnajökull, much use was made of available satellite imagery for identification of features and planning of routes. The best image available at that time for this purpose was taken on 31st January 1973 by Landsat-1. In this, a sun's elevation of only 7° threw into relief many slight features of the ice-cap. However coverage was restricted by cloud to the western half of Vatnajökull. Later in 1977, a computer enhanced image, covering the whole of Vatnajökull, was received from the United States Geological Survey and from this enlargements at 1:250,000 and 1:500,000 were made.

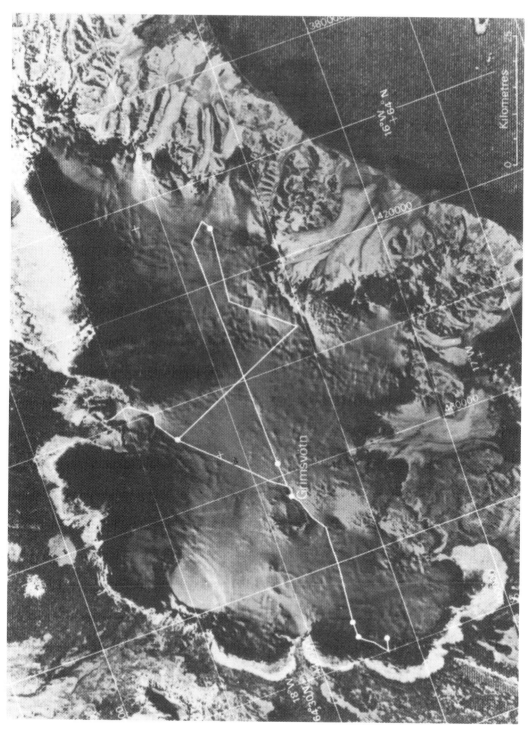

Fig. 4   Landsat-1 satellite photo of Vatnajökull from 920 km on 22 Sept.
1973.   Routes are shown and a 20 km Icelandic Lambert grid.

In order to transfer computed positions onto the enlarged image, it is first necessary to transpose the Icelandic grid in Lambert co-ordinates onto the image. No conventional projection fits these high level photographs which are rarely normal to the earth's surface along the axis of the camera. A Helmert transformation (5) of the type x' + iy' = f(x + iy), which has been used previously for this purpose in the polar regions, was adopted.

The method used enabled the grid to be plotted with great ease and accuracy. Firstly co-ordinated points in the Icelandic survey network were identified on the image and pricked through. Then the image was laid on a D-Mac digitising bed and the co-ordinates of these points measured to 0.1 mm relative to the bed. Without removing the image from the bed, these co-ordinates, together with their corresponding Lambert co-ordinates were fed into a computer program which computed a Helmert transformation between the two sets of points and then produced a list in terms of the digitising bed co-ordinates of points defining the Icelandic Lambert grid. These could then be transferred with an accuracy of 0.1 mm back onto the image. By interpolation between grid lines, all fixed points could be plotted directly onto the image and the antenna track related to natural features shown on the satellite image. The track and grid produced are shown in Fig. 4.

CONCLUSIONS

1. A satellite Doppler receiver, recording data on magnetic tapes, is a most rapid and secure means of fixing positions in remote regions.

2. This paper presents a grid transposed onto a satellite image of Vatnajökull from which future work and analysis can be based. Surface elevations provide control in determination of bedrock levels and will aid navigation on the ice-cap.

3. Recent developments based on the experience gained on Vatnajökull have provided a much lighter and more compact instrument based around the use of a microprocessor. This should lead the way to easier sounding of temperate ice sheets and the possibility of rapid measurements from aircraft or helicopters.

REFERENCES

1. Bishop, J.F., Cumming, A.D.G., Ferrari, R.L. and Miller, K.J. Cambridge University Vatnajökull Expedition, 1977. Polar Record 19 (1978) 51-54.

2. Miller, K.J. Under-ice volcanoes. Geog. J. 145 (1979) 36-55.

3. Bishop, J.F., Cumming, A.D.G., Ferrari, R.L. and Miller, K.J. Results of Impulse Radar Ice-Depth Sounding on the Vatnajökull Ice-Cap, Iceland. (These proceedings).

4. Böðvarsson, E.A. Greinargerd frá Landmaelingum Islands. Timarit Verkfraedingafelags Islands 2-3 Hefti 56 árg, 1971, 18-21.

5. Jordan, W., Otto, E. and Kneissl, M. Handbuch der Vermessungskunde. J.B. Metzlersche Verlagsbuchhandlung Stuttgart. Vol. II (1958) 66-73.

Electronic design and performance of an impulse radar ice-depth sounding system used on the Vatnajökull ice-cap, Iceland

A.D.G. Cumming, R.L. Ferrari, G. Owen

Department of Engineering, University of Cambridge, England

ABSTRACT

The design and performance of the equipment used to make an extensive impulse radar ice-depth survey on the Vatnajökull ice-cap in 1977 are described. The coupling between separate, resistively loaded dipoles, laid on the ice is analysed, leading to an estimate of 208 m/µs for the ice-air interface wave velocity. An "A"-mode run is analysed to yield a figure of 4.2 dB/100m for the attenuation of the radar pulse in temperate ice. The system overall performance is found to be 149 dB.

INTRODUCTION

Until recently, ice-depth sounding radar equipment has always taken a conventional form, using a pulse comprising of a sufficient number of high frequency carrier cycles to give a small fractional bandwidth about the central carrier frequency. Tuned circuit elements are exploited, resulting in relatively simple antenna design and electronic circuitry. A typical ice-depth sounding gear of this form might operate at 60 MHz with a 200 ns pulse length, comprising 12 cycles of the carrier. The velocity of electromagnetic waves through ice is 169 m/µs and such a device would give complete separation of the return from objects greater than 16.9m apart. However, the ice of temperate glaciers is electrically a very imperfect material and whereas the "conventional" ice radar apparatus is satisfactory in polar ice, it has failed to penetrate temperate ice of any substantial depth. Watts and England (1) observe that the imperfection of water-laden temperate ice becomes less apparent if frequencies considerably lower than 60 MHz can be used. To maintain even the coarse resolution quoted in the example above, lowering the frequency involves a reduction in the number of cycles per pulse to beyond the limit where the principles applying to a single-frequency carrier hold. Tuned circuits and resonant antennas can no longer be used. The radar pulse now carries a wide spectrum of frequencies and impulse radio echo techniques have to be employed. We give here details of the design and performance of the impulse radar apparatus which was the basis for the successful temperate glacier ice-depth measurements made on Vatnajökull described in an accompanying paper by Bishop et alia. Details of the routine operation of the apparatus are given in the latter. Here some of the specialised aspects of the performance are dealt with.

THE RECEIVING AND TRANSMITTING ANTENNAS

The basic antenna design used is the same as that described by Ferrari et alia (2). This is a resistively loaded linear array with parameters calculated from principles set down by Wu, Shen and King (3), (4) to give the wide bandwidth necessary to handle the unmodulated radar impulse. Essentially a centre-fed cylindrical dipole of length 2h and radius a is considered. The design problem is that of selecting suitable values of h and a and then calculating the resistive loading required by the theory of Wu, Shen and King to allow a travelling wave on the dipole without reflection from its extremities. The half length h was taken such that $hk_I = \pi/2$, where k_I represents the wave propagation number for quasi-plane TEM waves in the ice-air interface at the centre frequency value of the radar pulse spectrum. The design was worked out assuming the interface wave velocity to be $v_I = \omega/k_I = 225$ m/µs although subsequent measurements reported here suggest that 208 m/µs would have been a more accurate figure. An effective cylindrical radius a giving $k_I a = 0.021$ was assumed. In the light of the limited depth penetration of the 1976 apparatus described in (2) an antenna half-length h = 30.7m was chosen, approximately twice the length of the earlier apparatus. With $hk_I = \pi/2$ a central design frequency of $f = v_I/4h$ is implied, giving f = 1.83 MHz for $v_I = 225$ m/µs (or alternatively 1.69 MHz if $v_I = 208$ m/µs is taken). The antenna theory indicates that the resistively loaded array is broad band about the central frequency and thus suitable for accepting pulses with a wide frequency content spread about this central value. Because the array is necessarily open circuit to d.c., any pulse applied has net zero average current. A single period sinusoid effectively results from the voltage applied in the case here. Such a pulse of length T_o can be shown to have a maximum in its Fourier spectrum at the frequency $0.837/T_o$. Equating this to the centre frequencies obtained above leads to pulse lengths of 0.457 µs and 0.495 µs respectively. The receiving antenna response is an electric current which is scaled to the time derivation of the transmitter waveform, so that there is a different Fourier centre frequency. Thus the receiving antenna parameters should properly be a little different from those of the transmitting array associated with it. However, here the same parameters were used for both antennas.

The continuous resistive loading set by the above parameters has to be simulated by a finite number of steps. Five such steps per half antenna were taken, resulting in the total of 20 resistors shown deployed in Fig. 1. In the first instance the intention was to use the well-tried rule of thumb replacing the cylindrical geometry by parallel conductors spaced 4a apart, setting the lateral dimension given in Fig. 1.

However, the antenna design is not sensitive with respect to the breadth dimension a. There is much convenience to be gained if a single wire array can be used, although the theory is singular for a = 0. Furthermore great advantage is to be had if the receiver electronics can be housed within the shelter of the vehicle towing the antennas. The snow-vehicle becomes an electrical counterbalance behaving approximately like an earthed plane normal to the antenna axis. In the latter ideal, only a half-length unbalanced array is required. The transmitter is still conveniently placed

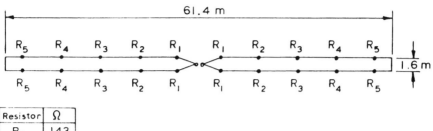

Resistor	Ω
R_1	143
R_2	184
R_3	257
R_4	428
R_5	1285

Fig. 1 Broad-Band Antenna

at the centre of a balanced array trailing yet some distance behind the receiver antenna. It was established during the 1976 expedition that the unbalanced receiver system could successfully be employed. Tests were carried out during the present work showing that the single wire antenna could also be used. The antenna configuration actually used thus corresponded to Fig. 1 with the lateral dimension closed to zero and the two parallel trains contained within a single, jointed plastic pipe. An optimum separation of 39.3m between the trailing end of the receiving array and the leading extremity of the transmitting antenna was established by experiment. The unbalanced terminal of the receiving array was connected to the display electronics in the "Weasel" snow vehicle by means of an 8.5m length of 50 Ω coaxial line. This was necessary to bypass the sledge which had always to be trailed immediately behind the Weasel.

THE TRANSMITTER

The transmitter was designed to produce a train of high voltage pulses across the antenna, each pulse being 0.5 μs in length. Its circuit is based on a delay line which is open circuited at one end and can be connected either to a high voltage supply or to its characteristic impedance at the other. The connection is effected with a transistorised switch as shown in Fig. 2. In operation, the line is first charged to a potential of 400 volts, then the switch is closed to discharge the line into the terminating network. A travelling wave of voltage then propagates down the line, is reflected at the open end and eventually absorbed by the matched terminating network. A rectangular voltage pulse is therefore coupled into the network which then transfers it to the antenna. The impedance of the antenna is fixed by the considerations in the Wu, Shen and King design (320 Ω here in fact) whereas the impedance required of the transmission line is determined by the voltage and current ratings of the transistorised switch. The matching network is then designed to couple these two impedances as efficiently as possible.

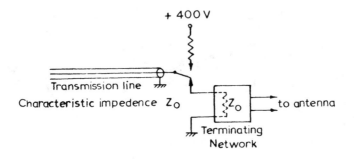

Fig. 2   Transmitter arrangement

THE RECEIVER

The receiver electronics has two functions.   Firstly, the signal from the
antenna is filtered to reduce electronic interference from external sources,
and gated to remove the primary transmitted pulse.   Secondly a trigger
pulse is generated to start the sweep of the display oscilloscope when the
primary pulse from the transmitter is received.   Thus two receivers are
used, a relatively insensitive one to generate the oscilloscope trigger pulse
from the primary received pulse, and a very sensitive one to amplify the
pulses reflected from the ice/rock interface.

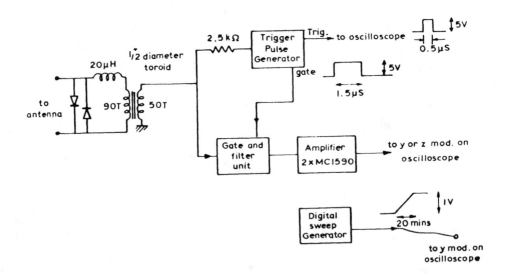

Fig. 3   Receiver block diagram

A block diagram of the receiver is shown in Fig. 3. The antenna feeds to a matching network which converts the impedence to 50 Ω, and removes any incoming signal which might be large enough to damage the electronics. The large primary received pulse then feeds into the trigger unit which gates it out of the delicate receiver circuit, and also triggers the oscilloscope trace. Secondary reflected signals are able to pass through the gate and into the filter. The filter is a sixth order Bessel bandpass type, with 3dB points at 0.5 MHz and 4.0 MHz. It effectively removes medium wave broadcast signals which lie outside the band and which can be a major source of interference, without introducing any "ringing" into the received pulses. The filtered signal is then amplified by the amplifier module, based on two cascaded MC1590 video amplifiers. This produces a signal of about 1 V peak amplitude, which is fed to the Y-plates of the oscilloscope for use in the A-mode. Alternatively this signal can be fed to the Z-modulation input for operation in the Z-mode. In the latter case, a slow generator is used to move the trace down the screen of the oscilloscope over a period of about thirty minutes. The sweep generator is based upon a ZN425E counter and a digital-to-analogue conversion circuit.

EXPERIMENT TO DETERMINE THE COUPLING CHARACTERISTICS BETWEEN TRANSMITTING AND RECEIVING ANTENNAS

In the apparatus used for the radio-echo sounding survey measurements the receiver and antenna were separate. The transmitter, at the centre of its balanced antenna, was some 100m distant from the receiver, located in the snow-vehicle at one end of an unbalanced array. The receiver and display electronics were triggered from the direct pulse picked up from the transmitter. Thus it is of interest to determine the coupling characteristics between the two systems, both amplitude and time-wise. Any theoretical treatment to predict such characteristics is difficult. The first order radiation field coupling between in-line antennae such as used here is zero. Determination of the delay between the two antennae requires knowledge of the characteristics of the quasi-TEM waves which propagate in a plane interface between two semi-infinite dielectrics. This problem is not to be found treated theoretically in the literature. In designing the antenna system before the experiment described here was undertaken, the ice-air interface wave velocity was guessed to be a value intermediate between the ice and air values.

Before the radio-echo sounder was dismantled prior to moving off the ice on 21st June, 1977, an experiment was performed in which the receiving antenna was moved successively in-line away from the stationary transmitter. Amplitude and delay characteristics were recorded with the in-line distance separating transmitter and receiver varying in 30m steps from 60m to 480m. The data recorded are given in Table 1.

The ice-air interface electromagnetic disturbance coupling transmitter and receiver may be assumed to be of damped spherical form:

$$V = \frac{A_1}{R} e^{-\alpha_1 R} \tag{1}$$

α_1 being the attenuation constant for plane waves in the ice-air interface.

This is confirmed by the plot of ln (VR) versus R in Fig. 4 to which a straight line of slope $\alpha_1 = 0.00169$ can be fitted. It corresponds to an

TABLE 1   Antenna coupling experiment data

Transmitter to receiver distance R m	Transmitter pulse amplitude at receiving antenna terminals V peak to trough millivolts	Delay between direct transmitter pulse and reflected pulse T μs
60	460	4.78
90	330	4.66
120	250	4.51
150	170	4.36
180	140	4.18
210	110	4.06
240	90	3.98
270	82	3.93
300	–	3.88
330	–	3.83
360	56	3.78
390	–	–
420	48	–
450	–	–
480	25	–

attenuation of 1.47dB/100m.   That this figure is considerably less than that for wave propagation in the bulk ice is not surprising.   A proportion of the signal energy connecting the receiver and transmitter directly is carried in the air above the ice, virtually without loss.

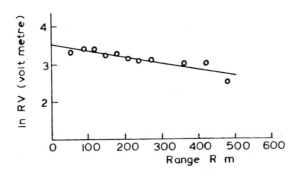

Fig. 4   Signal response.   Direct coupling
between transmitter and receiver.

The delay time T, observed at the receiver, between directly received signals and those reflected from the rock-ice interface may be analysed by considering Fig. 5.

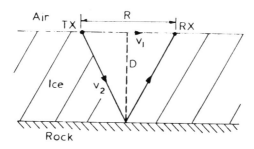

Fig. 5   Oblique reflection through ice from bedrock

It is seen that

$$\frac{T}{2} = \frac{[(R/2)^2 + D^2]^{\frac{1}{2}}}{v_2} - \frac{R/2}{v_1} \qquad (2)$$

where D is the depth of ice, R the distance between transmitter and receiver, v_2 is the velocity of TEM wave propagation through bulk ice and v_1 is the velocity at which electromagnetic energy propagates along the ice-air interface.   The parameters D, v_1 and v_2 corresponding to the experiment performed may be determined for a least-squares fit of expression (2) with the R and T data of Table 1. This calculation is readily done with a programmable pocket calculator and yields the values D = 426m, v_1 = 208 m/µs, v_2 = 170 m/µs.   Figure 6 shows the data compared with the theoretical curve of equation (1) fitted to it.   The dashed curves in the figure indicate the limiting relations between R and T assuming the ice-air interface velocity firstly to be 300 m/µs and finally to be the 170 m/µs predicted for bulk ice.

There is clearly some residual bias involved in associating the least-squares solid curve of Fig. 6 with the data.   A simple and plausible explanation is that the observed values of T corresponds to an ice-depth D which varies with range R.   Assuming that the delay time T corresponds to reflection of the electromagnetic wave energy from a point on the bedrock half-way between receiver and transmitter, the ice-depth D can be recalculated from the data as a function of R.   This results in the depth profile predicted in Fig. 7, D varying from 418m to 433m within a horizontal distance of 150m.

It may be noted that the figure of 170 m/µs for the bulk-ice wave velocity extracted from the data here agrees closely with the value 169 m/µs which is generally accepted and may be determined by more direct methods.

The bulk-ice relative dielectric constant is given by

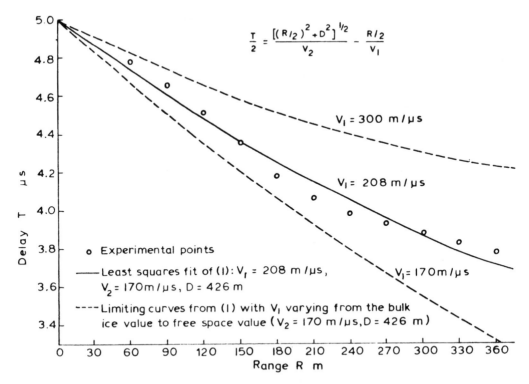

$$\frac{T}{2} = \frac{\left[\left(R/2\right)^2 + D^2\right]^{1/2}}{V_2} - \frac{R/2}{V_1}$$

$V_1 = 300\ m/\mu s$

$V_1 = 208\ m/\mu s$

o Experimental points

——Least squares fit of (1): $V_1 = 208\ m/\mu s$,
$V_2 = 170 m/\mu s$, D = 426 m

$V_1 = 170 m/\mu s$

- - - -Limiting curves from (1) with V_1 varying from the bulk
ice value to free space value ($V_2 = 170\ m/\mu s$, D = 426 m)

Range R m

Fig. 6   Delay T between reflected and direct pulses
with transmitter at range R.

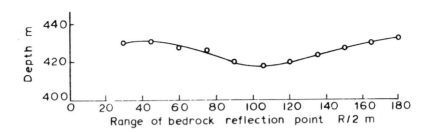

Range of bedrock reflection point   R/2 m

Fig. 7   Recalculated ice depth D from data,
using v_1 = 208 m/μs, v_2 = 170 m/μs.

$$\varepsilon_r = \frac{c^2}{v_2^2} \tag{3}$$

where c = 300 m/μs is the free-space TEM wave velocity. With v_2=169 m/μs
this yields ε_r = 3.151. It is plausible to assume that the interference wave

travels as if a TEM wave in a medium having relative dielectric constant the arithmetic mean between this value and $\varepsilon_r = 1$ for free space, viz. $\overline{\varepsilon}_r = 2.076$. Such an assumption yields the value $v_1 = 208.2$ m/µs, closely agreeing with the experimental figure obtained here of $v_1 = 208$ m/µ s.

THE "A"-MODE RECORD OF 20 JUNE 1977.

In normal operation the equipment was operated in its so-called "Z"-mode, the camera recording an intensity modulated trace swept across the oscilloscope screen at a constant rate. An "A"-mode record (x deflection being the echo delay time and the y deflection being signal amplitude) was only taken at the beginning and end of each 30 minute run, separated in distance by several kilometres. A section of the return track from Grimsfjall to the ice-edge base camp which had already been surveyed on the outward trip was selected for the purposes of taking more closely spaced "A"-mode records. The track began at Stake B8, continuing for some 5.5 km westwards, (see Fig. 3 of the accompanying paper by Bishop et alia) and corresponds to a region where the depth of ice increased from some 250m to greater than 600m. Recordings were taken every two minutes. On this occasion the terrain allowed speeds of at least 6 km/hour and it was attempted to hold this figure throughout the run. Thus, in general, the "A"-mode records were taken at distance intervals of about 200m.

TABLE 2   Data from the A-mode record of 20 June 1977

Distance along track km	Received echo delay T µs	Slant range $R_s = 169(T+0.48\ s)$ m	Ice-depth D m	Echo Signal amplitude peak-trough µV
0	2.91	573	282	627
0.210	2.72	541	266	804
0.406	2.57	515	253	1117
0.618	2.48	500	245	1090
1.044	3.14	612	302	981
1.302	3.44	662	327	409
1.440	3.44	662	327	409
1.640	3.61	691	342	381
1.862	3.81	725	359	150
2.264	4.62	862	428	99.80
2.48	4.92	913	454	62.30
3.040	5.62	1031	513	34.30
3.248	5.62	1031	513	37.40
3.460	5.62	1031	513	37.40
3.674	5.80	1061	528	74.80
3.882	5.38	990	493	21.80
4.082	5.20	960	477	37.40
4.226	5.78	1058	527	24.90
4.432	5.98	1092	544	37.40
4.860	6.22	1132	564	49.90
5.058	6.44	1169	583	34.30
5.264	6.80	1234	615	9.35
5.466	7.00	1264	630	9.35

The "A"-mode records were analysed to correlate the echo signal amplitude received (at constant transmitter power) with the ice-depth, as the equipment was moved over ice of steadily increasing thickness. The results are given in Table 2 and plotted in Fig. 8. It is clear that as well as the instrumental uncertainty in the signal amplitude, there is a considerable variation which must be ascribed to differences in reflectivity of the rock-ice interface and perhaps to inhomogeneity of the ice itself.

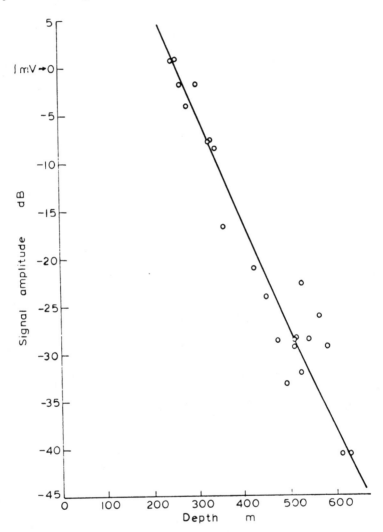

Fig. 8   Echo signal return to receiver

The depths D were calculated from the delay T as elsewhere in this paper, using equation (2), with the bulk-ice propagation velocity taken to be v_2 = 169 m/µs, the surface wave velocity as v_1 = 208 m/µs and for the transmitting and receiving antennae separated by 100.7m.

A damped spherical wave signal amplitude is expected once more. Here it may be written:

$$V = \frac{A}{R_s} \cdot e^{-\alpha R_s} \qquad (4)$$

where R_s is the total slant range from transmitter to receiver via reflection at the bedrock–ice interface. The receiver delay corresponding to the figures given above is 0.48 µs so that

$$R_s = 169(T + 0.48 \text{ µs}) \text{ m} \qquad (5)$$

From equation (4), we have

$$\ln (R_s V) = \ln A - \alpha R_s \qquad (6)$$

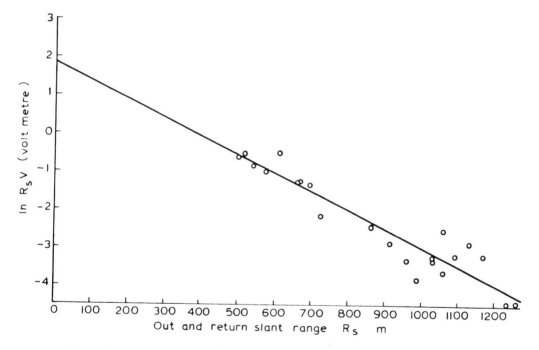

Fig. 9   Least-squares fit to the law of a linearly attenuated spherical wave.

Thus a plot of $\ln(R_s V)$ against R_s should be a straight line. Fig. 9 shows the experimental plot obtained and the least-squares linear regression straight line which was fitted by standard numerical methods. A slope $\alpha = 0.004845$ is found, corresponding to attenuation in the melting glacier ice of 4.21 dB/100 m. This figure may be compared with the results of Westphal (see Robin, Evans and Bailey, (5) and Robin (6)), which have been accepted hitherto as the best available, Westphal obtains, rather in-directly, a figure of 4.9 dB/100 m for melting glacier ice. We note that the value obtained in the work here has been obtained for glacier ice in situ. However attenuation figures are bound to be dependent to some extent upon the geological impurities present. These impurities may well be special in the volcanic Vatnajökull area.

The extrapolated values of ln(RV) at R = 0 for the direct and reflected signals (Figs. 4 and 9) are respectively 3.521 and 1.826 (natural logarithms relative to 1 volt metre). If this difference is entirely due to imperfect reflection at the bedrock-ice interface, then a voltage reflection coefficient of ρ = -0.1836 may be inferred for this plane. Simple theory for plane electromagnetic waves (see for instance Ferrari (7)) shows that the ratio of relative permittivities between ice and bedrock is given by

$$\frac{\varepsilon_{rock}}{\varepsilon_{ice}} = \left(\frac{1-\rho}{1+\rho}\right)^2 \tag{7}$$

Substituting in the figures here and assuming ε_{ice} = 3.151 gives ε_{rock} = 6.6. This value is, of course, an average figure for the 5.5 km of terrain traversed in the "A"-mode run, whereas Fig. 9 shows considerable fluctuations in the received signal compared with the linear regression. Some of this variation might well be ascribed to change in composition of the bedrock. A relative permittivity of 6.6 might be considered rather a low estimate for solid rock. However, the Vatnajökull bedrock may be inferred to be volcanic basalt, not at all a densely compacted material, explaining this low value.

THE TRANSMITTER PULSE

Figure 10 shows a record of the transmitter pulse developed in the balanced antenna under normal operating conditions, when the two parallel linear resistively loaded elements were closed together. The characteristic is of negative and positive current excursions of some 0.56 A approximating to a single sinusoid of duration about 0.6 μs. It is noted that the antenna was designed to accept such a waveform (but of duration 0.5 μs). A "ringing" of about 0.15 μs period, out to at least 2.5 μs is clearly discernable whilst it is observed that the time for radiation to travel the 30.7 m antenna half-length at 208 m/μs is 0.148 μs.

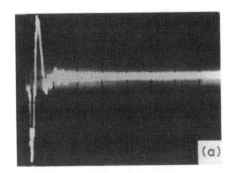 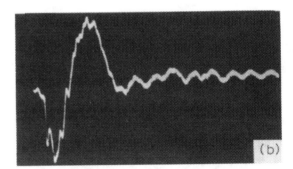

Fig. 10   Transmitter current pulse.   Vertical scale 0.2A per division.
Horizontal scale (a) 1 μs per division
(b) 0.2 μs per division

COMPARISON OF THE SINGLE LINEAR AND WIDE SPACED ARRAYS

Figure 11 shows an oscilloscope record of the pulse obtained in the balanced transmitter antenna when the parallel spacing of the two linear elements was restored to the 1.6m specified in Fig. 1. This record may be compared with that of the single linear array Fig. 10. The negative peak current appears to be somewhat greater than for the single array (0.81 A compared with 0.56 A) whilst the 0.15 μs ringing apparent in Fig. 10 is absent. Otherwise the single wire array appears to function satisfactorily and the observations recorded here justify the decision taken to exploit the logistic convenience of dispensing with the dual linear arrangement. A "fine" structure with periodicity about 42 ns is evident in both cases. This is probably associated with the fact that the measurements were made using equipment housed in the Weasel snow-vehicle which was drawn up alongside the antenna at its centre. This represented a substantial source of reflection only a few metres from the radiating array.

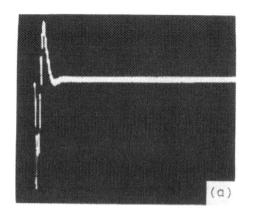 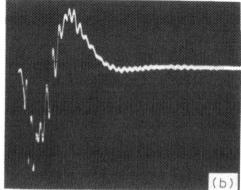

Fig. 11 Transmitter current pulse with wide-spaced (1.6m) parallel linear antenna elements. Scales as in Fig. 10.

THE ECHO WAVEFORM

Figures 12(a) and (b) show a typical received echo pulse, on the normal "A"-mode and expanded time scales respectively. Examination of the pulse in detail reveals that the two peak negative excursions are some 0.6 μs apart. This would correspond to maximum negative rates of change of transmitter current occurring at the transmitter time zero and 0.6 μs after this. Examination of Fig. 10 (b) confirms this. Figure 13 shows the received signal obtained when the interference suppressor, a 0.5 MHz to 4 MHz Bessel filter, was taken out of circuit. It illustrates the improvement to signal-to-noise ratio that is obtained by limiting the band of frequencies accepted to that most significant to the pulse employed.

THE SYSTEM OVERALL PERFORMANCE

An estimate of the minimum signal detectable in the apparatus used can be had from the data of Table 2. It corresponds to a signal excursion of some 4.67 μV going both positive and negative. Assume that a transmitter current of I_t is developed in an antenna of resistive impedance R_t and that a peak voltage V_r is developed from a matched receiving array of im-

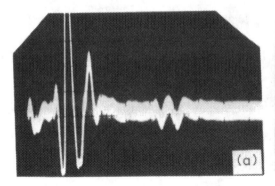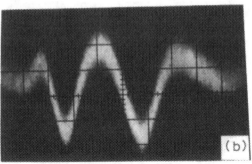

Fig. 12   Typical A-mode echo waveform.
Horizontal scale (a) 1 μs per division.
Vertical scales (a) 1 V per division
(b) 0.2 V per division

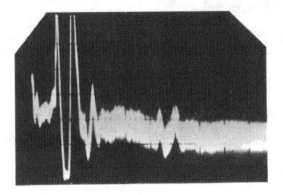

Fig. 13   The signal of Fig. 12 without interference suppression.

pedance R_r.   Then if we define a performance figure P as the ratio of the
transmitted and received peak powers we have

$$P = \frac{I_t^2 \, R_t R_r}{V_r^2} \qquad (8)$$

Here we have measured I_r and V_r and it is necessary to know R_t and R_r
in order to determine P.   The transmitter balanced antenna impedance was
designed to be 321 Ω, but it was not possible to measure this under operational conditions with the array lying on the ice.   The receiver array was
a half-length version of the transmitting antenna.   Reflected into a perfect
ground plane it would have half the latter's impedance, whilst the Weasel
snow-vehicle provided an approximation to the ground plane.   The receiver
antenna impedance was measured in the field.   The results revealed a complicated situation which was, however, not inconsistent with the design target

of 160 Ω. From equation (6) we have the performance figure in decibels to be

$$10 \log P = 20 \log \left[\frac{I_t R_t}{V_r} \right] - 3 \text{ dB} \qquad (9)$$

Substituting in the measured values for I_t and V_r and taking $R_t = 321 \Omega$ gives 149 dB for the overall system performance.

REFERENCES

(1)   Watts, R.D. and England, A.W.   Radio–echo sounding of temperate glaciers: Ice properties and sounder design criteria.   J.Glac.17, (1976), 39–49.

(2)   Ferrari, R.L., Miller, K.J., and Owen, G., The 1976 Cambridge-Reykjavik Universities Expedition to Vatnajökull, Iceland, Cambridge University Engineering Department Special Report 5, (1976).

(3)   Wu, T.T. and King, R.W.P.   The cylindrical antenna with non-reflecting resistive loading.   IEEE Trans. on Antennas and Propagation,   AP-13, (1965), 369–373.

(4)   Shen, L.C. and King, R.W.P.,   Correspondence, IEEE Trans. ibid, (1965), 998.

(5)   Robin, G. de Q., Evans, S. and Bailey, J.T.   Interpretation of radio–echo sounding in polar ice sheets.   Phil. Trans. Royal Soc. London,   A265, (1969), 437–505.

(6)   Robin, G. de Q.,   Radio–echo sounding: Glaciological Interpretations and Applications.   J. Glac., 15 (1975), 49–64.

(7)   Ferrari, R.L.   An introduction to electromagnetic fields.   Van Nostrand Reinhold, New York, (1975), 119.

Results of impulse radar ice-depth sounding
on the Vatnajökull ice-cap, Iceland

J.F. Bishop, A.D.G. Cumming,* R.L. Ferrari,* K.J. Miller.**

*Engineering Dept., University of Cambridge
**Faculty of Engineering, University of Sheffield

ABSTRACT

Impulse radar ice-depth soundings taken on the 1977 Cambridge University expedition to the Vatnajökull ice-cap, Iceland are described. The detailed results are presented for this first-ever major survey of a temperate glacier by radio-echo depth sounding. The data presented represent traverses approximately east-west and then north-south totalling some 295 km.

INTRODUCTION

The Vatnajökull ice-cap is the largest of the European glaciers, covering an area of some 8400 km^2 in south-east Iceland. There is much interest in obtaining detailed thickness measurements for this ice mass. It lies over a region of intense volcanic and geothermal activity whilst its melt water supplies .a substantial proportion of Icelandic hydro-electric power. Until recently only the very limited seismic ice-thickness data of Holtzscherer (1) and borehole results of Arnason et al (2) have been available. Radio echo sounding techniques have been used for some time to determine the thickness of polar ice covering Antarctica and Greenland. The early work is described by Robin et al (3) whilst Robin (4) reviews the more recent state of the art. However, Vatnajökull is a temperate glacier. The near-melting point temperature and substantial water content of such ice was observed by Ewen Smith and Evans (5) and Davis (6) to degrade its electrical properties to such an extent that equipment which works well sounding through several thousand metres of polar ice might not penetrate even a few hundred metres of a temperate glacier. Watts et al (7) reported the first successful radio echo sounding through temperate ice, on South Cascade Glacier, Washington. This was accomplished by using a centre frequency of 5 MHz, nearly an order of magnitude lower than had been used up to that time in radio-echo sounding work. Watts and England (8) published a paper throwing further light on the problem of radio-echo sounding through temperate ice. Following these reports, an equipment was designed and built in the Cambridge University Engineering Department and, with the co-operation of the University of Reykjavik and the Iceland Glaciological Society, tested on Vatnajökull during June/July, 1976. This work, in which a maximum penetration through 400m of temperate ice was attained with only a limited facility for making continuous recording, is described by Ferrari et al (9) and Björnsson et al

(10). As a result of the 1976 tests it was possible to design and build an improved radio-echo sounder in Cambridge during the year 1976/77. The apparatus was taken to Iceland in May 1977 and successfully put into operation during survey traverses of the Vatnajökull ice-cap totalling 295km carried out between 31st May and 22nd June 1977. The greatest ice depth recorded was 930 m, although it is likely that greater thicknesses occur on the route followed and elsewhere on Vatnajökull. This survey expedition has been described briefly by Bishop et al (11) and is the subject of the more detailed report here.

The members of the expedition, with their Cambridge affiliation, were:

K.J. Miller of Cambridge University Engineering Department, (C.U.E.D.) and Trinity College. Leader.

J.F. Bishop of British Antarctic Survey and St. John's College.

S. Chandler of C.U.E.D.

A.D.G. Cumming of C.U.E.D. and Emmanuel College.

R.L. Ferrari of C.U.E.D. and Trinity College.

C.M. Jonscher of Trinity College; and finally

G. Owen of C.U.E.D. and Trinity College, who worked with the team beforehand, designing the electronic equipment.

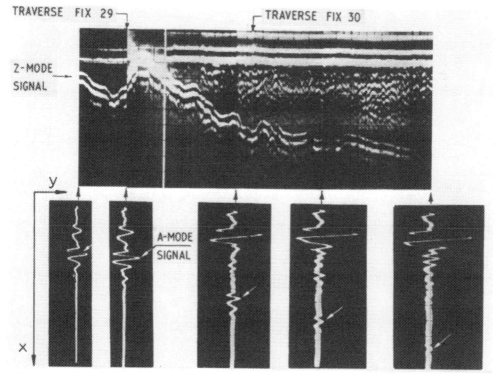

Fig. 1 Part of the Record of 17 June 1977, including Traverse Fixes 29 & 30. A-Mode Recordings taken at intermediate halts are included.

THE REGION SURVEYED

The Vatnajökull ice-cap is shown in relation to the rest of Iceland in Fig 1 of a companion paper (12) along with (their Fig. 2) a more detailed outline showing the expedition's track and some of the salient topographical features of the glacier. The approach to Vatnajökull on this occasion was from the west, using the rough-road route leading to the Jökulheimer glaciological hut some 3 km from the ice-edge. From here the glacier melt-stream is negotiable whilst the shallow, relatively lightly crevassed slopes of Tungnaárjökull allow convenient access onto the ice. The expedition moved over the ice-cap by means of a tracked snow-vehicle (the "Weasel") trailing a sledge loaded with two tons of stores and accompanied by a lighter "snow-mobile". The first portions of the track shown in Fig. 2 of reference (12) represent the expedition's approach over the well-tried route (of the Iceland Glaciological Society) to the Grimsfjall hut overlooking the Grimsvötn caldera.

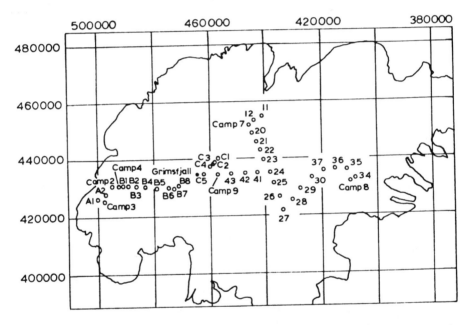

Fig. 2   Plot of Traverse Fixes (listed in Table 1 of Reference 9)

Camp 1 shows the point where progress was initially held up through the Weasel and sledge becoming bogged down in slush, whilst Camp 2 on higher ground is the point where the radio-echo sounding gear was first tested. In order to be able to check the apparatus over a known depth of ice a detour to camp 3 was made, a position where depth soundings had been made by the previous year's expedition. After Camp 3, continuous depth soundings were attempted whenever the expedition was in transit. From Grimsfjall (Camp 5) the expedition proceeded north-east to Kverkfjöll on the northern edge of the ice-cap, a geothermal area and where a hut (Camp 6) had recently been erected. The party retraced its track from Kverkfjöll in poor visibility until a suitable point (Camp 7) had been reached from where to strike out to the south-east. There, whilst waiting for clearer weather, the opportunity was taken to make test measurements on the antenna using laboratory equipment specially brought along for this purpose. On 17th June, in ideal weather, the expedition traversed three successive long, straight tracks, the first towards Esjufjäll, taking the party to its most southerly point on the glacier, then the north-west again until roughly the central latitude of the ice-cap had been reached and finally west to approach Breidabunga, the principal ice-dome in the eastern half of Vatna-jökull. The greatest ice-depth measurement made during the expedition, 930 m, was recorded during the first leg of this traverse between Camps 7 and 8. The next day, also in excellent weather, the ascent of Breidabunga was completed and a route heading back towards Grimsvötn taken. After the difficult ascent back to the Grimsfjall hut, the expedition's outward tracks, still mostly visible, were followed to take the party back to glacier-edge Base Camp. A final halt on the glacier at Camp 11 was made. It was here that the survey measurements finished, the unwieldy echo-sounding antenna being dismantled in order to allow safer manoeuvring over the by now (in the late season) extensively crevassed lower slopes of the glacier. However, before packing up the antenna arrays another session was occupied making more fundamental measurements on the electrical apparatus.

Throughout the expedition's traverse, navigation was carried out using the 1:250,000 survey maps of the Geodetic Institute, Copenhagen (published by Landmaelingar Islands, 1972 revision) sheets 5, 6, 8 and 9 which together cover the whole of Vatnajökull. Additionally the topographical information provided by the ERTS-1 satellite record of the Vatnajökull area on 31st January 1973 was taken into account in planning and executing the expedition's route, (see Fig. 5 of reference (12)). In clear weather it was always possible to discern geographical features external to the glacier, so that positions could be fixed with respect to the 1:250,000 map and ERTS-1 satellite record by means of theodolite and compass observations. A continuous record of barometer readings was kept to provide height information. At selected points along the route, accurate geodetic fixes of height and position were made using a JMR-1 satellite survey apparatus*, although the results of such observations became available only from computer analyses carried out after the expedition had returned home to the U.K. The general routine for setting the expedition's track was for two members to go ahead with theodolite, JMR and compass, using a snowmobile, marking out a safe and suitable track for the Weasel and sledge. The electronic receiving equipment was mounted in the Weasel and required three persons for operation. The radio-echo sounding results were correlated with position and height on the glacier by means of odometer and barometer readings taken every two minutes, the Weasel, sledge and antenna train proceeding on a constant bearing at typically 7 km/hr. Such readings were tied into the basic grid of points located by theodolite and JMR measurements. Extensive details of the surveys and subsequent analysis are given in the companion paper, reference (12), presented at this conference. Some more details of the nature of the project are also reported by Miller (13).

THE SURVEY RESULTS

The impulse radar ice-depth sounding techniques used in the work described here consisted essentially of measuring the time elapsed between initiation of a transmitted electrical impulse and its reception after reflection at the rock-ice interface immediately below, assuming an electromagnetic wave velocity of 169 m/μs in ice. Separate transmitting and receiving antennas were employed. These were necessarily long and unwieldy, having a total length of some 140 m trailing behind the Weasel snow-vehicle. The received signal voltage waveform corresponded roughly to 1½ cycles of a sine function extending over some 1.2 μs. Two possible modes were available for dis-

*Kindly loaned by Decca Survey Limited, Kingston Road, Leatherhead, Surrey

playing the rock-bed echo; see Fig. 1 of this paper. The A-mode record shows the signal amplitude (y) at a given location versus time (x). The Z-mode record shows the same signal, recorded along the oscilloscope x-direction, as an intensity (z) modulation, the y oscilloscope position being swept at a constant time rate as the radar equipment was hauled over the ice at constant speed. This gives a record corresponding to the rock-bed profile relative to the surface of the ice. Fig. 1 shows a typical set of records obtained by means of a polaroid film camera. The bright-dark-bright bands corresponding to the received signal can be seen. The true ice depth corresponds to the leading edge of these bands relative to the leading edge of the transmitter pulse which initiated the oscilloscope trace. (The large instrumental "ringing" signal near the trace x-origin must be ignored.) It was possible to determine the time elapse between these two instants to an accuracy of about ±0.05 μs corresponding to an ice-depth precision of some ±7 m. Details of the electronics used are given in an accompanying paper; see reference (14).

The collection of data amassed on polaroid photographs and in log books was analysed as described in references (12) and (15) and the results are presented here. The original somewhat arbitrary identification symbols for the route markers have been retained. Table 1 of reference (12) lists the Lambert co-ordinates (16) and heights established for these traverse fixes or "stakes". The marker positions are plotted on the Lambert grid of Fig. 2 of this paper, which also shows the glacier outline. Profiles representing the ice-surface and bed-rock elevations along the track joining the plotted points of Fig. 2 are shown in Figs 3-7.

DISCUSSION OF RESULTS

Figure 3 represents the first successful radio-echo sounding traverse after the equipment had been set up at Camp 3, a point near the 1976 expedition's static camp. Approaching Camp 4, the bedrock profile shows ridges and a trough which are considered to be the subglacial continuation of the north-east orientated surface features to be seen at F 2n of sheet 6 of the 1: 250,000 survey maps. Figure 4 shows more of this feature on leaving Camp 4, then the steady climb both of the surface ice and bedrock as Stake B8 and the Grimsfjall peak were approached.

Figure 5 shows the track from Grimsfjall to Kverkfjöll. The intermittent record up to some 12 km from the start is a result of technical difficulties with the apparatus. The sharp "precipice" of some 340 m at 10 km was a very definite feature which can be associated with a continuation of the volcanic caldera edge constituting the three Sviahnukar (Grimsfjall) peaks. However the gap left in the record between 15 and 21 km represents a situation where with the apparatus working well, the echo returns nevertheless became exceedingly confused. The 6 km-wide region corresponding to this confused pattern is, in fact, to the lee (for southwesterly winds) of the active volcano under the ice at Grimsvötn. Until 1936 it had erupted regularly every 10 years, so that large amounts of volcanic ash might be expected to lie in the ice. Thus the confused pattern of echo returns is tentatively ascribed to the presence of ash-laden ice. The gaps in the record of Fig. 6 from Camp 7 to Stake 27 also represent similar cases of confused echoes in regions where the ice-depth was probably no deeper than at other places where perfectly good returns were obtained. Again, the region is to the lee of the volcano. The greatest ice-depth recorded of some 930 m was obtained on this leg, between Stakes 24 and 25. Figure 7 includes soundings made at the summit of the ice-dome Breidabunga, where 150 m of ice cover was found.

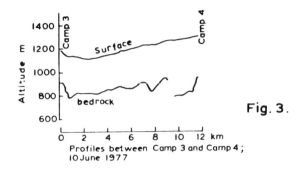

Fig. 3.

Profiles between Camp 3 and Camp 4;
10 June 1977

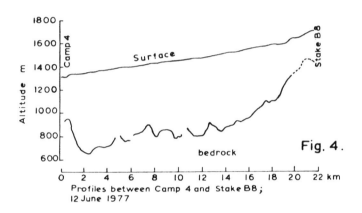

Fig. 4.

Profiles between Camp 4 and Stake B8;
12 June 1977

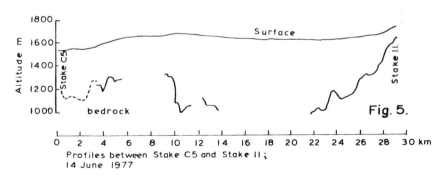

Fig. 5.

Profiles between Stake C5 and Stake II;
14 June 1977

Figs. 3, 4, 5   Profiles of Surface and Bedrock
Vatnajökull, 10–14 June, 1977.

ACKNOWLEDGEMENTS

The authors wish to acknowledge financial support from the Royal Society of London and the Science Institute, University of Reykjavik (1976), the Natural Environment Research Council (1977) and Trinity College, Cambridge (1976 and 1977) enabling them to carry out the work described here. They are grateful for the co-operation of the Department of Engineering, University of Cambridge and the Science Institute, University of Reykjavik. Both

132

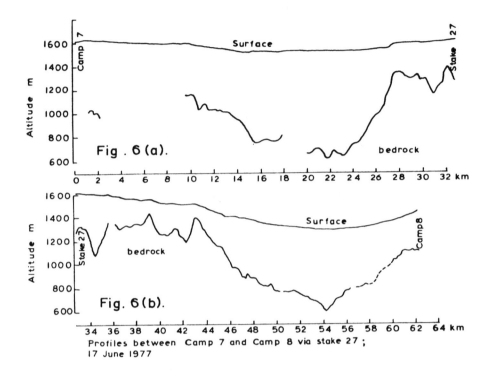

Fig. 6 (a).

Fig. 6 (b).

Profiles between Camp 7 and Camp 8 via stake 27;
17 June 1977

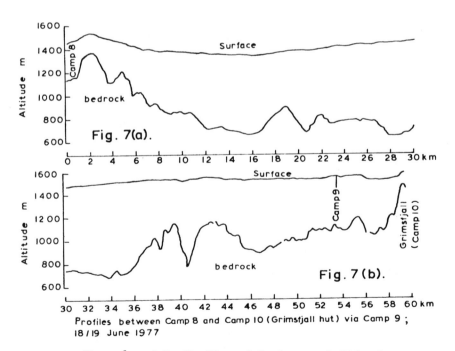

Fig. 7(a).

Fig. 7 (b).

Profiles between Camp 8 and Camp 10 (Grimstjall hut) via Camp 9;
18/19 June 1977

Figs. 6 and 7 Profiles of Surface and Bedrock
Vatnajökull, 17–19 June, 1977.

establishments provided facilities and personnel contributing substantially to the success of the expedition. The authors extend their thanks to the many other organisations, too numerous to name here, which helped the expedition in one way or another.

REFERENCES

(1) Holtzscherer, J.J. Expedition Franco-Islandaise au Vatnajökull mars-avril 1951. Resultats des sondages seismiques. Jökull (Iceland Glaciological Society). 4 (1954), 1–33.

(2) Arnason, B., Björnsson, H. and Theodorsson, P. Mechanical drill for deep coring in temperate ice. J. Glac. 13 (1974), 133.

(3) Robin, G. de Q., Evans, S. and Bailey, J.T. Interpretation of radio-echo sounding in polar ice sheets. Phil. Trans. Royal Society London, A265 (1969), 437–505.

(4) Robin, G. de Q., Radio-echo sounding: Glaciological interpretations and applications. J. Glac. 15 (1975), 49–64.

(5) Ewen Smith, B.M. and Evans, S. Radio Echo sounding: Absorption and scattering by water inclusions and ice lenses. J. Glac. 11 (1972), 133–146.

(6) Davis, J.L. The problem of depth sounding temperate glaciers. M.Sc. Thesis, University of Cambridge, England, (1973).

(7) Watts, R.D., England, A.W., Vickers, R.S. and Meier, M.F. Radio-echo sounding on South Cascade Glacier, Washington, using a long-wave-length, monopole source. J. Glac. 15 (1975), 459.

(8) Watts, R.D. and England, A.W. Radio-echo sounding of temperate glaciers: Ice properties and sounder design criteria. J. Glac. 17 (1976) 39–49.

(9) Ferrari, R.L., Miller, K.J. and Owen, G. The 1976 Cambridge-Reykjavik Universities Expedition to Vatnajökull, Iceland. Cambridge University Engineering Department Special Report 5 (1976).

(10) Björnsson, H., Ferrari, R.L., Miller, K.J. and Owen, G. A 1976 radio-echo sounding expedition to the Vatnajökull ice-cap, Iceland. Polar Record 18 (1977), 375–377.

(11) Bishop, J.F., Cumming, A.D.G., Ferrari, R.L. and Miller, K.J. Cambridge University Vatnajökull Expedition, 1977. Polar Record 19 (1978), 51–54.

(12) Bishop, J.F. and Miller, K.J. Survey systems for the Vatnajökull ice depth sounding expedition 1977. (These proceedings).

(13) Miller, K.J. Under-ice volcanoes. Geogr. J. 145 (1979), 36–55.

(14) Cumming, A.D.G., Ferrari, R.L. and Owen, G. Electronic design and performance of an impulse radar ice-depth sounding system on the Vatnajökull ice-cap, Iceland. (These proceedings).

(15) Bishop, J.F., Cumming, A.D.G., Ferrari, R.L., Miller, K.J., Owen, G.
The 1977 Cambridge University Expedition to Vatnajökull, Iceland.
C.U.E.D. Special Report 6 (1979).

(16) Bomford, G. Geodesy. 3rd Ed., Clarendon Press, Oxford (1971), 212–215

The mechanics of fracture applied to ice

K.J. Miller

University of Sheffield, England

ABSTRACT

In order to assess the safety and integrity of engineering plant, the design engineer has to assume that defects exist in the materials used and/or are created during construction and operation. Assessment of structural relia- bility from fatigue failure is but one aspect of the requirement to under- stand the behaviour of cracked bodies.

This paper reviews current parameters used by engineers to predict failure conditions of engineering constructions. Their relevance to the fracture of ice is then outlined from both a theoretical and an experimental view- point. Some recent results on the fracture of laboratory-prepared bubble- free ice are presented.

INTRODUCTION

The design of safe structures in the aeronautical, chemical, and nuclear engineering industries today requires a theoretical and experimental know- ledge of the mechanics of fracture. Calculations can be made, for example, for allowable stress limits of detectable defects or fatigue lifetime predic- tions for growing but sometimes unobservable defects. Indeed all structures are today assessed on the basis that they inherently contain defects. Such defects may only be parts of a millimetre in length prior to fracture. Some of nature's largest defects occur in ice crevasses. This paper focuses attention to the application of engineering mechanics to the study of frac- ture of ice and was stimulated from the author's unintentional confined study of numerous crevasses and an urgent desire to have information con- cerning their depth and distribution. (1)

FRACTURE MECHANICS

1. Linear Elastic Fracture Mechanics (LEFM)

If materials behave in a purely elastic manner Griffith (2) showed that unstable crack propagation takes place when an increment of crack growth

136

occurs which results in more stored energy being released than is absorbed by the creation of the new crack surfaces. However, in real materials considerable plastic and/or time-dependent deformation can take place which presents great difficulties in evaluating the energy release term. Nevertheless, Griffiths's equation

$$\sigma_{crit} = \sqrt{\frac{2E\gamma}{\pi a}} \tag{1}$$

laid the basis of fracture mechanics.

Here σ_{crit} is the critical stress for unstable fracture, E is Young's modulus of elasticity, γ is surface energy and a is crack length. Griffith showed that the product of σ_{crit} and $\sqrt{a}$ was constant for a given material, and so it is possible to analyse large scale structures by performing tests on smaller models in a laboratory. He also showed that stresses σ_x, see Fig 1, had no effect on σ_{crit} $\sqrt{a}$. Note that central cracks in wide plates have a length 2a to correspond with edge cracks of length a in equivalent half plates. Irwin (3) and Orowan (4) suggested that (i) γ_p should replace γ in equation (1) since γ_p, the energy absorbed by plastic deformation, can

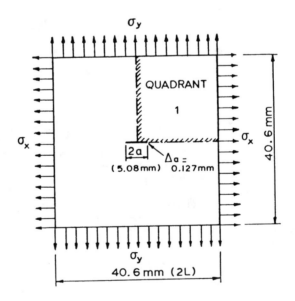

Fig. 1. Configuration used for Elastic-Plastic Analyses of Crack Growth in Biaxially Stressed Plates.

be three orders of magnitude greater than γ and (ii) provided the plastic zone is small in comparison to the crack length and body dimensions the energy released could still be calculated from elastic analyses with sufficient accuracy. Irwin and Kies (5) then proposed that if the fracture process was similar for different specimens and loading conditions, fracture occurred when the strain energy release rate attained a critical value, G_c, a material property; here

$$G = \frac{\partial U}{\partial a} = \frac{F^2}{2} \frac{\partial C}{\partial a} \qquad (2)$$

where U is the strain energy in an elastic body, F = force and C = compliance. Whilst this approach has much appeal, the evaluation of C is frequently difficult in complex shaped cracked components with complex stress-strain distributions. An alternative approach developed by Irwin (6) was to consider the spatial stress distribution at the tip of a crack. Solutions show that

$$\sigma = \frac{K}{\sqrt{2\pi r}} \cdot f(\theta) + \ldots \text{ series} \qquad (3)$$

where r and θ are polar co-ordinates with the origin at the crack tip. Thus the stress intensity factor K helps describe the mechanical environment of the crack tip and is equivalent to G, i.e.

$$GE = K^2 \text{ for plane stress conditions}$$

$$GE = K^2 (1-\nu^2) \text{ for plane strain conditions} \qquad (4)$$

where ν is Poisson's ratio. There are three modes of extending a crack (7) see Fig. 2, but here we shall confine ourselves to the tensile opening mode, designated I, and the critical value of the Mode I stress intensity factor K_{1c}. The value of K is dependent on loading conditions and specimen shape

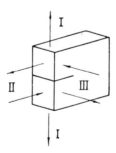

Fig. 2. The Three Modes of Crack Growth.

and these effects are embodied in the Y term of the K calibration, i.e.

$$K = Y\sigma \sqrt{\pi a} \qquad (5)$$

For the specimen shown in Fig. 3a

$$K = \frac{P\sqrt{a}}{BW} \left[29.6 - 185.5 \left(\frac{a}{W}\right) + 655.7 \left(\frac{a}{W}\right)^2 - 1017 \left(\frac{a}{W}\right)^3 + 638.9 \left(\frac{a}{W}\right)^4 \right] \qquad (6)$$

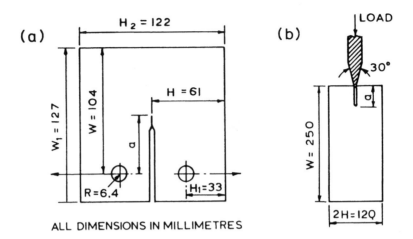

ALL DIMENSIONS IN MILLIMETRES

Fig. 3   Typical Fracture Toughness Specimens.
(a) Compact Tension Specimen and
(b) Wedge Opening Specimen for Crack Arrest Studies

The K calibration for the specimen illustrated in Fig. 3b is given in Ref.(8).
In equation (6), P = applied force and B = specimen thickness.   Unlike G
the parameter K is relatively easy to determine and now may be obtained
from reference books, e.g. (9).

2.   Elastic-Plastic Fracture Mechanics (EPFM)

The difference between LEFM and elastic-plastic fracture mechanics is difficult to define because even brittle-like materials, such as glass, exhibit
some degree of permanent deformation at the crack tip prior to fracture.
Indeed it was Liu (10) who pointed out that it is not the elastic stress and
strains outside the plastic zone which cause fracture, but rather, the mechanical state within the plastic zone.   Therefore, as in LEFM, an EPFM single
parameter characterisation of fracture was next attempted.   This was required
because LEFM was too conservative since it could not quantify the energy
absorbed at crack tips that required an increase in stress before criticality
occurred.   The first attempt by Dugdale (11) and Barenblatt (12) considered
a small plastic strip zone developing ahead of the crack tip thus allowing
the crack tip to blunt by an extent later termed the crack tip opening displacement CTOD (or δ) by Wells (13) who proposed that fracture occurred
at a critical value, δ_c.   Burdekin and Stone (14) then evaluated CTOD as

$$\delta = \frac{8\sigma}{\pi} y \frac{a}{E} \log \sec \left(\frac{\pi\sigma}{2\sigma_y} \right) \tag{7}$$

and so after expanding the log sec term, it was found that

$$\delta = \frac{\pi\sigma^2 a}{E\sigma_y} \left[1 + \frac{\pi^2}{24} \left(\frac{\sigma}{\sigma_y} \right)^2 + \cdots \right] \tag{8}$$

It follows that $G = \alpha\delta\sigma_y$, with $\alpha = 1$ to 2.

Whilst the CTOD approach is a valuable attempt to realistically quantify events at the crack tip it is difficult to apply in design situations for ductile materials. This is why the K approach is still often preferred. Nevertheless, the K approach requires modification to permit non-linear deformation at the crack tip. Thus the J contour integral was used by Rice (15) where

$$J = \int_\Gamma \left(wdy - T_i \frac{du_i}{dx} \right) ds \qquad (9)$$

Here J is a contour from the lower crack face anticlockwise around the crack tip to the upper face, whilst s is the path along the contour and w is the strain energy density of the form $\int_0^e \sigma_{ij} d\epsilon_{ij}$. Finally $T_i du_i$ are work terms when components of surface tractions T_i (on the contour path) move through displacements du_i. The benefits of the J approach are (i) the result is independent of the path when the crack faces are stress free and so the path can be chosen to be remote from the crack tip, (ii) non-linear behaviour of crack tip material can be accounted for albeit that the stress and strain functions derived by Hutchinson (16) and Rice and Rosengren (17) are not based on incremental plasticity considerations. It should be noted that the contour integral had previously been developed separately by Eshelby (18) and Cherepanov (19). The approach is conceptually the same as that of Griffith and equivalence is maintained by G, J, and K^2/E.

A further development was to consider the growth of cracks due to creep. Some researchers adopted a K approach whilst others claimed better correlation with the nett section stress across the ligament or the use of a reference stress. Understandably the first approach is suitable for brittle materials whilst the second approach favours ductile materials. Recently some experiments (20, 21, 22) have attempted to introduce non-linear behaviour, characterised by J, to account for creep crack growth. The creep equivalent of J has been called C*.

Thus if J equals $-\frac{1}{B}\frac{dU}{da}$ where B is the plate thickness (crack width) and U is potential energy, then

$$C^* = \frac{1}{B} \cdot \frac{d\dot{U}}{da} \qquad (10)$$

This approach is now being actively pursued but will probably suffer the same limitations as the J approach and, like LEFM, will be geometry dependent. It should be noted that none of the parameters yet mentioned consider growth, only initiation and none take true plasticity into consideration.

3. A Crack Growth Parameter

All the previous parameters do not consider real material events at the instant of crack growth and what occurs immediately after the start of growth. With reference to Fig. 4, if the crack growth step Δa is much greater than the plastic zone size, r_p, which is an alternative way of saying the latter term is exceedingly small, then the growth step is adequately characterised

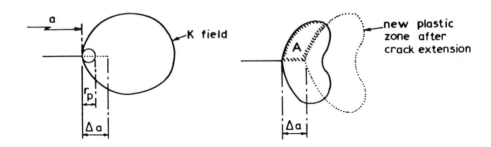

Fig. 4 Crack Tip Zones

by the elastic stress field and LEFM applies. If Δa is smaller than r_p then LEFM does not apply and growth must be determined from EPFM parameters. In this case, immediately crack growth commences, elastic recovery takes place in zone A which helps to close the crack which thus requires more force to open it for propagation to continue. It is this energy redistribution within the fracture process zone Δa that provides the complication. At once an anomaly is explained. In a conventional elastic energy balance approach $\partial w/\partial a$ (=G) is defined as the limit of $\delta W/\delta a$ as δa tends to zero. The energy balance is therefore calculated over a distance δa which is always infinitely smaller than r_p, however small the latter. Thus it is not surprising that Rice (23) showed the fracture separation energy cannot be accounted for from a continuum mechanics energy balance. Thus there is a requirement for a process fracture zone Δa which requires ΔW, the work absorbed during separation of the crack surfaces over the distance Δa.

The growth parameter G^{Δ} is defined as ΔW/Δa and if G^{Δ} exceeds the cohesive strength of the material extension will occur even if $\partial W/\partial a$ is zero. To appreciate the significance of these events, an elastic-plastic finite element analysis was performed on the plate shown in Fig. 1 (24). Different states of biaxiality were introduced since K and possibly J cannot account for any biaxial effect at the singularity. The effect of biaxiality is shown in Fig. 5. For $\lambda = 0$ (uniaxial loading ; $\sigma_x = 0$) an intermediate plastic zone is created in comparison with $\lambda = +1$ (equi-biaxial tension) and $\lambda = -1$ (shear loading); the latter producing a very large plastic zone. In these cases K is nominally constant. In Fig. 5 A is a crack-tip plastic zone size factor which is usually related to the applied load and yield stress by

$$r_p = A \left(\frac{K_1}{\sigma_y} \right)^2 \tag{11}$$

The value of A is assumed to be 0.175 (25) but it is seen to be a function of load, ψ, and biaxiality except for the case of the smallest plastic zone ($\lambda = +1$). Here $\psi = G/G_o$ where G = Griffith's energy release rate and G_o is the value of G at incipient yielding (26).

Figure 6 shows what happens when plasticity occurs in relation to purely

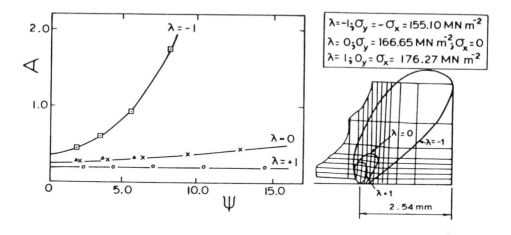

Fig. 5 Plastic Zone Sizes

elastic behaviour at the crack tip.   It is obvious that as the crack begins
to propagate, there is a partial self-healing effect that helps close the crack
and so a larger force is required to maintain a constant CTOD.   Meanwhile
the plastic zone acts not only as an energy sink limiting the amount of
energy required for crack separation, but also blunts the crack and causes
a stress redistribution in the crack tip zone.   Fig. 6b indicates that under
elastic-plastic conditions there is a lack of linearity in the nodal force
vs node displacement characteristic from which ΔW and hence G^Δ $(=\Delta W/\Delta a)$
is determined.

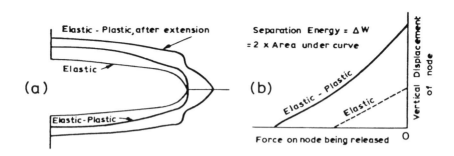

Fig. 6   Elastic and Elastic-Plastic Effects on Crack Tip Shape

Figure 7 shows the increasing divergence of parameters J and G^Δ as load
increases (27).   Note the tendency of G^Δ to zero for large loads (greater
crack tip plasticity) as Rice predicted (23).   Using the Dugdale-Barenblatt
model, Kfouri and Rice (28) established the relationship between G^Δ and
J uniaxial curves but this relationship is also a function of stress biaxiality

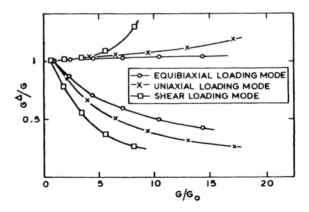

Fig. 7   A Comparison of J and G^Δ Parameters for
Different Biaxial Stress States.

since J does not take plastic zone size into account. Finally, Fig. 8 shows
that if the ratio of process zone size to plastic zone size is (a) constant,
then G^Δ is constant irrespective of state of stress biaxiality in the plate
of Fig. 1 and (b) when r_p is small or Δa large, then G^Δ = G (=J) and LEFM
applies.   Using this approach, Kfouri and Miller (24) accurately predicted
the variation of K_{1c} with temperature for three materials simply by accounting
for variations in plastic zone size as temperature increased.   It should be
noted that since r_p is a function of both the mode 1 opening stress and

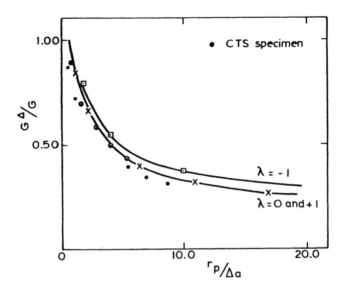

Fig. 8   Normalization of G^Δ with $r_p/\Delta a$.

the state of biaxiality λ a single parameter characterization for fracture is not possible. Similarly, in multiaxial fatigue studies, Miller (29) and co-workers have shown that in complex stress situations it is necessary to describe fracture by two parameters, the maximum shear strain (or stress) and the strain normal to the plane of maximum shear strain. The first controls the direction of crack growth and the second affects the speed of growth.

To conclude it should be recorded that G^Δ has also been used successfully to explain the reduction of fracture toughness in steels due to hydrogen embrittlement (30) and is now being used to quantify fatigue crack growth.

FRACTURE TOUGHNESS OF ICE

Very little work has been published to date concerning the fracture toughness of ice. Liu and Miller (8) reported some 63 CTS tests on bubble-free fresh-water ice at temperatures in the range -1 to -46°C at loading rates in the range 0.5 to 480 mm/min. Eleven other tests were also reported concerning the effect of speed of freezing of samples, crack arrest and environmental effects. Goodman and Tabor (31) reported 5 fracture toughness calculations from indentation tests and 3 point bend tests. Gold (32) also deduced the toughness characteristics of ice under thermal shock conditions. These latter results are difficult to interpret if fracture toughness is related to crack tip opening events since maximum COD does not necessarily occur at the maximum K value due to yield variations with temperature throughout a cycle; an identical problem to thermal shock studies of gas turbine blades. More than 100 tests have recently been conducted on glacier ice samples and these results are published in a companion paper (33).

The application of fracture mechanics theory to the problem of crevasse penetration was first introduced by Smith (34) who assumed that K_{1c} was negligibly small. The stress intensity factor was a superposition of three effects namely the longitudinal stress in the glacier σ_t, the overburden pressure and the wedge effect due to water in the crevasse. From the general equation,

$$K = 1.12 \sigma_t \sqrt{a} + 0.683 \rho_w g d^{3/2} - 0.683 \rho_i g a^{3/2} \tag{12}$$

where a is the depth of the crevasse, d the depth of water, and ρ_i , ρ_w the density of ice and water respectively, he simply deduced the depth of dry crevasses as 36m as well as the effects of crevasse spacing and water filling.

The problems of testing ice samples are many, especially at temperatures close to melting conditions. Grips can cause pressure melting which is a particularly difficult problem when testing at low strain rates. Another problem is obtaining suitable ice samples for making specimens especially from depths. From both these considerations the pressurization of annular discs containing edge slits would appear to be the best compromise. External radial slits have a K calibration strongly dependent on crack length as well as the internal and external radii but internal slits are not so strongly influenced (9) although pressure fluctuations due to seeping into the crack could be a problem.

Fracture Toughness of Ice Results

Figure 9 shows some unusual trends (8), e.g. as temperature decreases K_{1c} increases indicating more energy is absorbed at the crack tip prior to fracture. The reverse happens in metals when a decrease in temperature causes a decrease in plastic zone size because yield stress increases. Toughness is also strongly dependent on crack tip strain rate, characterized by $\acute{K}$, much more than in metals. For ice at these relatively high temperatures creep can take place causing stress redistribution at the crack tip and hence a greater force is required to cause fracture. This effect will be enhanced the slower the straining rate.

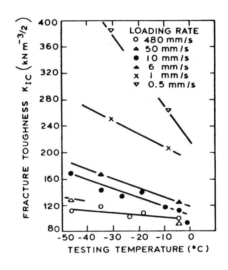

Fig. 9   Fracture Toughness of Laboratory Prepared Ice

Another effect is that of recrystallization to produce stress free material at the crack tip. None of these effects, however, account for the decreasing temperature effect which can only be explained by an increase of cohesive strength. Fletcher (35) has shown that a transition from a crystalline to a quasi-liquid surface exists at a temperature between -2 and -10°C. This thin liquid film will lower the total free energy of the ice surface. Tests carried out in a water + antifreeze mix indicated a 50% reduction in fracture toughness (8).

CONCLUSIONS

1. Only in the past few years has fracture mechanics been applied to ice. In the now expected multiplication of studies, care should be exercised when selecting a relevant fracture toughness parameter.

2. Studies of ice, particularly those concerned with the effects of temperature and strain rate, will help elucidate the role of creep, stress redistribution, recrystallization and surface energy.

3.   Don't look for dislocations at the bottom of crevasses.

REFERENCES

(1) Miller, K.J.   Traverse of the Staunings Alps, North-East Greenland. Alpine Journal 81 (1976), 143-152.

(2) Griffith, A.A.   The phenomena of rupture and flow of solids.   Phil. Trans. Roy. Soc. 1921 A221, 163-198.

(3) Irwin, G.R.   Fracture dynamics.   Fracturing of metals.   A.S.M. Cleveland 1948.

(4) Orowan, E.   Fracture and strength of solids.   Rep. Prog. Phys. 12 (1949) 185-232.

(5) Irwin, G.R. and Lies, J.A.   Fracturing and fracture dynamics.   Weld. J. Res. Suppl. 17 (1952) 95s-103s.

(6) Irwin, G.R.   Analysis of stresses and strains near the end of a crack traversing a plate.   J. App. Mech. 24 (1957) 361-364.

(7) Knott, J.F.   Fundamentals of Fracture Mechanics (1973) Butterworths.

(8) Liu, H.W. and Miller, K.J.   Fracture toughness of fresh-water ice. J. Glac. 22. (1979) 135-143.

(9) Rooke, D.P. and Cartwright, D.J.   Compendium of Stress Intensity Factors.   HMSO 1976.

(10) Liu, H.W.   Discussion in fracture toughness testing and its applications. ASTM (1965) S.T.P. 381. 23-29.

(11) Dugdale, D.S.   Yielding of steel sheets containing slits.   J. Mech. Phys. Solids 8 (1960) 100-104.

(12) Barenblatt, G.I.   The mathematical theory of equilibrium cracks in brittle fracture.   Advances in Applied Mechanics 7 (1962) 55-111.

(13) Wells, A.A.   Unstable crack propagation in metals; cleavage and fast fracture.   Crack Propagation Symposium, Cranfield 1961.

(14) Burdekin, F.M. and Stone, D.E.W.   The crack opening displacement approach to fracture mechanics in yielding materials.   J. Strain Analysis 1 (1966) 145-153.

(15) Rice, J.R.   A path independent integral and the approximate analysis of strain concentration by notches and cracks.   J. Appl. Mechanics. (1968) 379-386.

(16) Hutchinson, J.W.   Singular behaviour at the end of a tensile crack in a hardening material.   J. Mech. Phys. Solids. 16 (1968) 13-31.

(17) Rice, J.R. and Rosengren, G.F.   Plane strain deformation near a crack tip in a power-law hardening material.   J. Mech. Phys. Solids 16 (1968) 1-12.

146

(18)   Eshelby, J.D.   Solid State Physics 3, Ed Seitz F. & Turnbull D.
        New York (1956) 79–144.

(19)   Cherepanov, G.P.   J. App. Math. Mech. 31   (1967) 503–512.

(20)   Landes, J.D. and Begley, J.A.   A fracture mechanics approach to creep
        crack growth.   Westinghouse Res. Lab. Rep. (1974) 74–1E7–FESGT–P1.

(21)   Nikbin, K.M.   Webster, G.A. and Turner, C.E.   A comparison of methods
        of correlating creep crack growth.   Proc. ICF 4 (1977) 627–634.

(22)   Ellison, E.G. and Walton, D.   Fatigue, creep and cyclic creep crack
        propagation in a 1% Cr–Mo–V Steel.   Instn. Mech. Engrs. (1973/74).
        173.1 – 173.12.

(23)   Rice, J.R.   An examination of the fracture mechanics energy balance
        from the point of view of continuum mechanics ICF 1 (1966).   309–340.

(24)   Kfouri, A.P. and Miller, K.J.   Crack Separation Energy Rates in
        Elastic–Plastic Fracture Mechanics.   Proc. Instn. Mech. Engrs. 190
        (1976), 571–584.   First published as a Cambridge University Report
        (1974) CUED/C–Mat/Tr18.

(25)   Rice, J.R. and Johnson, M.A.   Inelastic Behaviour of Solids.   Ed.
        M.F. Kanninen et al.   McGraw–Hill, New York (1960) 641–672.

(26)   Miller, K.J. and Kfouri, A.P.   A Comparison of Elastic–Plastic Fracture
        Parameters in Biaxial Stress–States.   ASTM STP 668 (1979)   214–228.

(27)   Kfouri, A.P. and Miller K.J.   The effect of loading biaxiality on the
        fracture toughness parameters J and G^{Δ}.   Proc. ICF 4 (1977) 241–245.

(28)   Kfouri, A.P. and Rice, J.R. Elastic–Plastic Separation Energy Rate for
        Crack Advance in Finite Growth Steps Proc. ICF 4, 1, (1977) 43–59.

(29)   Miller, K.J.   Fatigue under complex stress.   Metal Science Journal,
        August/September 1977   432–438.

(30)   Howard, I.C.   Models of the reduction of fracture toughness due to
        hydrogen in strong steels.   Proc. ICM 3 2 (1979) 463–474.

(31)   Goodman, D.J. and Tabor, D.   Fracture toughness of ice; a preliminary
        account of some new experiments. J. Glac. 21 (1978) 651–660.

(32)   Gold. L.W. Crack Formation in Ice Plates by Thermal Shock.
        Canadian J. Physics 41 (1963) 1712–1738.

(33)   Andrews, R.M., McGregor, A.R. and Miller, K.J.   Fracture Toughness
        of Glacier Ice.   (These proceedings).

(34)   Smith, R.A.   The application of fracture mechanics to the problem of
        crevasse penetration.   J. Glac.   17 (1976) 223–228.

(35)   Fletcher, N.H.   The Chemical Physics of Ice.   Cambridge University
        Press 1970.

Fracture toughness of glacier ice

R.M. Andrews, A.R. McGregor and K.J. Miller

University of Sheffield

ABSTRACT

This paper describes a simple test rig for evaluating the fracture toughness, K_{IC}, of glacier ice. Results of K_{IC} tests on Roslin glacier ice (North-East Greenland) using an easily portable test rig are compared with previous data obtained on bubble-free ice manufactured and tested under laboratory conditions.

The fracture toughness of ice is approximately 125kN m$^{-3/2}$, i.e. approximately three orders of magnitude less than common structural steels.

INTRODUCTION

In 1976 an international symposium on applied glaciology (1) was largely concerned with discussions on the strength, hardness, creep resistance, flow and deformation behaviour, and finally fracture criteria of snow and ice. Three papers were directly concerned with engineering properties of snow and ice (2) (3) (4). However, recent developments in the mechanics of fracture of metals had not then been applied to the studies of crack formation and propagation in snow and ice, a new development that will certainly lead to a greater appreciation of avalanches and crevasses.

Fundamental questions as to how crevasses form and why they stop propagating at a certain depth are two initial and simple questions that require serious study. Glaciologists, alpinists, and mountaineers are those primarily concerned with understanding crevasses, particularly their formation and, hopefully, their stable behaviour, see Fig. 1, whilst skiers and walkers will undoubtedly acknowledge a greater understanding of crack initiation in snow slabs that can so easily initiate avalanches, see Fig. 2.

A companion paper in this conference (5) outlines the common parameters in use concerning the analysis of cracked bodies and so this paper will confine itself to describing a test rig, suitable for field work, in which the fracture toughness of glacier ice can be evaluated. This rig was used during the 1978 University of Sheffield expedition to North East Greenland and is of light construction and easily transported. Finally, the results obtained from the Roslin glacier of the Staunings Alps, see Fig. 3, are presented which are compared with previously obtained laboratory data.

Fig. 1   Traversing a Karakoram Glacier Crevasse

FRACTURE TOUGHNESS TESTING

Since ice is a relatively brittle material linear elastic fracture mechanics (LEFM) can be invoked, see reference (5). In this approach only the stress intensity factor, K, at the tip of a crack need by analysed. Methods of measuring the fracture toughness of metallic materials are defined in British

Fig. 2   An Avalanche in a Karakoram Glacier Valley

Standards (6) and these procedures were considered adequate for testing pre-notched ice specimens for which the fracture toughness is given by:

$$K_{IC} = F_c Y \qquad (1)$$

Here K_{IC} is the critical mode I value of the plane strain fracture toughness and is calculated from the load to cause fracture, F_c, and Y a tabulated function of pre-cracked specimen dimensions.

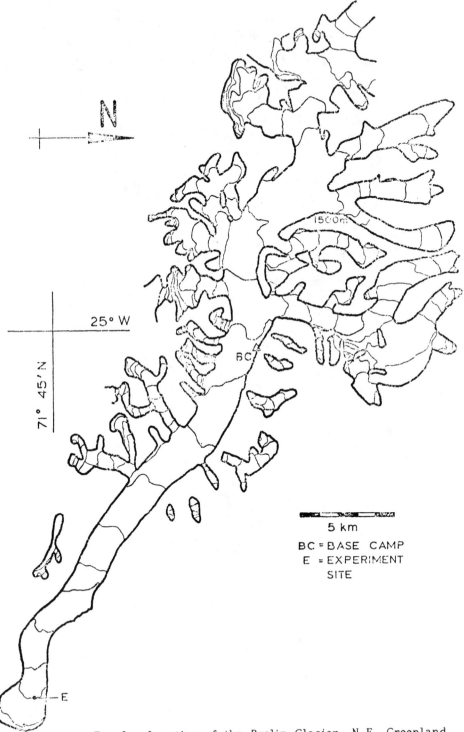

Fig 3.   Location of the Roslin Glacier, N.E. Greenland

In previous experiments by Liu and Miller a CTS specimen was used for laboratory tests, see Fig. 4, and the Y values for this geometry are given in reference (7). These tests showed that the fracture toughness was not affected by the method used to form the starter crack at the root of the notch; they standardised by forcing a razor blade into the notch root.

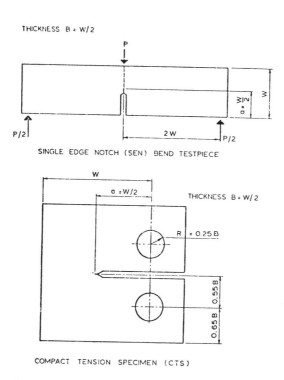

THICKNESS B = W/2

P

P/2

2 W

P/2

SINGLE EDGE NOTCH (SEN) BEND TESTPIECE

W

a = W/2

THICKNESS B = W/2

R = 0.25 B

0.55 B

0.65 B

COMPACT TENSION SPECIMEN (CTS)

Fig. 4 Geometry of Standard Testpieces

For tests on ice samples in remote areas, it is essential to simplify ice extraction methods, shaping of specimens, and test procedures. Specimens from the snout zone of the Roslin glacier were cut directly from ice hummocks which had previously been cut into a rectangular box shape. Specimens were of the SEN form shown in Fig. 4. These specimens were fractured by simple three point bending after a crack had been formed by drawing a 0.3mm wire across the bottom of a saw cut notch.

THE TEST APPARATUS

A schematic diagram of the rig is shown in Fig. 5 and illustrated in use in Fig. 6. The specimen, when located in its grips, is loaded via a wire cable and drum. The other grip is anchored to the rig frame via the load cell. This cell carries strain gauges, which are connected to a bridge circuit. The output from the bridge circuit is amplified and used as input to a chart recorder, which records the applied force.

The frame of the rig, which should be rigid, is made from 19mm square mild

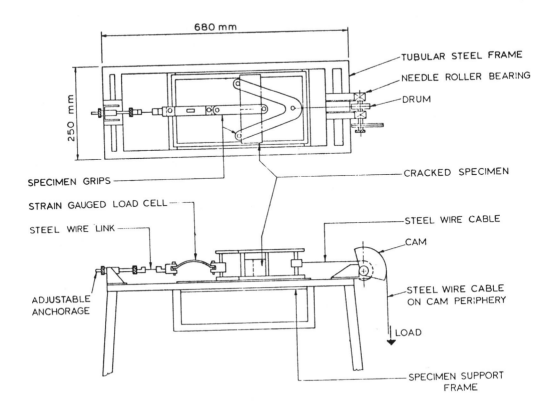

Fig. 5 Layout of Fracture Toughness Test Rig

Fig. 6 Test Rig on Roslin Glacier

steel tube.   One end provides an adjustable anchorage for the load cell.
This adjustment permits variation in specimen size.   A short flexible wire
link connects the load cell to the anchorage.   Interchangeable load cells
are easily fitted to allow a range of applied forces.

Several types of specimen can be tested in this apparatus including the
Single Edge Notch (SEN) specimen which is loaded as a simply supported
beam and the Compact Tension Specimen (CTS) used by Liu and Miller.

In this latter case, the tension load is applied using pins fitted through
the holes of the specimen.   A jig should be used to ensure that the notch
of the CTS specimen is cut in the correct position and orientation relative
to the holes.

The specimen and grips are supported by a low friction and flexible base
of elastic cord stretched across a light alloy frame.   The height of this
support frame may be altered to suit the specimen shape and size.   The
elastic cord eliminates the need for precise adjustment, and also has a lower
thermal conductivity than a metal support.

Loading is applied by filling a bucket hanging from the light alloy cam on
the drum spindle.   Controlling the rate of water flow into the bucket gives
direct control of the loading rate and hence the rate of stress intensification
$\dot{K}$.   Various load ranges are obtained by using cams of different radii.

The force measuring system is shown in block diagram form in Fig. 7.   Electric resistance strain gauges mounted on curved brass bars are used as
load cells.   This method eliminates errors due to friction in the bearings
of the drum spindle.   Different sensitivities are obtained by using load cells
of different size.   The strain gauges are connected to a terminal box which
contains the bridge network.   This bridge network also eliminates temperature effects.   The output from the bridge circuit is amplified before input
to a chart recorder which produces the force-time curves.   The amplifier
unit permits adjustments for the calibration of the system.   Dry cells provide
the amplifiers' power supply during field operations.

TEST PROCEDURE AND FIELD EXPERIENCE

Tests were carried out daily over a period of 3 weeks in July and August,
1978.   For convenience, experiments were conducted close to the glacier snout
well below the firn line.   Ice samples were cut by hand from approximately
150mm below the glacier surface.   The ice temperature was -1°C and the
air temperature was generally around 6°C.   With practice, tests could be
carried out at a rate of 6 per hour.   Two operators were required; one returned the water to the top container and re-set the loading system, the
other placed the specimen in the grips.   Gradually increasing load was
applied, until fracture occurred, by control of the water supply to the bucket
via a rubber hose and clip gauge.

Only SEN bend specimens were tested, as trials showed that drilling accurate
holes in the CTS specimen without cracking was both difficult and time consuming.   Specimens were cut about 200mm long and between 30mm and 70mm
wide.   The thickness was in the range of 25-50mm, and the crack length
in the range of 7-30mm.   Overall dimensions of each specimen were measured
by steel rule after fracture, to allow for any pressure melting at the loading
points.   This was only significant during long tests at low loading rates.
The important dimensions were L, a and the remaining ligament depth.   These

154

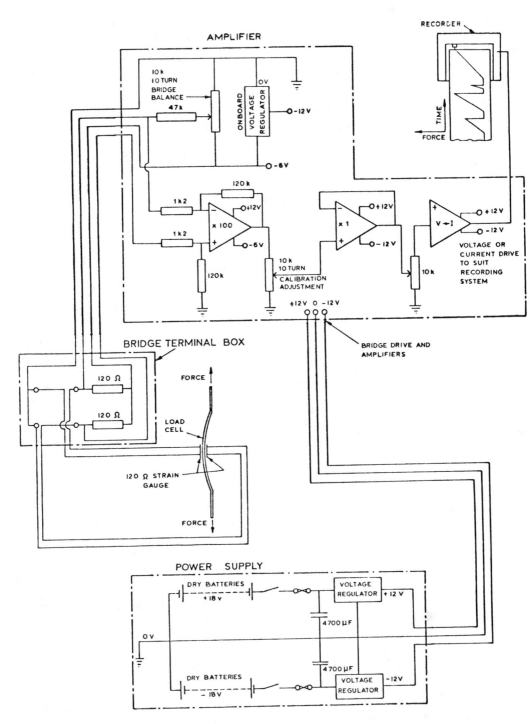

Fig. 7   Force Measurement System

latter two dimensions being taken directly from the fracture surfaces.

Some problems were encountered during the tests. Guides to hold the mid-specimen grip central in relation to the "U" shaped grip would have simplified testing. Any slight loss of accuracy due to friction at such guides would be offset by the consistent alignment and the ease of use that the guides would provide. The loading system was simple to use and adjust. However it needs water and so a totally mechanical system would be required for sub-zero conditions. The design of the force measuring system is critical since any small movement of the load cell in its holders can cause a serious error. A load cell using a constant stress cantilever beam which is fixed to the rig would remove this error and also make the rig more compact.

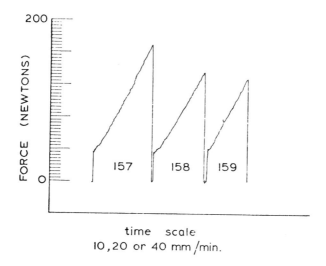

Fig. 8   Example of Force v Time Curve when Fracturing Ice Specimens

RESULTS

The field tests involved 194 experiments. An example of a force time curve is given in Fig. 8. The maximum force and the loading rate prior to fracture were measured from these curves on returning to the U.K. Values of K_{IC} and and $\dot{K}_I$ were calculated using the University of Sheffield's ICL 1906S computer. Some twenty tests were eliminated from K_{IC} calculations since the specimen proportions were outside the range of the Y-calibrations available. The K_{IC} values for the range of specimen dimensions used were checked for validity (8) using the criteria

$$ a \geq 2.5 \left(\frac{K_{IC}}{\sigma_y}\right)^2 \text{ and } B \geq 2.5 \left(\frac{K_{IC}}{\sigma_y}\right)^2 \text{ and } W-a \geq 2.5 \left(\frac{K_{IC}}{\sigma_y}\right)^2 \quad (2) $$

where σ_y is the yield stress of ice which depends on the strain rate. The crack tip strain rate was estimated to be in the range 1×10^{-6} s^{-1} to 1×10^{-5} s^{-1}. Hence σ_y was taken as 21×10^{5} Pa (9). Thirty-five K_{IC} values were rejected as invalid. The valid results are shown in Fig. 9, as a function of $\dot{K}_I$. The mean value of K_{IC} is 125 kN m$^{-3/2}$; there is no statistically significant correlation between K_{IC} and $\dot{K}_I$.

DISCUSSION

Figure 9 shows the currently available results of laboratory fracture toughness tests on ice. The results of Liu and Miller are for their tests at –4°C and –8°C only. The results of Goodman and Tabor (10) are only those for 3 point bend tests at –14°C; the loading time in these tests was less than 10 seconds. Hence $\dot{K}_I$ can be estimated as approximately 10 kN m$^{-3/2}$ s^{-1}, which is comparable to the present work. The results of Gold's work (11) were derived from the arrest of cracks induced by thermal shock at temperatures of approximately –10°C. It is not possible to estimate $\dot{K}_I$ for these tests. These results are also dependent on the size of the segments formed on cracking.

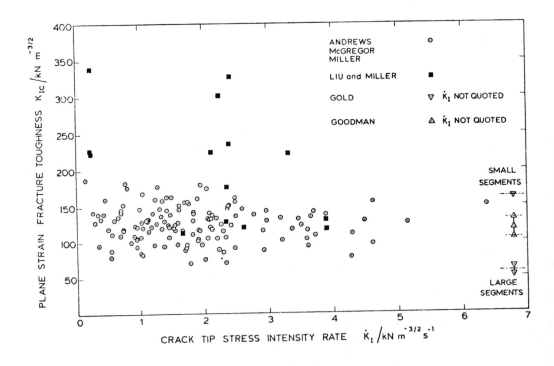

Fig. 9   Fracture Toughness as a Function of Stress Intensity Rate

The present results are in general agreement with those of earlier workers, although they are less than those of Liu and Miller at low $\dot{K}_1$. Over a $\dot{K}_I$ range of 0.2 kN m$^{-3/2}$ s^{-1} to 2000 kN m$^{-3/2}$ s^{-1}, Liu and Miller found a decrease in K_{IC} from approximately 250 kN m$^{-3/2}$ to 100 kN m$^{-3/2}$. The present work only covers the lower part of this $\dot{K}_I$ range, and so good agreement is not to be expected. The present results agree with those of Goodman and Tabor.

The discrepancy between Gold's "large segments" and "small segments" which describes wide and narrow thermal crack spacing respectively can be partly explained as follows. Gold assumed the cracks to be single isolated edge cracks when calculating K_{IC}. For the "small segments", the average segment size was about three times the crack depth. If the cracked surface is approximated as an array of parallel equi-spaced cracks in a semi-infinite plane, the analysis of Bentham and Koiter (12) can be used. When the crack spacing is equal to three times the depth, the Y factor of equation (1) reduces to 64% of the value for an isolated crack. Furthermore the cracked surface is a 3-dimensional stress problem, and the true reduction in the Y factor would be greater than this 2-dimensional approximation suggests. Finally Gold's results are for the arrest of a propagating crack, not K_{IC}. Liu and Miller found that K_{IC} at crack arrest was 30% lower than K_{IC} at temperatures of -4°C to -12°C. For these reasons the discrepancy between the present results and Gold's is not significant.

The analysis of crevasse penetration by Smith (13) is worth consideration. He derived an equation for the depth of an isolated, dry, crevasse:

$$1.12 \; \sigma_t \; \sqrt{\pi a} \; - \; 0.683 \; \rho_i ga \; \sqrt{\pi a} \; - \; K_{IC} = 0 \qquad (3)$$

Here σ_t is the longitudinal stress, a the crevasse depth and ρ_i the density of ice. Smith assumed that ice was so brittle that K_{IC} was zero.

The above equation was solved for a range of values of K_{IC}, assuming the same values of σ_t = 2 x 10^5 Nm^{-2}; ρ_i = 920 kg m^{-3}; g = 9.81 m s^{-2} as Smith used. The results, shown in Fig. 10, indicate that the depth of a crevasse is not sensitive to the value of K_{IC}. For K_{IC} = 200 kN m$^{-3/2}$, the depth is 8% less than that predicted assuming the fracture toughness of ice to be zero. Using K_{IC} = 125 kN m$^{-3/2}$, as measured for Roslin glacier ice, the depth of a dry crevasse predicted by equation (3) is 34.4 metres.

CONCLUSIONS

(1) It is possible to measure the fracture toughness of glacier ice in the field using apparatus of the type described.

(2) The fracture toughness of ice obtained from the lower part of the Roslin glacier is 125 kN m$^{-3/2}$. This value is in agreement with laboratory test data.

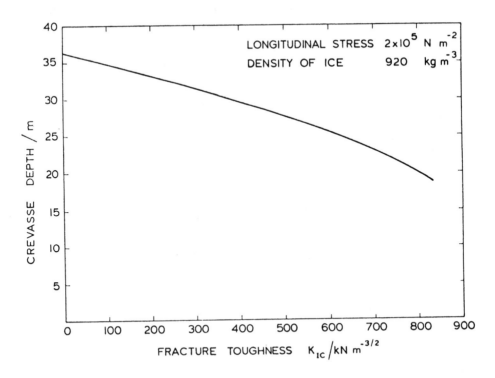

Fig. 10   Variation of Crevasse Depth with Fracture Toughness

(3)   The prediction of crevasse depths using LEFM theory is not sensitive to the value of fracture toughness.   If K_{IC} is assumed to be 125 kN m$^{-3/2}$ the predicted depths will be within 5% of those obtained using previously reported fracture toughness data.

ACKNOWLEDGEMENTS

The authors gratefully acknowledge the assistance given by members and supporters of the Sheffield University N.E. Greenland expedition 1978.   They also wish to thank the technicians of the Mechanical Engineering Department of the University of Sheffield; Mr. Stephen Grimes, for his help throughout the field work, and Dr. A.P. Kfouri for supplying some of the Y-calibrations used in the analyses.

REFERENCES

(1)   Journal of Glaciology, (1977) Symposium on applied glaciology, Cambridge, September, 1976, J. Glac. 19.

(2)   Mellor, M. (1977)   Engineering properties of snow. ibid 15–66

(3)   Gold. L.W.   (1977)   Engineering properties of fresh water ice.   ibid 197–212

(4) Schwarz, J. and Weeks, W.F. (1977) Engineering properties of sea ice. ibid 499–532.

(5) Miller, K.J. (1980) The Mechanics of Fracture applied to Ice. (this conference).

(6) British Standards Institute (1977) BS 5447, Methods of test for the plane strain fracture toughness of metallic materials. London

(7) Liu, H.W. and Miller, K.J. (1979) Fracture toughness of fresh water ice. J. Glac. $\underline{22}$, 135–143.

(8) Brown, W.F. and Strawley, J.E. (1966) Plane strain crack toughness testing of high strength metallic materials. ASTM STP 410, Philadelphia.

(9) Hawkes, I. and Mellor, J. (1972) Deformation and fracture of ice under uniaxial stress. J. Glac. $\underline{11}$ 103–131.

(10) Goodman, D.J. and Tabor, D. (1978) Fracture toughness of ice – a preliminary account of some new experiments. J. Glac. $\underline{21}$ 651–660.

(11) Gold, L.W. (1963) Crack formation in ice plates by thermal shock. Canadian J. Physics, $\underline{41}$ 1712–1728.

(12) Bentham, J.P. and Koiter, W.T. (1973) Asymptotic solutions to crack problems. IN Sih, G.C. (Ed), Methods of analysis and solution of crack problems, 131–178, Groningen, Noordhoff.

(13) Smith, R.A. (1976) The application of fracture mechanics to the problem of crevasse penetration. J. Glac. $\underline{17}$ 223–228.

Indo-Asian convergence and the 1913 survey line connecting the Indian and Russian triangulation surveys

R. Bilham and D. Simpson
Lamont-Doherty Geological Observatory
of Columbia University
Palisades, New York, 10964.

ABSTRACT

In 1913 a 480km long north-south triangulation survey was conducted across the Karakoram and Pamirs near the 73rd meridian. It spans more than one third of the zone of convergence characterising the collision of the Eurasian and Indo-Australian plates and its remeasurement after more than 65 years is expected to reveal considerable crustal deformation.

This article discusses the errors associated with the 1913 survey and attempts to predict the magnitude and style of deformation that may have occurred since then. It is shown that the original survey accuracy is associated with errors greater than $\pm 10^{-6}$ strain and that, unless a convergence rate of more than 6 cm/year has occurred, uniform compression of the survey line will be undetectable. At 73°E the Indo-Asia convergence is expected to be 4 cm/year. However, it is clear from historical and continuing seismicity that the convergence is not uniform and that locally large deformations may have occurred. Two areas appear particularly promising. These are the area just north of Gilgit and the area north of Lake Karakul in the Soviet Pamirs.

INTRODUCTION

In 1913, after four years of discussion and planning, a route was finalised that would link the extensive geodetic networks of Russia to those of India. The primary incentive for connecting the two surveys was to complete a line from Madras (9°N) to Siberia (60°N), defining almost a quarter of the Earth's surface in a north/south line, enabling a direct measure of the shape of the Earth.

The survey was conducted several years before Wegener suggested that continental movements may be occurring. By the greatest good fortune, the geodetic link may provide an opportunity to monitor the velocity of convergence at the continental collision boundary between the Indo-Australian and Eurasian plate. It is difficult to monitor directly the relative motions of plates by conventional surveying techniques, because the distances between exposed land areas on the plates are usually large or because geodetic measurements of sufficient antiquity do not exist. Relative plate motions are monitored geodetically in Iceland (1,2), East Africa, and California (3). Very different processes are involved in each case. In the Himalayas, shortening and thickening of the mountain ranges are known to be occurring, although the rate of convergence between the Indian and Eurasian plates is subject to dispute. Estimates range from less than 1cm/year to more than 6cm/year in an approximately N/S direction.

The 1913 survey line starts at a point south of Gilgit and extends 160 km to the Russian/Chinese border. A Russian survey, completed at the same time, extends a further 320 km northward to Osh in the USSR (4). Both ends of the combined lines (latitudes 36°–41°N) are connected to first-order survey networks that extend to locations well outside the Himalayan Mountain zone. The 480 km line represents more than one-third of the width of the zone of deformation characterising the India/Asia collision; this article attempts to assess the possible magnitude and style of crustal deformation that may have occurred along this line in the last 67 years.

SURVEY ACCURACY

A detailed study of errors associated with parts of the 1913 survey is presented elsewhere (5). It is important, however, to recognise two types of error assessment for the purposes of discussing crustal deformation in the region: the errors associated with individual readings and triangulation figures, and the much larger cumulative errors that are associated with locating end-point positions relative to each other. The identification of deformation in a single triangulation polygon or adjacent pair of polygons is of great importance since strain across the Himalaya is unlikely to be uniform and may locally concentrate in narrow deformation zones.

The survey line follows river valleys for most of its route and the survey marks consist of points located on the summits of 5000 m mountains bordering the valleys. Inevitably, the resulting geometry has a large number of poorly-shaped triangulation figures; elongated triangles and polygons that prevent optimum survey accuracies from being attained. The survey theodolites had to be small to facilitate frequent perilous mountain ascents and were thus inferior to those usually used for first-order triangulation. The details of the measurement of the Russian line are not documented but judging from comments made by Mason (4) the accuracy of individual measurements varied considerably, depending on line lengths and optical conditions.

Table 1 summarises closure error information for the Pakistan side of the survey (4).

TABLE 1

Latitude °N	No. of Triangles	Closure Error
35.8 – 36.8	23	2.2" (3.8 x 10^{-6} rad)
36.8 – 37.0	6	7.3" (12.7 x 10^{-6} rad)
37.0 – 37.5	11	3.6" (6.3 x 10^{-6} rad)

The closure error is the modulus of the difference between the sum of the measured angles of a triangle and 180°. Angular closure errors result in lateral displacement errors that are proportional to distance. The least error in lateral location accuracy in the Karakoram survey is approximately ± 1 cm at a distance of 3 km (the shortest line in the survey) and the worst error in relative location may be ± 40 cm in 30 km (the longest line in the survey). The mean line length is 13.9 km. The standard mean error for an individual angle in the 1913 Karakoram survey was computed to be 5×10^{-6} radians (4). This value will be used to provide a rough estimate of overall survey accuracy in order to assess the feasibility of detecting gross crustal shortening. If we assume that sixteen 14 km polygons comprise the Pakistan survey segment and that each polygon is associated with random (Gaussian) errors of the order of 5×10^{-6} we obtain a cumulative error from Gilgit to the Russian border of approximately ± 30 cm. Assuming that the Russian link is of comparable accuracy, or better, since the triangulation polygons are better constructed, the cumulative error will be larger than ± 1 m. Unfortunately, the north-south error will be larger than the east-west "shear" error since there is a north/south elongation bias in most of the triangulation figures. A conservative estimate of the cumulative error (Gilgit-Osh) is closer to ± 3.0 m (± 6×10^{-6} strain).

PLATE CONVERGENCE RATES AND DEFORMATION

Although the closure velocity of India relative to Asia was of the order of 15 cm/year from 70 million years ago until about 40 million years ago (6), the continental collision is thought to have slowed the rate to less than 6 cm/year; see Figs. 1 and 2.

Fig. 1   Positions of India, corresponding to magnetic anomalies in the oceans (6)

Fig. 2   Movement of the northeast** and northwest* tips of India (6)

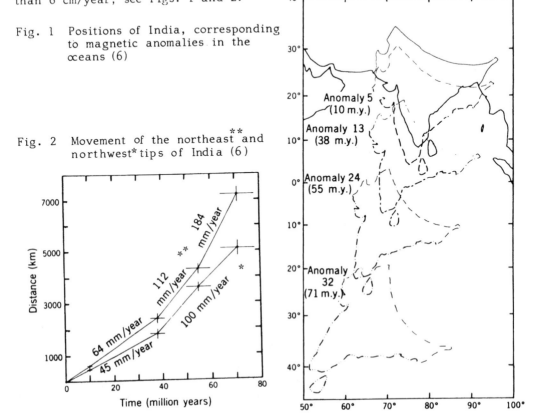

The value for the closure velocity is closer to 4 cm/year at 73° longitude due to the location of the pole of rotation of the Indian continent.

A significant problem arises in determining the closure velocity across the Himalaya since to do so presupposes a clear understanding of the width of the deformation zone involved in the continental collision. At the 73rd meridian the zone of fold mountains separating the relatively undisturbed continental areas to the North and South is approximately 1200 km wide (from the Punjab in the south to the northern edge of the Tian Shan). To the west and more so to the east, the width of the zone of deformation increases. Hence it is reasonable to assume that at least 25% and possibly as much as 95% of the 4 cm/year of continental closure at 73°E will appear as deformation within the 1200 km mountain ranges of the Karakoram, Pamir, and Tian Shan. In Figure 3 the location of the 1913 survey line is shown on

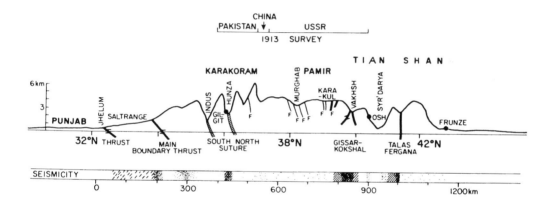

Fig. 3   Cross section along 73°E showing the total region of deformation. Present microseismicity is also shown. The dashed lines show major historical events and continuous lines recent large events.

a cross section between latitude 32°N and 42°N. The 1913 survey line crosses almost half of the fold-belt and joins with other survey lines at its southern and northern extremities.

If we assume that the closure rate across the 1200 km zone is 4 $\pm$ 2 cm/year, this corresponds to a strain of approximately (3 $\pm$ 2) x 10^{-8}/year. The cumulative strain in 67 years amounts to (2 $\pm$ 1.3) x 10^{-6} strain or about 2 $\pm$ 1.3 m of convergence.

The significance of this estimate is that it is smaller than the error estimate derived earlier for the survey line ($\pm$ 5 x 10^{-6}) and is thus unobservable with these data. Even the measurement precision in the best observed polygons is barely adequate to observe convergence if it is uniformly distributed across the region. However, considering the number of assumptions in the derivation of the expected convergence rate, the result is encouraging, particularly since we know that deformation within the region is not uniform. The major tectonic units are bounded by fault systems that are known to

164

be active from seismic and geomorphological evidence, and by geological boundaries that may or may not be zones of weakness. Thus, although ubiquitous folding and thickening of the Himalayan Mountains may appear to result in homogeneous strain, much of the plate tectonic convergence is clearly being absorbed as sporadic slip on extensive thrust faults and strike-slip faults. If one or more of these faults have moved in the last several decades, there may be some locations in which strains will be well above the noise level of the survey observations.

TECTONICS AND SEISMICITY

In Figs. 3 and 4, the major geological structures are shown. The 1913 survey line is shown stippled. Faults south of Karakul and North of Kilik are poorly documented. Detectable deformation is expected North of Karakul and North of Gilgit following major earthquakes in these regions in the last few decades. The location of the survey line is in an area of enormous geological complexity. The region is characterised by dramatic changes in the azimuth of structural trends involving thrust faults, grabens, strike-slip faults and folding and is clearly undergoing severe distortion (7).

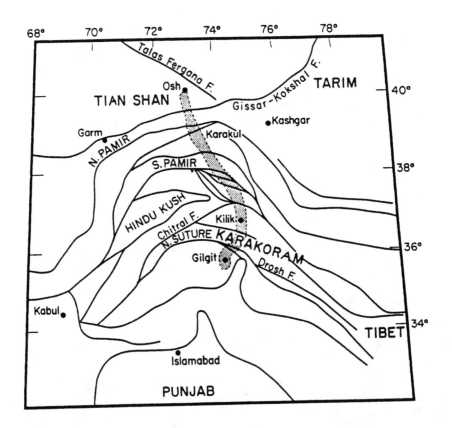

Fig. 4   Structural map of the Karakoram   (7)

The Karakoram and the Talas-Fergana faults are right-lateral strike-slip faults with hundreds of km offset (8,9,10), at least part of which has occurred during recent (Alpine) deformation. Neither of these faults crosses the region of the survey line, although their linear projections pass very close to the line. Several well developed thrust faults in the Pamirs cross the Russian survey near Lake Karakul (11,12,13,14), but become less prominent to the east of the line where the tectonics appears to have a larger right-lateral component (14). To the south, two sutures are crossed north and south of Gilgit, respectively. The main Himalayan thrusts are well to the south and are crossed by other first-order triangulation networks (Gilgit Principal Series).

In the last few decades, several magnitude 6 earthquakes have occurred in the survey region, especially near 39.5°N, 73.5°E. Larger events have also occurred (Chatkal, 1946; Alai, 1974) and nearby regions have experienced magnitude 8 events in recent history (Kashgar, 1902). The larger historical earthquakes and recent seismicity are shown in Fig. 5. Most of the activity is concentrated along the Gissar-Kokshal fault zone between the Pamirs and Tien Shan. There is a relatively low level of crustal seismicity in the Pamirs. What activity there is, may be mostly sub-crustal and related to the northeastern end of the Hindu Kush zone of intermediate depth earthquakes. The Pamirs appear to act as a semi-rigid crustal block, with little internal deformation, bounded on the north (Gissar-Kokshal) and south (Indus suture) by narrow zones of more intense deformation. The more concentrated these zones of deformation are, the more likely it is that measurable displacement has occurred along shorter segments of the survey line since 1913.

The Soviet part of the survey line crosses the Gissar-Kokshal zone. Seismic energy release is higher here than elsewhere along the line (Fig. 3), and at least three large earthquakes (1955, 1974, 1978, Fig. 5) have occurred in this region since the first survey. Large and measurable displacements are most likely in this part of the survey line. The survey polygons in this area north of Lake Karakul are geometrically favourable for detecting line contraction, see Fig. 6. As part of a major Soviet program in earthquake prediction, extensive geophysical measurements are being made at Garm, 250 km west of where the survey line crosses the Gissar-Kokshal fault (Fig 4). Geodetic studies in the Garm area show high strain rates (15,16).

Near the Chinese/Pakistan border it is possible that shear strains have accumulated since this region lies close to the northward extension of the Karakoram fault. Survey polygons north of the Kilik pass are well-braced and of moderate accuracy so that even in the absence of surface faulting, it is reasonable to be optimistic that deformation will be identified if it has occurred.

Although the two sutures near Gilgit mark the original collision boundaries between India and Asia and were not thought to be currently active, a M_b = 6.3 earthquake may have occurred on the northern suture to the south-west of Gilgit in 1973. Resurvey of the Gilgit quadrilaterals will provide important information. The seismicity of the region south of 36°N does not appear to follow any of the major structural trends evident in the Himalayas between 73°E and 75°E. However, Seeber et al. (17) indicate a striking correlation between seismicity and topography along the Himalayan arc. They hypothesize that since a linear belt of thrust earthquakes follows the mean 4,000 m contour, there may be a causal relationship between the two. In this hypothesis, the high Himalaya lie above a basement thrust that extends southward as a more gently dipping detachment thrust. In

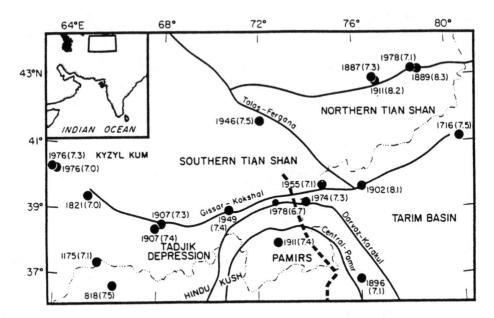

Fig. 5.(a)　Large historical earthquakes
with dates and magnitudes (23).
Approximate location of 1913 survey
line is shown by dashed line.

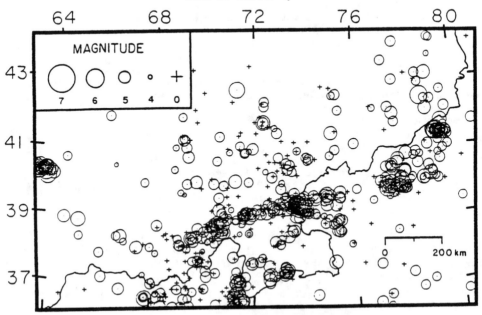

Fig. 5.(b)　Recent shallow seismicity (1964–1978)
from International Seismological Center
Catalog (see (13))

the NW Himalaya, this detachment emerges at the surface as the Salt Range thrust. Thus, a significant amount of plate-tectonic convergence could occur far to the south of the 1913 survey line.

Similarly, the right-lateral Talas-Fergana fault is known to be active and at least one of its associated branches has moved since the turn of the century (Chaktal, 1946; 18,19). Unfortunately, this fault zone, which is a striking feature on LANDSAT imagery, is just north of the survey line. It is not clear whether the Russian survey extension to the north crossed it earlier this century. Displacement on this fault is being currently monitored in connection with studies related to the Toktogul reservoir (18,20).

DISCUSSION

Resurvey of the Indo-Russian survey connection of 1913 is not expected to reveal relative motion between India and Asia unless the velocity of approach of the two continents is significantly larger than 4 cm/year, or unless tectonic movements are non-uniformly distributed in space and/or time. The occurrence of major earthquakes near part of the survey line suggests that strain rates, at least in some locations, are sufficiently high to be detected. Strain release to the north and south of the Indo-Russian connection may be significant on known active faults, although no major earthquakes have occurred on these faults near the survey line since 1913.

The estimates of survey accuracy used in this article do not take into account the problem of location of the original triangulation points. To optimise survey accuracy, it is vital that as many of these be used as possible. If new survey marks are introduced, they should be sufficiently robust to last many years. It is also assumed that the new survey, using modern triangulation and triangulation equipment, will be more accurate than the original survey by at least a factor of five. Thus, one of the important objectives of resurvey will be to provide an accurate baseline for future generations of geodesists. Major earthquakes in the region have been estimated to occur every 100-200 years and the possibility of one occurring in the near future is considerable.

The estimation of vertical movements from the survey data is too imprecise to be of use. Yet the region is known to be rising in response to horizontal compression (19). It would be of great value to introduce a first-order levelling line along the survey route starting from south of the Frontal Himalayan Thrust and extending as far north up the Hunza Valley as is feasible. Levelling data have the advantage of being inherently more precise than trilateration and triangulation, and can be performed in the valley floors along existing roads. The disadvantages of levelling are that it takes longer to perform and bench-marks, being closer to frequented paths, can be inadvertently destroyed.

The major benefits of re-survey of the complete 1913 line (including its extension to Osh in the USSR) are that, since it crosses at least one third of the zone of continental deformation between two major plates, a conservative measurement of instantaneous plate motions can be obtained. This value would provide constraints on present estimates of plate motions and gross seismic energy release. Resurvey of part of the line is less satisfactory. The Gilgit to Kilik Pass survey line crosses only one tenth of the total zone of deformation and although it crosses several tectonic features, none of them is as prominently active as other features to the north. Understanding the style of deformation in the south is important for the inter-

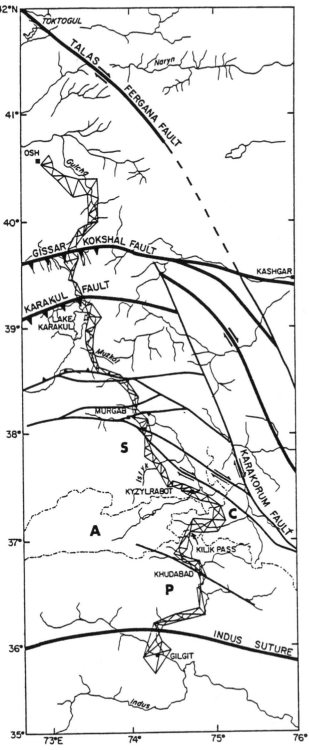

Fig. 6 The 1913 survey line.

In this figure, major faults are shown, although their locations are sometimes inaccurate due to the inadequate quality of existing maps of the region (4,7,11, 12,13). The various authors relate some of the major faults in different ways. Minor lineations on LANDSAT imagery suggest numerous other faults in the region. It is likely that geodetic resurvey, together with local microseismic studies will clarify structural relationships in the region. Segments of international borders are shown to indicate Pakistan (P), China (C), and USSR (S) sections of the survey line and Afghanistan (A).

pretation of the curious discrepancy between surface structure and known seismicity (17,21). The measurement of a significant amount of deformation (10^{-4} strain) within the Karakoram part of the survey since 1913 cannot be excluded and might be indicative of impending seismicity. A cumulative strain of 10^{-4} to 10^{-5} would require accelerated preseismic strain rates well above the strain rates predicted from plate motions, perhaps involving unloading of adjacent regions and strain concentration within a future rupture zone.

Finally, the remeasurement and reinforcement of old survey marks must be regarded as an important responsibility that this generation owes to the next. With improvements in satellite geodesy it is likely that new techniques will be available in the next century when similar magnitudes of deformation will have occurred. A century of previous deformation history will be invaluable to future geodesists.

ACKNOWLEDGEMENTS

The authors wish to thank Leonardo Seeber, John Armbruster and John Beavan who have contributed significantly to this article. The work was supported by the National Aeronautics and Space Administration contract NGR 33-008-146, by the U.S. Department of the Interior, Geological Survey under Contract Number 14-08-0001-16844 and by an NSF travel award INT 80-16497; see also Observatory publication number 3010.

REFERENCES

1. Neimczyk, O. (1943) Fissures in Iceland. Stuttgart: Verlag von Konrad Wittwer.

2. Gerke, Von K., et al. Geodetic measurements in the determination of crustal movements in N.E. Iceland. Walter Hopke Memorial Volume, Scientific Works of the Department of Geodesy, Photogrammetry and Cartography, Hannover Technical University, 83, 23-33 (1978).

3. Prescott, W.H., Savage, J.C. and Kinoshita W.T. (1979) Strain accumulation rates in the Western United States between 1970 and 1978. J. Geophys. Res., 84, 5423-5436.

4. Burrard, S.G. (editor) (1914) Completion of the link connecting the triangulations of India and Russia 1913. Records of the Survey of India, 6, pp. 116.

5. Crompton, T.O. (1980). The Pakistan to Russia Triangulation Connection: Past and Projected Error Analyses. This volume.

6. Molnar, P., and Tapponnier, P. (1975) Cenozoic tectonics of Asia, effects of continental drift. Science, 189, 419-426.

7. Desio, A (1979) Geologic evolution of the Karakoram. Geodynamics of Pakistan, Geol. Surv. Pakistan, (A. Farah and K.A. DeJong, eds), Quetta, pp. 361.

8. Molnar, P., and Tapponnier, P. (1978). Active tectonics of Tibet. J. Geophys. Res., 83, 5361-5375.

9. Tapponnier, P., and Molnar P. (1979) Active faulting and Cenozoic tectonics of the Tien Shan, Mongolia and Baykal regions. J. Geophys. Res. 84, 3425-3459.

10. Ni, J. (1978). Contemporary tectonics in the Tien Shan region. Earth Planet. Sci. Lett., 41, 347-354.

11. Burtman, V.S., Peive, A.V. and Rushentsev, S.V. (1963). The main strike slip faults and horizontal movements of the Earth's crust. (in Russian) Trans. Geol. Inst. Acad. Sci. USSR, 80, 152-172.

12. Ruzhentsev, S.V. (1963) Transcurrent faults in the S.E. Pamir, faults and horizontal movements of the Earth's crust (in Russian) Trans. Geol. Inst. Acad. Sci. USSR, 80, 113-127.

13. Kristy, M., and Simpson, D. (1980) Seismicity changes preceding two recent Central Asian earthquakes. J. Geophys. Res., 85, 4829-4837.

14. Jackson, J., Molnar, P., Patton, H., and Fitch T. (1974) Seismotectonic aspects of the Markana Valley, Tadjikistan, earthquake of Aug. 11, 1974 J. Geophys. Res. 84, 6157-6167.

15. Kuchay, V.K., Pevnev, A.K. and Guseva T.V. (1979) Deformation of the near surface parts of the Earth's crust from geodetic measurements. Izvestiya, Earth Physics, 15, 8.

16. Bulanger, Yu. D., Guseva, T.V., Dem'yanova, T.E., Pevnev, A.K. and Shevchenko, V.I. (1979) Morphology and origin of the Vakhsh Thrust Fault in Tadzhikistan from geodetic and gravimetric data. Izvestiya, Earth Physics, 4.

17. Seeber, L., Armbruster, J., and Quittmeyer R. (1981). Seismicity and continental subduction in the Himalayan arc. I.C.G. Working Group 6 Report, (H. Gupta, ed.).

18. Simpson, D.W., Hamburger, M.W. Pavlov, V.D. and Nersesov I.L. (1981) Tectonics and seismicity of the Tiktogul Reservoir region, Kirgisia, USSR submitted to J. Geophys. Res.

19. Molnar, P., Chen, W.F., Fitch, T.J. Tapponnier, P., Warsi, W.E.K. and Wu, F.T. (1977). Structure and tectonics of the Himalaya: a brief summary of relevant geophysical observations. Colloques Internationaux du CNRS, 268, 269-294. Ecologia et Geologie de L'Himalaya.

20. Latynina, L.A. and Karmaleeva R.M. (1978) Extensometer measurements: The measurement of crustal strain. Nauka: Moscow, pp. 152.

21. Seeber, L., and Armbruster J. (1979) Seismicity of the Hazara Arc in Northern Pakistan - decollement versus basement folding. In Geodynamics of Pakistan, (A. Farah and K. DeJong, eds.) Geol. Surv. Pakistan, Quetta, pp. 361.

22. Chugh, R.S. (1974) Study of recent crustal movements in India and future programs. Int. Symp. Recent Crustal Movements, Zurich.

23. Kondorskaya, N.V., and Shebalia N.V. (1977). New Catalogue of Strong Earthquakes in USSR from Ancient Times to 1975 (in Russian), Moscow, Nauka, 535 pp.

The Pakistan to Russia triangulation connexion: past and projected error analyses

T.O. Crompton
University College, London

ABSTRACT

Using those data available, an error analysis of the triangulation connexion from Gilgit to the Russian border, which was observed by Kenneth Mason's 1913 expedition, is performed. This analysis is compared with an error analysis of the projected re-observation of this connexion. The latter analysis may be treated, in part, as an optimisation process enabling crucial measurements for the control of scale, to be identified, based on a priori estimates of measuring accuracies.

The paper continues with some comments on how the calculation of the observations should proceed with a view to abstracting tectonic movements by comparison of the two epochs of measurement. The difficulties of obtaining such results and their statistical significance are described. Finally, looking to the future, the relevance of the 1980 observations is discussed in the context of possible repeat measurements within, say, the next ten years.

INTRODUCTION

Whenever a relative movement of the earth is taking place, for example crustal compression, shear, doming or subsidence, then conventional field survey methods may be employed in an attempt to quantify these movements. The problem is analogous to surveying on ice where moving stations, reflecting variable ice velocities, are combined with fixed rock stations. The survey measurements may be of several types involving various combinations of horizontal and vertical angles, distances and height differences. Such measurements may be supplemented by specialised measurements using tiltmeters and extensometers in the immediate vicinity of the movement. Only conventional survey methods are considered here applied to the detection of horizontal movements.

The principle on which the detection of such tectonic displacements is based is easily understood. A network of permanently marked stations is observed at two epochs separated by a known interval of time. The calculation of the observations from both epochs results in each station having two distinct sets of co-ordinates. From these it is an easy matter to calculate vector displacements. In practice, there are considerable difficulties in interpreting the observations. All survey measurements contain errors which can never be eliminated completely, either by proper observing procedure or by the application of adjustment techniques. Hence, the co-ordinates derived from the observations at each epoch will contain errors. Their magnitude depends upon the actual errors in the observations which are, of course, never known. The error in a deduced vector may be larger than the errors in the co-ordinates at either epoch. Although it is impossible to quantify these errors exactly, they may be estimated in terms of certain statistical parameters. It is vital that observational and computational procedures are designed to give vector displacement values that are statistically significant. Obviously the anticipated magnitude of displacements is important in this context. For example, suppose it was possible to deduce a vector displacement with a standard error of + 200 mm: then such results would give significant estimates of movements of the order of 1m, whereas estimated movements of say 100mm would be meaningless. There is a grey area where one must be less pedantic and unfortunately many results fall into this category. Under these circumstances, significance can be enhanced by considering the broad pattern of movements over a wider area and by basing results on more than two epochs of measurement.

One of the proposed activities of the International Karakoram Project 1980 is to perform repeat measurements on part of the triangulation connecting Gilgit to the Russian border. Such a project has exciting possibilities.

(i)   The network is situated in a highly active seismic region.

(ii)   Estimated movements are large and almost 70 years have elapsed since the first epoch of measurements. Armbruster et al. (1) quote a value of 37mm per year for the northwards movement of the Indian plate relative to the Eurasian plate, which represents a displacement of approximately 2.5m between epochs.

(iii) The results can be used for the modelling of seismic activities and may thus lead to a greater understanding of earthquake prediction.

(iv)   It is hoped that the proposed work will provide the foundation and impetus for a continuing programme of such measurements resulting in a regular output of information, produced largely by Pakistan scientists, which would be of worldwide interest.

The main aim of this paper is to examine the first epoch of measurements, to describe the proposed measurements and to estimate the quality of the results that will be obtained.

HISTORICAL BACKGROUND

Figure 1 shows the triangulation connexion which was observed by Kenneth Mason's expedition in 1913. This is a classical triangulation chain consisting of triangles, braced quadrilaterals and centre point figures. The chain links two stations of the primary triangulation of Pakistan, Yashochish and Dinewar, with two stations of the Russian triangulation, Kukhtek and Sarblock. (Henceforth in the text, and in all diagrams, stations are identified numerically. Table 1 gives the required station names, numbers and co-ordinates.)

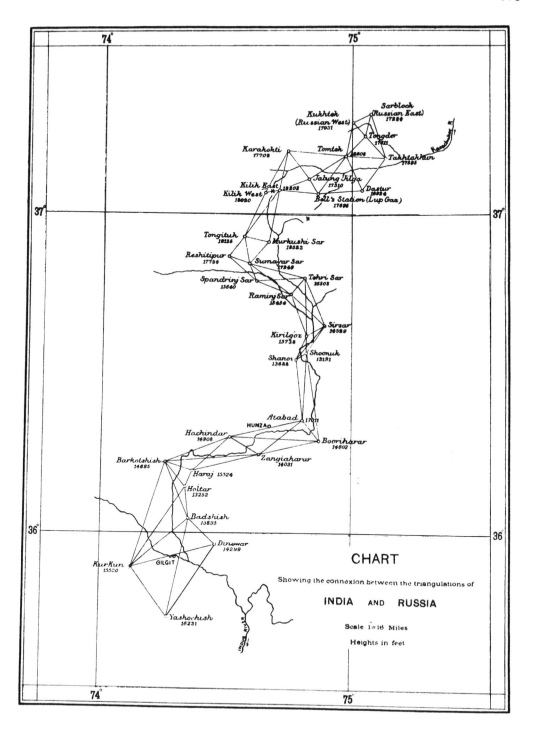

Fig. 1   Mason's 1913 Triangulation Connexion, Reproduced from (2)

TABLE 1   Station Names, Numbers and Co-ordinates for
the Southern Section of the Connexion.

No.	Name	Latitude	Longitude	Eastings	Northings
1	Yashochish	35°44'04".14	74°16'31".43	14 608.9	1 551.3
2	Kurkun	35 53 30 .81	74 07 42 .89	1 424.9	19 077.9
3	Dinewar	35 57 42 .37	74 28 02 .72	32 029.3	26 715.2
4	Badshish	36 02 43 .32	74 21 34 .81	22 339.8	36 015.4
5	Holtar	36 08 40 .41	74 20 32 .48	20 816.5	47 025.7
6	Barkotshish	36 13 22 .01	74 15 42 .50	13 604.4	55 731.9
7	Haraj	36 11 41 .67	74 22 19 .85	23 517.3	52 603.6
8	Hachindar	36 18 03 .40	74 31 19 .34	37 011.4	64 338.1
9	Zangiaharar	36 14 57 .07	74 38 13 .06	47 330.5	58 586.1
10	Booriharar	36 17 30 .30	74 52 25 .04	68 588.3	63 328.1
11	Atabad	36 21 24 .87	74 48 22 .66	62 530.6	70 546.8
12	Shanoz	36 32 34 .87	74 46 41 .50	59 984.9	91 193.7
13	Shoonuk	36 33 44 .44	74 49 09 .69	63 666.9	93 343.0

Scale is introduced from the calculated side length of the Pakistan primary triangulation and may be checked by comparison at the two Russian stations. All angles were observed using Troughton and Simms six-inch theodolites of the type shown in Fig. 2. It was impossible to carry larger, more accurate instruments to the stations and consequently the connexion is deemed to be of second order accuracy. Observations were made on six zeros, normally to helios, but in a few cases to opaque signals.

Mason (2) quotes figures for the mean triangular errors of various sections of the chain which may be used to give a useful estimate of the quality of the observations. Figure 3 shows the southern portion of the link for which an average triangular error of 2.18" was achieved. It is this part of the triangulation that will be considered in the following analyses and which forms the priority area for re-observation. Using Mason's figures, a value of ±0.9" may be derived for the standard error of a direction which is a remarkable achievement. Any uncertainty in the distance from 1 to 3 constitutes an additional source of error. Such a scale error will result from several sources, namely:

(i)   round off in the available co-ordinates of 1 and 3;

(ii) inherent scale error in the measured base-lines of the primary triangulation of Pakistan, and

(iii) accumulation of scale error due to observational errors in the primary triangulation.

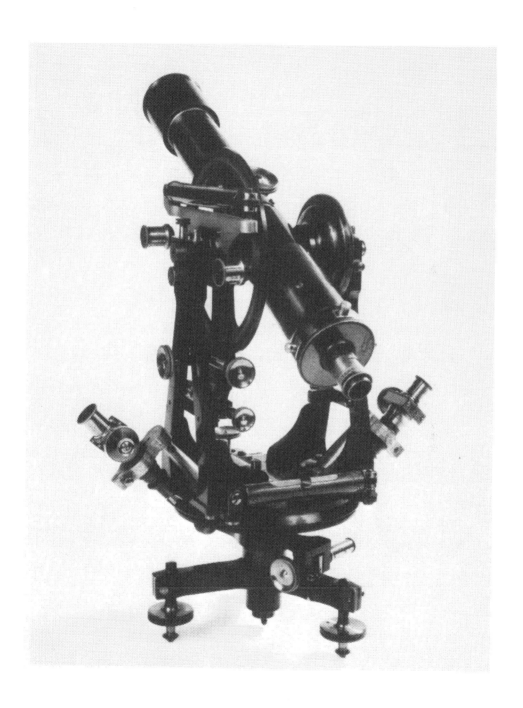

Fig. 2 A Troughton and Simms Theodolite of the type used by Mason

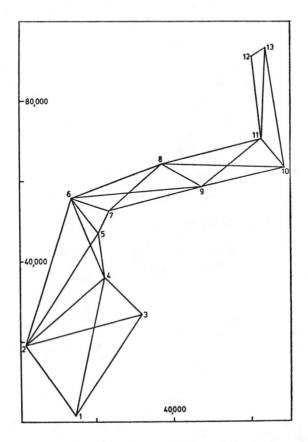

Fig. 3   The priority area for resurvey on a transverse Mercator
projection, scale 1:750,000 approximately.

The co-ordinates of 1 and 3 are quoted to two decimal places of a second of
arc.   Consequently, the latitude and longitude of these stations may exhibit
round off errors of as much as ±0.005".   Appendix I shows that these round
off errors may, in the worst case, cause a distance error of ±13ppm or an
azimuth error of ±2.6" in the line 1 to 3.   Bibby and Walcott (3) warn
against problems of orientation and Crompton (4) gives similar advice regarding scale when displacements are calculated directly from co-ordinates.

The second two sources of error may be taken together though, unlike round
off, neither may be formally quantified.   A base-line, measured in catenary
with properly standardised equipment, would be unlikely to contain errors
of more than 1 or 2 ppm.   However, the degeneration of scale through an
apparently first-class triangulation can be extremely serious.   The retriangulation of Great Britain has areas where scale error has accumulated
to 40 ppm (5).   With modern instrumentation it is relatively simple to obtain
these figures by comparing computed and directly observed side lengths of
the triangulation.   The conclusion is that the line 1-3 of Mason's work will
contain an unknown scale error, the magnitude of which is almost certain
to be intolerable in the context of tectonic displacement measurements.

The resurvey will include distance measurements, but they can only be used to maximum effect if they are able to be reconciled with the scale of Mason's work. This may be achieved by experimentally determining a scaling factor for the proposed observations by remeasurement of the line 1-3. Ideally this scaling factor should be determined by several observations which include other lines in that figure of the primary network from which the connexion springs. The success of this technique depends upon the line 1-3 remaining unchanged in length between epochs and upon finding suitably precise permanent marks at 1 and 3.

THE PRINCIPLES OF NETWORK ANALYSIS

The attempted analysis represents, in part, the second and third order aspects of the more general problem of network design (6). The problem is simplified by knowing the configuration of the network and having a posteriori knowledge of the quality of Mason's work. A priori estimates must be made of the quality of the proposed work but a certain amount of optimisation may take place by testing suitable combinations of proposed observations. The techniques used are well understood but the theory is outlined here for completeness.

Consider the observations made at a specific epoch. They will contain errors and, because redundancies exist in the network, inconsistencies will arise if all the available information is used. The usual practice is to perform some form of adjustment, often by the method of variation of co-ordinates. Each observation is used to construct an observation equation which results in a set of such equations of the form $Ax=\ell$. This set of equations will be inconsistent because of the redundancies in the network and the unavoidable observing errors. Using a criterion of minimum variance, an estimate $\hat{x}$ of x is made from the equations

$$A^T pAx = A^T p\ell . \tag{1}$$

Now the solution $\hat{x}$ will contain errors e as a result of errors ε in the observations. Hence:

$$A(\hat{x} + e) = (\ell + \varepsilon) \tag{2}$$

and

$$A^T pA(\hat{x} + e) = A^T p (\ell + \varepsilon) \tag{3}$$

Expanding gives:-

$$A^T pA\hat{x} + A^T pAe = A^T p\ell + A^T p\varepsilon \tag{4}$$

Therefore

$$A^T pAe = Ap\varepsilon \tag{5}$$

and

$$e = (A^T pA)^{-1} A^T p\varepsilon \tag{6}$$

Hence

$$ee^T = (A^T pA)^{-1} A^T p\varepsilon\varepsilon^T pA(A^T pA)^{-1} \tag{7}$$

since $\quad p = p^T \quad$ and

$$E\{ee^T\} = (A^TpA)^{-1}A^TpE\{\varepsilon\varepsilon^T\}\ pA(A^TpA)^{-1} \qquad (8)$$

Now, since the observations are assumed to be independent,

$$E\{\varepsilon\varepsilon^T\} = \sigma_o^2\ p^{-1} \qquad (9)$$

where σ_o^2 is the coefficient of variance or the variance of unit weight.

Hence

$$
\begin{aligned}
E\{ee^T\} &= (A^TpA)^{-1}A^Tp\sigma_o^2 p^{-1}pA(A^TpA)^{-1} \\
&= \sigma_o^2(A^TpA)^{-1}A^TpA(A^TpA)^{-1} \\
&= \sigma_o^2(A^TpA)^{-1} \\
&= \sigma_o^2 N^{-1} \qquad (10)
\end{aligned}
$$

The term $E\{ee^T\}$ is a symmetric matrix known as the variance-covariance matrix which contains the variances of the solution on the diagonal and the covariances at the off-diagonal locations. Hence, the quality of the solution may be estimated by knowing σ_o^2 and N^{-1}. The term $N^{-1} = (A^TpA)^{-1}$ may be computed for a given network. The coefficients of A depend only upon the relative positions of the network stations and the elements of p are the weights attached to the observations. Usually σ_o^2 is estimated from $\sigma_o^2 = (v^Tpv) / (m-n)$ where $v = A\hat{x}-\ell$, m is the number of observations and n is the number of unknowns. A value of σ_o^2 close to 1 would indicate that the a priori weights of the observations (i.e. their standard errors) had been estimated reasonably accurately. In the case of proposed observations, or indeed for Mason's work, where the actual observations are not immediately available, a priori standard errors may be allotted, and assuming that they are achieved (i.e. that $\sigma_o^2 = 1$), the variance-covariance matrix of the solution may be computed before an observation is even made. The calculations require moderately powerful computers and the validity of the results depends largely upon the sensible choice of a priori standard errors for the observations.

It is often convenient to depict the error estimates for a point as an ellipse. This is known as an error ellipse of position and is drawn centred on the accepted position of the point with its semi-major axis oriented in the direction of maximum error. Figure 4 shows this representation with the pedal to the ellipse which more correctly describes the error distribution. Allan (7) derives the formulae from which the ellipse may be drawn. These ellipses generally have the disadvantage of being dependent upon the choice of network origin. However, this is precisely the situation in movement surveys as one point, the origin, is assumed not to have moved between epochs and hence displacements are determined relative to that origin.

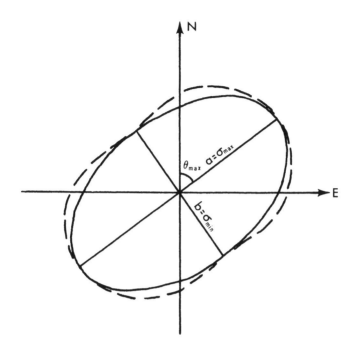

Fig. 4 The error ellipse of position and its pedal curve

ALLOCATION OF STANDARD ERRORS

It has been shown that the validity of network testing depends largely upon a sensible choice of standard errors for the observations. The following assumptions have been made for this work.

(i) Station 1 is held fixed as the origin at both epochs.

(ii) The standard error of a direction for both Mason's work and the proposed work is $\pm$ 0.9".

(iii) Distances will have a standard error of $[\pm 0.02^2 + (5.10^{-6} \cdot L)^2]^{0.5}$ m.

(iv) The base-line 1-3 is permanently marked and has not moved. This implies that no a priori azimuth standard error is required (i.e. that line 1-3 has maintained its azimuth between epochs) and that a scaling factor can be obtained to bring the proposed distance observations in line with the scale of Mason's work. This scaling factor will itself have a standard error, probably of the order of $\pm$5ppm, and this is best accounted for by increasing the value given at (iii) to $[\pm 0.03^2 + (7.10^{-6} \cdot L)^2]^{0.5}$ m and holding 3 fixed. If this assumption cannot be made, then computations would have to proceed along the lines described in (3) which detects displacements by analysing the changes in horizontal angles between two epochs. Measured distances would still be of importance as they help to control the errors in the angle observations.

THE ANALYSES

All analyses were performed using the same program running on an IBM 360/65 machine. With the options used, the input data consist of the approximate co-ordinates of the fixed and other network stations together with a list of the type and standard error of each proposed observation. The program is designed to operate in terms of grid co-ordinates and approximate projection grid co-ordinates were computed from the geodetic co-ordinates on a traverse Mercator projection, central meridian $74°40'E$, origin $35°00'N$. (See Table 1). From these data the variance-covariance matrix is computed. The elements of A are derived from the approximate station co-ordinates and the elements of p from the a priori standard errors. Various forms of output are available but the most appropriate to this problem consists of the parameters for error ellipses of position, relative error ellipses and the standard errors of azimuth and distance.

Five analyses have been performed.

(i) Mason's work, holding 1 and 3 fixed with $\pm0.9''$ as the standard error of a direction.

(ii) The proposed observations, holding 1 and 3 fixed with $\pm0.9''$ as the standard error of a direction and all distances with a standard error of $(\pm0.03+7ppm)m$.

(iii) As for (ii) except 9 of the shorter distances are not included.

(iv) As for (i) but including 10 of the longer distances.

(v) As for (ii) except 7 distances longer than 30km are not included.

The results of the five analyses are summarised in Table 2. The semi-axes $(\sigma_{max}$ and $\sigma_{min})$ and the orientation (θ_{max}) of the error ellipses of position

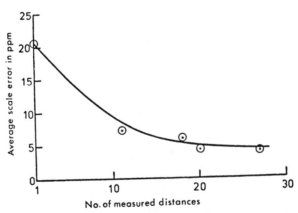

Fig. 5 Graph of average scale error against number of measured distances

STN	2	4	5	6	7	8	9	10	11	12	13	ANALYSIS
σ max	0·124	0·134	0·197	0·253	0·219	0·520	0·667	1·236	1·142	1·502	1·634	1
σ min	0·099	0·061	0·146	0·181	0·191	0·296	0·317	0·434	0·422	0·716	0·717	
θ max	349 04	304 11	335 23	325 12	344 35	51 01	76 33	77 26	65 17	33 52	35 40	
$\bar{\sigma}_L$							20·6 ppm					
σ_Δ	0·158	0·155	0·219	0·275	0·245	0·537	0·685	1·248	1·154	1·510	1·642	
σ max	0·099	0·078	0·112	0·145	0·134	0·207	0·218	0·324	0·319	0·406	0·429	2
σ min	0·067	0·054	0·083	0·091	0·088	0·109	0·121	0·151	0·142	0·161	0·159	
θ max	354 29	294 11	281 23	80 08	282 25	295 53	315 11	330 16	318 59	303 48	305 27	
$\bar{\sigma}_L$							4·0 ppm					
σ_Δ	0·158	0·155	0·219	0·275	0·245	0·537	0·685	1·248	1·154	1·510	1·642	
σ max	0·121	0·126	0·164	0·187	0·186	0·269	0·287	0·401	0·394	0·483	0·507	3
σ min	0·081	0·058	0·132	0·150	0·141	0·151	0·153	0·175	0·170	0·198	0·195	
θ max	344 12	302 59	314 59	276 58	301 38	304 26	319 22	329 30	320 27	305 27	307 10	
$\bar{\sigma}_L$							6·0 ppm					
σ_Δ	0·173	0·184	0·255	0·305	0·281	0·547	0·694	1·254	1·158	1·515	1·645	
σ max	0·122	0·126	0·165	0·193	0·186	0·273	0·295	0·407	0·399	0·492	0·517	4
σ min	0·084	0·058	0·132	0·158	0·152	0·173	0·181	0·221	0·213	0·240	0·233	
θ max	345 41	303 10	315 13	287 06	299 58	302 21	316 40	328 01	319 22	305 01	306 41	
$\bar{\sigma}_L$							7·2 ppm					
σ_Δ	0·174	0·184	0.255	0·313	0·281	0·553	0·702	1·261	1·166	1·521	1·650	
σ max	0·102	0·080	0·113	0·146	0·135	0·210	0·221	0·326	0·321	0·410	0·433	5
σ min	0·073	0·056	0·089	0·097	0·093	0·115	0·127	0·160	0·151	0·172	0·171	
θ max	359 01	295 28	283 02	79 33	283 14	295 35	314 20	329 34	318 21	303 49	305 31	
$\bar{\sigma}_L$							4·3 ppm					
σ_Δ	0·161	0·156	0·221	0·276	0·247	0·539	0·687	1·249	1·155	1·512	1·643	

TABLE 2   Results of the Five Analyses

are given for all stations except those held fixed. Also, the average scale error $(\bar{\sigma}_L)$ of each network is given in ppm and the standard error of a displacement (σ_Δ) for each station between epochs. Figure 5 plots average scale error against number of measured distances using the technique described in (8) and demonstrates the diminishing returns that accrue from the measurement of all possible distances. The theory for the calculation of σ_Δ is developed in Appendix II and σ_Δ may be regarded as the most pessimistic estimate of the standard error of the calculated displacement of a station between the two epochs.

CONCLUSIONS

In the realms of statistical estimation, great caution must be exercised in drawing conclusions, a situation which is reinforced by the fact that the second epoch of measurements has not yet been attempted. A greater degree of confidence will be possible once the second epoch of measurements is complete and Mason's original work has been unearthed. However, certain remarks can be made. The quality of the original work stands in tribute to Kenneth Mason and any improvement in the proposed work is only made possible by virtue of technological advances. The accuracy of Mason's work, coupled with the 67 year time interval and large rates of movement enable one to predict that useful results will be obtained and that smaller scale networks will be established in the immediate vicinity of movement. Even so, a uniform distribution of displacement throughout the area (9) or a high destruction rate for the original station markers would seriously hamper the programme. Whatever the outcome of the 1980 observations, it is hoped that they will define a network which could profitably be remeasured at shorter intervals and act as a regular source of data for international geophysical research.

ACKNOWLEDGEMENTS

The author is indebted to his colleagues in the Department of Photogrammetry and Surveying at University College London, for their help and advice.

APPENDIX I

Figure 6 shows a typical situation depicted on the plane for the line joining 1 and 3 on the spheroid. From the diagram it can be seen that errors in length and azimuth, $\delta \ell$ and $\delta \alpha$, are given by:

$$\delta \ell_{13} = \rho_1 \delta \phi_1 \cos \alpha_{13} + \upsilon_1 \cos \phi_1 \delta \lambda_1 \sin \alpha_{13} + \rho_3 \delta \phi_3 \cos \alpha_{31} +$$
$$\upsilon_3 \cos \phi_3 \delta \lambda_3 \sin \alpha_{31} \tag{A1}$$

and

$$\delta \alpha''_{13} = (\rho_1 \delta \phi_1 \sin \alpha_{13} - \upsilon_1 \cos \phi_1 \delta \lambda_1 \cos \alpha_{13} + \rho_3 \delta \phi_3 \sin \alpha_{31}$$
$$- \upsilon_3 \cos \phi_3 \delta \lambda_3 \cos \alpha_{31}) / \ell \sin 1''. \tag{A2}$$

Now displacements are small
$$(|\delta \phi_1| = |\delta \phi_3| = |\delta \lambda_1| = |\delta \lambda_3| < 0''.005) \tag{A3}$$

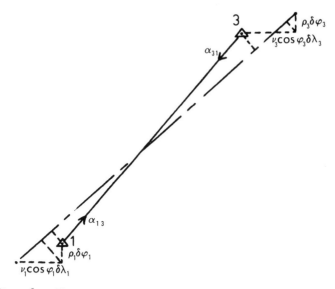

Fig. 6.   The Effect of Round Off on Azimuth and Distance

and line lengths are moderately short, hence some simplifications may be made. Putting

$$\rho_1 = \rho_3 = \upsilon_1 = \upsilon_3 = R \tag{A4}$$

and

$$\alpha_{31} = \alpha_{13} \pm 180° \tag{A5}$$

gives

$$\delta \ell_{13} = R \cos \alpha_{13} (\delta \phi_1 - \delta \phi_3) + R \cos \phi \sin \alpha_{13} (\delta \lambda_1 - \delta \lambda_3) \tag{A6}$$

$$\delta \alpha_{13} = R \sin \alpha_{13} (\delta \phi_1 - \delta \phi_3) - R \cos \phi \cos \alpha_{13} (\delta \lambda_1 - \delta \lambda_3) \tag{A7}$$

These give error bounds of 0.40m for $|\delta \ell_{13}|$ and 2.58" for $|\delta \alpha''_{13}|$.

APPENDIX II

Any station will have parameters for the error ellipses of position at two different epochs. The problem is to use this information to estimate the maximum standard error that might occur in a computed displacement. Ideally one requires the covariances between the two epochs. These cannot be evaluated and, although logical arguments exist that would indicate cor-relation, the results from the two epochs must be considered independent, i.e. uncorrelated. For the first epoch, the standard error of position in any direction θ is given by:-

$$2\sigma_1{}^2{}_{,\,\theta} = a_1{}^2 + b_1{}^2 + (a_1{}^2 - b_1{}^2) \cos 2 \,(\theta_{1,\,max} - \theta) \qquad\qquad (A8)$$

and for the second epoch by:

$$2\sigma_2{}^2{}_{,\,\theta} = a_2{}^2 + b_2{}^2 + (a_2{}^2 - b_2{}^2) \cos 2 \,(\theta_{2,\,max} - \theta) \qquad\qquad (A9)$$

Combining these as uncorrelated quantities gives:

$$2\sigma_\Delta{}^2 = 2\sigma_1{}^2{}_{,\,\theta} + 2\sigma_2{}^2{}_{,\,\theta} = a_1{}^2 + b_1{}^2 + (a_1{}^2 - b_1{}^2) \cos 2(\theta_{1,\,max} - \theta)$$

$$+ \; a_2{}^2 + b_2{}^2 + (a_2{}^2 - b_2{}^2) \cos 2 \,(\theta_{2,\,max} - \theta) \qquad\qquad (A10)$$

By putting $2 \,(\theta_{1,\,max} - \theta) = \alpha$ and $2(\theta_{2,\,max} - \theta) = \alpha + \Delta\alpha$ a value of α which maximises $\sigma_\Delta{}^2$ may be found from:

$$\cot \alpha = \frac{- \,(a_1{}^2 - b_1{}^2)}{(a_2{}^2 - b_2{}^2)} \; \mathrm{cosec}\,\Delta\alpha \; - \; \cot \Delta\alpha \qquad\qquad (A11)$$

Hence the value of θ which maximises the function is obtained and substitution then gives $\sigma_\Delta{}^2$.

REFERENCES

(1) Armbruster, J., Seeber, L. and Jacob, K.H.   The northwestern termination of the Himalayan mountain front: active tectonics from micro–earthquakes. Journal of Geophysical Research, 83: 269–282 (1978).

(2) Mason, K.,   A note on the degree of accuracy of the secondary link series, triangular errors, mean errors, etc.   Appendix B in Records of the Survey of India, Volume VI.   Completion of the link connecting the Triangulations of India and Russia 1913.   Survey of India, Dehra Dun. 121 pages; 114–115 (1914).

(3) Bibby, H.W. and Walcott, R.I.   Earth deformation and triangulation in New Zealand.   New Zealand Surveyor, XXVIII(6): 741–762 (1977).

(4) Crompton, T.O.   Some aspects of tectonic movement survey in Southern California.   Published   in – Chartered Land Surveyor/Chartered Minerals Surveyor, Vol. 2   No. 4: 19–37 (1981).

(5) Ordnance Survey.   The history of the retriangulation of Great Britain 1935–1962.   H.M.S.O., London.   395 + xix pages.   See pages 207–208 (1967).

(6) Cross, P.A. and Thapa, K.   The optimal design of levelling networks. Survey Review, XXV(192): 68–79   (1979).

(7) Allan, A.L.   The error ellipse: a further note.   Survey Review, XXI(166): 387–390 (1972).

(8) Ashkenazi, V. and Cross, P.A.   Strength analysis of block VI of the European triangulation.   Bulletin Géodésique, 103: 5–24 (1972).

(9) Bilham, R. and Simpson, D.   Indo–Asian convergence and the 1913 survey line connecting the Indian and Russian triangulation surveys. This volume pp. 160–170.   Paper read at the Islamabad Conference; Recent Advances in Earth Sciences. (June, 1980).

Special techniques for surveying on moving terrain

J.L.W. Walton
University College, London

ABSTRACT

At some stage in almost all land surveying projects a network of fixed or "control" points has to be set up. When the area being studied involves moving terrain, such as a glacier, special survey techniques have to be used either to eliminate errors incurred by the moving control points, or to actually determine their velocity. This paper contains a review of a number of these special techniques ranging from simple "by eye" approximations to complex methods requiring extensive computer facilities. Some of these techniques may be applied to problems involving infinitesimal velocities such as the tectonic plate movement studies planned for the 1980 International Karakoram Project.

INTRODUCTION

Glaciology is the study of ice. This sounds straightforward but in fact this single subject concerns scientists of many disciplines with varied interests and skills. These may include electronic engineers, chemists, geophysicists, geographers or theoretical physicists. Not the least important discipline is surveying, the science of measurement.

In its traditional form, land surveying is used for map making. It can indeed be used in this way for glaciological studies; a map of a glacier or associated feature is often essential before detailed field work can proceed.

However, in the study of ice dynamics concerned with how and why ice masses move and deform, land surveying methods have long been used as a straightforward way of obtaining quantitative field data which can be used to help form, verify, or even refute hypotheses concerning this highly complex medium.

Use of land surveying in this context dates back as far as 1846 (1) when some of the earliest glaciological field work was carried out on the Mer-de-Glace in France. This study used traditional methods and indeed it is only over the last 30 years or so that it has been found necessary to modify these methods for glaciological work.

These modifications initially became necessary as the glaciological studies became more detailed and approximations invoked in the computing had to be minimised so as to increase the accuracy of the results. Initially there were severe limitations as to the refinements that could be introduced because all computations had to be carried out by hand. However, since the advent of electronic computing aids it has become possible to eliminate virtually all approximations and to produce meaningful results from observations that would previously have been considered incomplete.

BASIC PROBLEMS OF GLACIER SURVEYING

In most land surveying problems, one is dealing with a number of rock stations whose positions relative to one another must be measured. These stations do not move, so only three dimensions need be considered; the coordinates in three orthogonal axes. The axes commonly used are North, East and a vertical height. Observations between these stations do not vary with time.

The glaciologist is faced with a different situation. Consider a number of marked stations on a glacier. The glacier is continuously moving and deforming so all observations made to or from these stations will be time dependent.

To measure glacier movement, the usual observing technique is to measure the positions of these ice stations relative to co-ordinated rock stations at one instant in time, then allow a suitable time to elapse before repeating the measurements. Two sets of co-ordinates for the ice stations result and the difference between the two sets is a measure of the movement.

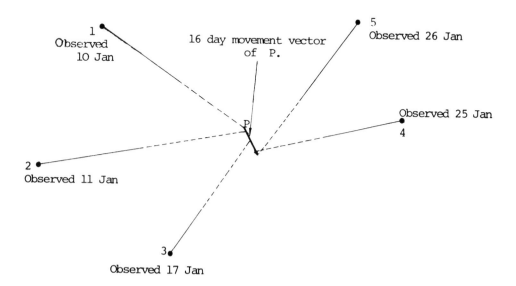

Fig. 1   Movement of ice station P during period of observation
from rock stations 1 to 5.

Consider the case illustrated in Fig. 1, where a single measurement has been made to an ice station from each of a number of rock stations. Because of accessibility, weather and other problems, some 16 days have elapsed between the first and last of these observations. Thus the point sighted from station 1 does not have the same co-ordinates as the point sighted from station 5. To use these five observations to compute a single value for the co-ordinates of "P" is hardly appropriate. If the movement vector of "P" during the 16 day period is such as to subtend an angle of say "10x" seconds of arc at station 5, with accuracy of observation being "x" seconds, then neglecting to take account of the 16 days elapsed time could clearly result in observation equation residuals of as much as "5x" seconds. Thus best use is not being made of the available data.

SPECIAL TECHNIQUES

Reduction to Epoch of Complex Networks

The first attempt to overcome the problem mentioned above was made in 1951, (2). Two complete sets of observations, about a year apart but with each set spread over 1 to 2 months were used to produce values for ice deformation. No values for absolute movement were obtainable as the nearest exposed rock was some 250 kilometres distant! A triangulation network was used with one baseline, measured twice in successive years; see Fig. 2. A "reduction to epoch" technique was devised to eliminate the time variation of observations Each angle was observed twice, thus a rate of change of angle could be calculated assuming that all angles varied linearly with time. Each observation was reduced to what it would have been at a selected instant of time, this instant chosen to lie somewhere near the actual time of observation. Thus an "initial" and a "final" value was computed for each angle and the procedure repeated for all observations, reducing each to the same two instants. The mechanics of this procedure is illustrated in Table 1 for data related to Fig. 1.

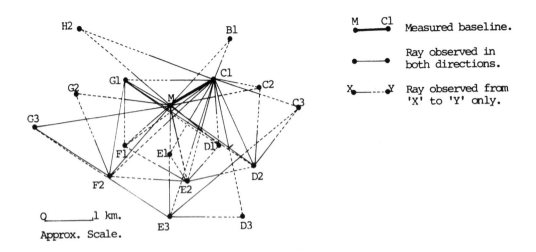

Fig. 2   Maudheim Survey April–June 1950 and May–Aug. 1951.

TABLE 1   Example of Reduction-to-Epoch of Angle and Distance Observations

	OBSERVATIONS					Period	Rate of	REDUCED QUANTITIES	
	Initial	Date	Final	Date	Change	Days	Change	to 14 Jan	to 27 Oct
Angle at 1 to P	41°27'36"	10 Jan	41°37'32"	31 Oct	+ 596"	294	+ 2.03" per day	41°27'44"	41°37'24"
Angle at 3 to P	37°26'14"	17 Jan	37°15'56"	24 Oct	− 618"	280	− 2.21" per day	37°26'21"	37°15'49"
Distance 5 to P	4753.71 m	26 Jan	4746.68 m	20 Oct	−7.03m	267	−0.026m per day	4753.45 m	4746.51 m

Since the advent of electromagnetic distance measuring devices, trilateration or mixed networks have become much more widely used. This observation reduction technique can and has been used for many of these networks, a few of which are illustrated in Fig. 3. This method makes one assumption, that all observations vary linearly with time.

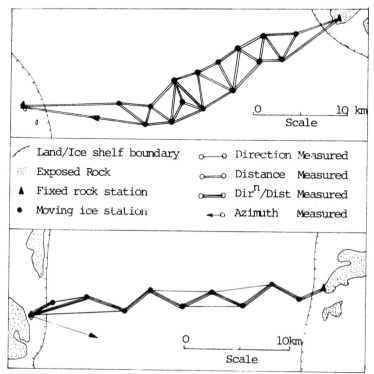

Fig. 3.

Sketch Maps of two survey networks set up for ice movement studies on George VI Ice Shelf, Antarctica

An alternative method, albeit similar in concept, was evolved in the early 1960's (3). This involves a time reduction of <u>positions</u> instead of observations, with the assumption that the velocity of any station on the ice is independent of time. Consider the start of a traverse as shown in Fig. 4.

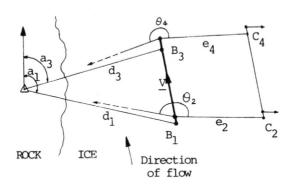

Fig. 4   Start of a traverse which will be computed by
the position reduction method.

The initial azimuth of the traverse, from rock station "A" to ice station "B" was observed at two points in time (t_1 and t_3), with values a_1 and a_3. The distances (d_1 and d_3) are also measured so two sets of co-ordinates can be computed for "B". Thus the movement is known (vector $\underline{V}$). The computed sets of co-ordinates are now adjusted by applying the appropriate velocity vectors so as to produce values (B_2 and B_4) corresponding to the two instants of observation (t_2 and t_4). Angles (θ_2 and θ_4) and distances (e_2 and e_4) as observed are applied to these new co-ordinates and a value for the movement of "C" is found. This process is repeated along the traverse.

Clearly, this position reduction method will involve much computing and the resulting accuracy is only as good as the assumption made. As with the observation reduction method, it could be refined with respect to acceleration and curvature of the ice movement but computing would then become very complex. It is not easy to apply the position reduction method to other than simple traverses.

Resection of a Single Station

Both the above techniques discuss large networks tied into fixed rock stations that must be visited. Access to these rock stations can often pose a serious problem; severe crevassing is all too widespread near the edge of glaciers and even when the surveyor has reached terra firma he may have a long walk or climb, burdened by heavy equipment, to reach the station. One way around this is to use a resection survey method. This enables the observer to co-ordinate a single station accurately with only one instrument set-up, thus obviating all the problems of adjustment of time dependent quantities. Repeat the resection some time later and a movement vector can be computed.

An accurate resection requires at least three accurately co-ordinated rock stations. Glaciers frequently inhabit the more inaccessible parts of this planet where maps are rudimentary and of small scale while co-ordinated and visible rock stations are rarely available.

This impasse can be overcome. Firstly consider what information is being sought by the glaciologist. The answer is movement, unlike the co-ordinates required in traditional surveying. Usually a movement vector is expressed as the difference between two co-ordinates separated by a period of time. Yet it is possible to measure the movement directly and accurately and only obtain approximate absolute co-ordinates.

Conventional resection formulae require one to know the geometry of the figure formed by the control points. If the co-ordinates of these points are in error, the computed position of the unknown station will also be in error. However, this error will be the same if the resection is repeated after a period of time provided that (a) the computations are carried out assuming the same co-ordinates for the control points, and (b) that the movement during the period is small compared with the size of the control triangle. The computed movement vector will be accurate and the erroneous co-ordinates still good enough for plotting on a map.

Consider the network shown in Fig. 5. Stations "1", "2", "3" are prominent but unco-ordinated rock features. Approximate co-ordinates for these relative to "P" can be found by setting out a short baseline P – P' and intersecting the features. The triangles are ill-conditioned but this does not matter. If the accuracy of observation is "x" then as long as the angle subtended by the baseline at any of the rock features is more than 100x, then the distances P to 1, P to 2 etc. can be found to better than 1%. If the azimuth of the baseline is measured by compass to say +1° then it would be possible to compute movements accurate to +1% and 1° limits. For most glaciological studies these error limits are completely acceptable.

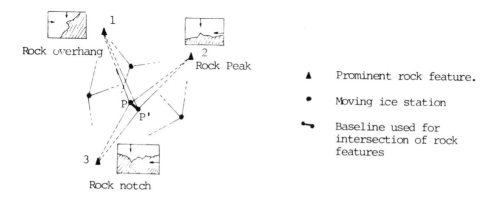

Fig. 5   Network of typical rock features used for resection of a moving ice station.

This method can be extended to cover a number of other ice stations on the same glacier. An accurate movement can be computed at each ice station as before by observation to the same rock stations without having to set up another base-line.

A technique similar to the one above involves using a map of the area, selecting features for observing as before, then co-ordinating them by placing a grid over the map. This has been used extensively in Antarctica (4) where the only maps available are from 1:250000 satellite imagery with a graticule overdrawn. Up to seven control features were used, between 15 and 80 kilometres distant and movement vectors were measured directly by using a modified semi-graphic resection method. The initial assumed position corresponds to the time of the first resection whilst the "computed-minus-observed" values were replaced by "final-minus-initial" values; a measure of the angular change between resections.

This method is used to measure velocities of isolated points and it can also be used to provide control for larger networks. Fig. 6 shows a net-work measured in Antarctica (4). This was a triangulation chain with four baselines. Three stations were used for control, the velocity for each was measured as above. The scheme was remeasured and both sets of measure-ments were computed by an observation reduction method. One of the control stations was used as origin while the other two provided checks for the computed velocities. Complete results were obtained without the observer ever having to leave the ice. The results of these velocity checks are shown in Table 2.

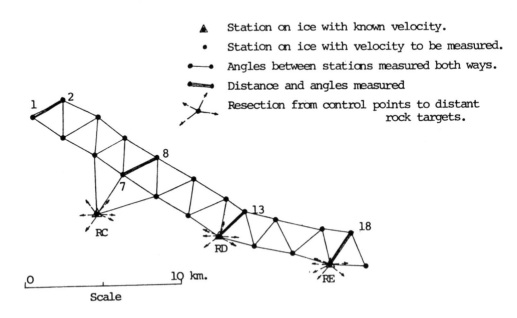

Fig. 6   Network of stations for ice movement study.
Control for the network is supplied from ice
stations of previously measured velocity.

TABLE 2   Comparison between measured and computed quantities
assuming Station RE of Fig. 6 as control velocity.

Distances and Velocities	Baseline RD to 13	Baseline 7 to 8	Baseline 1 to 2	Station RC	Station RD
Measured	2213.48m	2460.58m	2262.16m	158.5m/year 311.9°True	196.2m/year 300.8°True
Computed	2213.41m	2460.35m	2361.48m	162.4m/year 309.7°True	195.8m/year 304.1°True

Rigorous network analysis using time dependent co-ordinates

During the last few years a method has been evolved which eliminates all erroneous assumptions, can compute a network with incomplete observations, and can accept any combinations of rock and ice stations with known or unknown co-ordinates and velocities (5).   To date this is probably the most powerful and adaptable method available to the glaciological surveyor.

The basis of the method is the least squares solution of a set of observation equations.   This is the standard technique for obtaining the most probable values of station co-ordinates in a conventional static network and is discussed in detail elsewhere (6, 7).   The extension of this technique, to allow for moving stations, increases the number of parameters to be adjusted for each station but does not increase the mathematical complexity.

Instead of two unknowns for each station, all co-ordinates are considered to be time-dependent, the order n of the dependence polynomial varying with the example under investigation.   Each station now has $n + 2$ unknowns; X, Y, dX/dt, dY/dt, $d_2X/dt^2 \ldots d_nX/dt^n$.

To carry out the subsequent computation it is essential to have access to a sizeable computer and indeed the original development of this technique was carried out on the Cambridge University IBM 370/165 machine.

The programme currently in use accepts only up to second order polynomials but this is sufficient for almost all glaciological studies.   Control stations are specified with all parameters fixed.   Approximate co-ordinates are input for all other stations, controlling azimuths are specified, all other observations are listed, together with their time of observation and the system then uses standard software packages to carry out a rigorous least squares analysis.

The output consists of adjusted values of the unknowns at the time of the earliest observation with their estimated standard errors and the residual error of each observation.   Care must be taken in interpreting the output to allow for the fitting procedure giving values of velocities and accelerations unjustified by the accuracy of the observations.   For example, a static network badly observed, would give a lower residual variance by this method if each station was allowed to have a velocity associated with it; the velocity would, of course, be meaningless but the extra degrees of freedom would convert position errors to velocity values.

The method was first used to analyse data from Spartan Glacier in Antarctica. This data were incomplete and using traditional methods it would not have been possible to obtain results (Fig. 7). A single baseline was measured only once and only 16 days spanned the first and last observations of the network. The velocities of the moving stations have subsequently been measured over a one year time span. A comparison of the short and long term results is shown in Table 3.

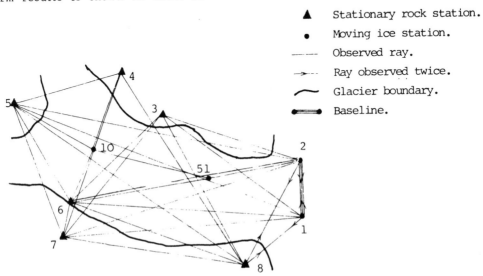

▲　Stationary rock station.

●　Moving ice station.

------　Observed ray.

→---　Ray observed twice.

⌒　Glacier boundary.

●══●　Baseline.

Fig. 7　Spartan Glacier Survey. (5)

The comparison is disappointing for stations 10 and 51 until one realises that velocities have been measured over a 3-day time span only, during which time the stations have moved only 70mm and 130mm respectively. Clearly, if a network were specifically designed for this computing method, then a single measure of each observation spread out over say three weeks will produce good results. To obtain velocities from a single survey is unprecedented.

TABLE 3　Comparison of short and long term velocity results from Spartan Glacier.

Station	VELOCITY			VELOCITY		
	Magnitude m/year	Azimuth °True	Period days	Magnitude m/year	Azimuth °True	Period days
1	109.7	284	17	105.3	286	350
10	5.5	114	3	8.8	120	350
51	15.8	68	3	15.0	96	350

The great strength of the method is its adaptability. It can be used for computing a single traverse (though with no great advantage over other methods) or for computing a complex network of stationary and moving stations that may have been measured many times. All observations can be used, even those not repeated.

This method could possibly be applied to a problem concerning tectonic deformation. It is possible that the crustal movements in the 1980 study will be very small and scarcely detectable above random survey errors. This is analogous to the short term survey at Spartan Glacier and as has been stated, care must be taken in the interpretation of results obtained by using this method.

CONCLUSIONS

This paper discusses several techniques for determining glacier velocities using standard survey instruments. The methods described attempt to overcome problems concerned with difficult and limited access to stations, and limited computing facilities without affecting the accuracy of results.

ACKNOWLEDGEMENTS

As with all review compilations, the sources from which ideas and data have been drawn are numerous. In particular the author would like to thank all his glaciological colleagues at the British Antarctic Survey who have so patiently helped and guided him in his association with glacier movement studies over the past seven years.

REFERENCES

1. Forbes, W.S.G. (1846) Illustration of the viscous theory of glacier motion. Philosophical transactions of The Royal Society of London. 136, 143-210.

2. Swithinbank, C.W.M. (1958) The movement of the ice shelf at Maudheim. Norwegian-British-Swedish Antarctic Expedition 1949-52. Scientific results III(C).

3. Dorrer, E., Hoffman, W.R. and Seufert, W. (1969) Geodetic results of the Ross Ice Shelf Survey Expeditions, 1962-63 and 1965-66. Journal of Glaciology, 8(52), 67-90.

4. Walton, J.L.W. (1979) Resection on moving ice. Survey Review, XXV, 191, 33-44.

5. Wager, A.C., Doake, C.S.M., Paren, J.D. and Walton, J.L.W. (1980) Survey reduction for glacier movement studies. Survey Review, XXV, 96, 1-13.

6. Rainsford, H.F. (1957) Survey adjustments and least squares. Constable London.

7. Allan, A.L., Hollway, J.R. and Maynes, J.H.B. (1968) Practical field surveying and computations. Heinemann, London.

Variations of the Batura glacier's surface from repeated surveys

Chen Jianming

Institute of Glaciology & Cryopedology, Lanzhou
People's Republic of China

ABSTRACT

Repeated terrestrial stereophotogrammetric surveys have been used to make various maps. The glacier example quoted shows basic data requirements and presents standard techniques to study height variations of the surface, the terminus mass and ice movement. The results show that the system is a good method to determine glacial variations. However, when applying this technique the orientation of each photo should be identical. In order to obtain reliable data, a combination of methods is necessary.

INTRODUCTION

The main purposes of glacier surveys are to determine the state of a glacier at a certain definite time and to measure time dependence of glacial surfaces in both horizontal and vertical directions. Such studies advance our knowledge of glaciers and assist predictions of future behaviour. It must be pointed out that the glacial behaviour is a function of locality, height and orientation all of which have distinctive effects. Thus differences in relief, direction of slope, thickness of moraine, ice grain size etc cause distinctive variations in behaviour even between close neighbouring points. It follows that when we study the characteristics of a certain glacier, in order to derive meaningful conclusions, a vast quantity of survey data including time, climate, hydrology and topographical relief is required. It is difficult to carry out a substantial theodolite intersection survey and simultaneously make the other essential observations. However terrestrial stereophotogrammetry permits rapid collecting of survey data. Furthermore for glacial variation studies it is more accurate and economical than an aerophotogrammetric survey. Hence, a repeated terrestrial stereophotogrammetric survey is an ideal method for measuring glacial variations.

Using repeated terrestrial stereophotogrammetric surveys of the Batura Glacier in 1974, 1975 and 1978 as an example, the application of this technique for making maps, and measuring glacial variation permits an assessment of its accuracy. The example besides providing the basic requirements also provides data for standard surveying techniques to study basic requirements and height variations of the glacial surface, the terminus ice mass, glacial movement and terminal advance.

BASIC REQUIREMENTS OF A REPEATED TERRESTRIAL STEREOPHOTOGRAMMETRIC SURVEY

Repeated terrestrial stereophotogrammetric survey requires the taking of photographs at a certain interval on the same fixed base line, and from such photographs to make topographic maps, as well as to compare and take measurements from them by stereocomparator to analyse glacial variations. In consequence, besides the basic requirements of common photography, for repeated terrestrial stereophotogrammetric survey the following are required. For periodic photography external conditions of the photostation must be maintained constant. The photostation must be positioned at a stable place, not on sliding ground, and a markstone must be buried with its centre accurately marked for long-term use. Near the photostation it is advisable to set up a witness mark so that the station can easily be found afterwards, e.g., a moraine cairn 1 m high. The photogrammetric mapping scale should be greater than 1:10000 and the accuracy of the photo-base length should not exceed 1:2000; the most suitable length of the photo-base line is about 1/5 to 1/15 of the glacier average width. In order to ensure intersection accuracy and stereoscopic effect, when the glacier width exceeds 2 km, the base line must be laid out on both sides of the glacier. Naturally, "dead space" must be avoided or at least be minimized.

Depending upon the accuracy of the measuring method, the extent of glacial variation and the required accuracy of its measurement, the interval of repeated surveys can be either several years or only days. Generally speaking, the longer the interval, the higher the relative accuracy of the collected data, but long intervals cannot also provide glacial variation over different periods of days, months, seasons or years. Furthermore, glacial surface points may disappear, and hence consecutive data can not always be collected. If the interval is too short, the reliability of data cannot be ensured. If conditions allow, the repeated surveying may be conducted at the beginning and the end of the ablation period. Under these conditions the reliability of data can be ensured and the annual glacial variation from ablation and accumulation can be obtained. The least time interval Δ between two repeated surveys is

$$\Delta t = \frac{\Delta D}{V \cdot n} \qquad (1)$$

where ΔD is the error of measurement, n the required accuracy of data and V is velocity of the point being examined.

THE MEASUREMENT METHOD OF GLACIAL VARIATIONS

1. Glacier movement in the horizontal direction

(a) The coordinate comparison method

From photographic prints taken at different times, the homologous image points must be carefully identified so that their coordinates can be determined. If the differences of the coordinate points are ΔY and ΔX then the movement D and the direction of glacier motion can be calculated.

$$D = \Delta Y^2 + \Delta X^2 \qquad (2)$$

Generally, if the photo-distance is 500 to 1500 m, the mean square error of movement by this method is $m_d = \pm 0.3$ to ± 1.8 m, $m_h = \pm 0.1$ to ± 0.4 m, the latter being the height error, whilst m_d is the horizontal movement error.

(b) The pseudo-parallax method

Since the glacier always has some variation over the time interval, the photopair (they are not a true stereoscopic pair) will have a so-called "pseudo-parallax" which includes both a horizontal and a vertical displacement parallax that can be measured by stereocomparator. The principle of the pseudo-parallax method is shown in Fig. 1.

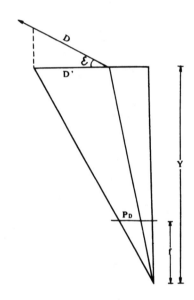

FIG. 1.   DIMENSIONS FOR THE PSEUDO-PARALLAX METHOD.

The glacial movement distance which is vertical to the optical axis direction of the photograph is given by

$$D' = \frac{Y}{f} P_D \qquad (3)$$

In the same way the variation in glacial height can be shown to be

$$\Delta h = \frac{Y}{f} Q \qquad (4)$$

where f is the focal length, Y is the photo-distance and P_D is the horizontal displacement parallax.

Figure 1 shows that the photographic direction is vertical to the glacial movement direction. In fact, they cannot be vertical to each other and there is an angle of deflection between them, so the actual distance D of glacial movement can be shown as follows:

$$D = \frac{D}{\cos \varepsilon} \qquad (5)$$

As a general rule, the direction of movement of a point is determined by calculation from coordinates or by measurement on a topographic map. The photo-distance Y is determined by a bridging method. It will be seen from this that in order to measure glacial variations the pseudo-parallax must be combined with terrestrial stereophotogrammetry.

To measure the accuracy of glacial variations by pseudo-parallax methods, differentiate equations (3) and (4). Neglecting the micro-increments of higher orders and rearranging, we obtain

$$m_D' = \pm \sqrt{(\frac{Y}{f}m_{P_D})^2 + (\frac{P}{f}m_Y)^2}$$

$$m_{\Delta h} = \pm \sqrt{(\frac{Y}{f}m_Q)^2 + (\frac{Q}{f}m_Y)^2}$$

When $m_{P_D} = m_Q = 0.01$ mm, Y = 1500 m, f = 200 mm, m = 1.2 m

we obtain, from equation (6) $m_D = \pm 0.14$ m, $m_{\Delta h} = 0.08$ m.

Therefore the accuracy of the pseudo-parallax method is higher than the accuracy of the coordinate comparison method. At present, it is the main measuring method for assessing glacial variations.

2. Glacier movement in the vertical direction

(a) The comparative method using topographic maps

This method uses the same coordinate system and scale as in the aerial photogrammetric surveying method. The topographic maps are drawn by using a Zeiss Stereoautograph 1318 and each map has its own distinctive colour. In order to avoid errors during mapping, each photo of the photopair from the same photostation must use the same orientation. Ground objects and geomorphology (e.g., superglacial moraine, lake, ice crevasse, ice hummock, boundary of ice tongue, homologous contour and man-made mark) are carefully plotted onto the topographic map. If displacements of these marks on the map are found, they represent glacial variation. Generally, the use of transparent thin film is a much simpler way of determining variations of a glacial surface. Thus all variation elements are traced on the thin film map and over-laying onto the other map permits a measure of glacial variations by direct comparison.

(b) The square grid method

First of all, the topography is plotted onto a map divided by a small square grid whose side length is 1 cm (or according to accuracy and scale required). Then the height of each square corner is measured by the stereoautograph 1318 instrument and the difference of height at each square corner is $h_2 - h_1 = \Delta h$.

Hence variation of the glacial surface height is obtained. The height of each square corner point can also be derived by contour interpolation.

The small square grid can be made to agree with the coordinate grid of the topographic map and be directly plotted by a plotting machine. The square grid can also be plotted on the transparent thin film instead of on the topographic map. The grid can have various directions and forms to allow for different positions of the glacier.

The effort required by the square method is excessive and it is only applicable to detailed glacial investigations.

(c) Profile method

This method is a standard one for glacial investigations. Generally, the profile is laid out at typical positions, such as the upper, middle and lower sections of a glacier, terminus, turns and junctions, and the places where surface slopes are obviously changing.

Generally speaking, variations of surface height are small and a profile map is drawn with a vertical scale ten times larger than the horizontal scale.

The variation value of the glacial height h is a vertical direction value; the perpendicular value to the surface is

$$h = \Delta h \cos\beta$$

where β is the slope angle of the ice surface.

RESULTS OF A REPEATED TERRESTRIAL STEREOPHOTOGRAMMETRIC SURVEY AT THE BATURA GLACIER TERMINUS

A repeated terrestrial stereophotogrammetric survey at the Batura Glacier was mainly within 4 km of the terminus. Five photo-bases were laid out at the 2 km wide terminus and a repeated terrestrial stereophotogrammetric survey was taken at the beginning and at the end of the ablation period during 1974, 1975 and 1978. The Zeiss Stereoautograph 1318 and the Stereocomparator 1818 were then used back at the office in Lanzhou. With all the methods mentioned above, we obtained the results of terminus variations of the Batura Glacier shown in the following tables and figures.

From the data given overleaf it is clear that if we compare the 1978 data with those of 1975, the height of the terminus evidently dropped and much ice mass was lost but the ice surface of 1978 was higher than that of 1966. It follows that there is a considerable amount of ice advancing at the glacier terminus. Thus, it is quite possible that the Batura Glacier

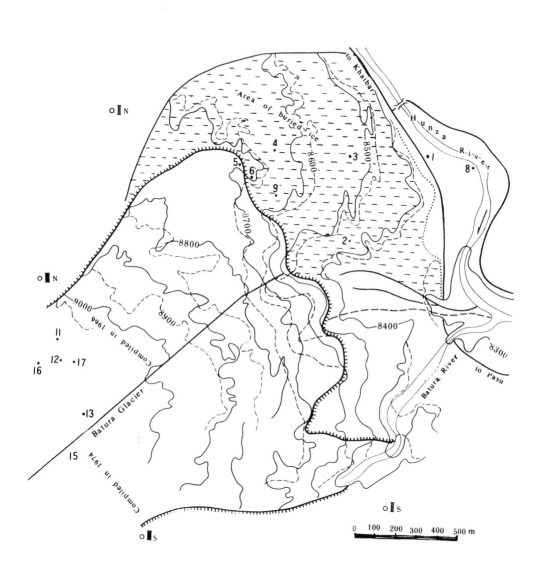

FIG. 2. SPATIAL CHANGES ON THE TERMINAL SURFACE OF THE
BATURA GLACIER, 1966 - 1974.

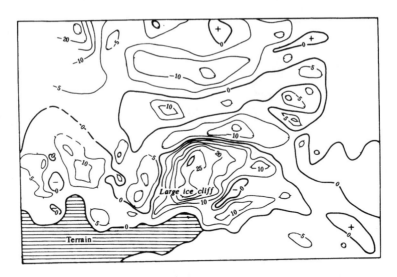

FIG. 3.   ISOLETH MAP OF HEIGHT VARIATION IN THE LARGE ICE CLIFF
AT THE TERMINUS OF THE BATURA GLACIER, 1975 – 1978 (m).

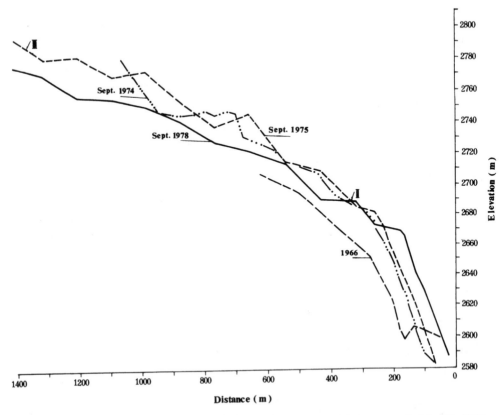

FIG. 4.   HEIGHT VARIATIONS OF CENTRAL LONGITUDINAL PROFILE OF THE
BATURA GLACIER'S TERMINUS, IN 1966 – 1978.

Time	Advancing Value (m)	V(m/yr.)
1966* - 1974	90.0	11.0
1974 - 1975	9.6	9.6
1975 - 1978	5.5	2.8

* Pakistan survey, aerial map 1:10.000
1966

Notes:

In comparison with maps

General surveying

From homologous contour displacements the greatest speed point was 37.5 during 1975 - 78.

Contour (m)	Variation of Height (m)	Glacial Area ($10^4 m^2$)	Variation of Ice Mass ($10^4 m^3$)
2650 - 2700	-3.3	73	-241
2700 - 2750	-7.6	60	-456
2750 - 2800	-10.9	39	-425
2800 - 2850	-11.8	59	-696
2850 - 2900	-10.9	128	-1395
South Cliff	3.8		142
North Cliff	8.6		

Note:

Square grid method with about 1500 points

will continue to advance in future years but with decreasing force; see reference (1).

The glacier variation results obtained from the above mentioned surveying and mapping techniques shows that the repeated terrestrial stereophotogrammetric survey is a good method to determine glacial variations. It must be particularly noted however that when applied the orientation of each photo should be identical. In order to obtain reliable data, a combination of these methods should be employed.

REFERENCES

1) SHI YAFENG, ZHANG XIANGSONG and BAI ZHONGYUAN (Eds.). Professional Papers on the Batura Glacier, Karakoram Mountains. Science Press, Beijing, China. (In Chinese with English Abstract).

BIBLIOGRAPHY

WANG WENYING, CHEN JIANMING and WANG MINGYUAN, (1973). 'Terrestrial stereophotogrammetry and its application', Science Press, Beijing, (in Chinese).

KONECNY, G., (1964).   'Glacial Surveys in Western Canada', Photo-grammetric Engineering, Vol. 1.

KICK, W., (1966).   'Measuring and mapping of glacier variations', Canadian Journal of Earth Sciences, Vol. 3.

Point positioning by Doppler satellite

J.P. Allen

Mechanical Engineering Department,
University of Sheffield

ABSTRACT

The Doppler satellite system is an all-weather, global survey system enabling the user to introduce survey control into remote and mountainous areas with a minimal requirement of personnel and time. With the present technology, single-point positioning with accuracies in the range 2-5m is possible. Future developments may bring this down to 1-1½m for absolute positioning, and to 10cm over 300km for relative positioning. With the continually improving accuracy, deficiencies in the existing reference ellipsoids and gravitational models are revealed which can only be corrected by extensive and meticulous doppler measurements. The system can also be used to investigate crustal mass distribution and long-term crustal movements. Two new systems offer shorter term crustal monitoring, with one of them potentially achieving relative positioning accuracies of a few centimetres.

INTRODUCTION

Details of the U.S. Navy Navigation Satellite System, usually called Transit, were first released to the general public in 1967. This system allows the user to determine his position on a global co-ordinate system, to better than ± 5 metres, in less than 48 hours. Stations located using the system do not have to be intervisible, nor do they have to be tied into any local or regional survey network. The receiver can be operated in all lighting and weather conditions, except electrical storm, without significant reductions in accuracy occurring. As such it is ideal for survey use in remote and mountainous regions, allowing substantial savings in time and manpower.

SYSTEM DESCRIPTION

The system consists of 5 (in 1979) satellites in circular, polar orbits around the earth. With an altitude of about 1075 km, each satellite completes one orbit every 107 minutes. Transmitted from each satellite are two very stable signals (the frequencies of which change less than one part in 10^{11} during a pass). These are at approximately 150 and 400 MHz. Encoded by phase modulation into these signals is a navigation message, which contains the details of the satellite's position as a function of time. This message is programmed to start and end at the instant of every even minute, thus also providing precise timing marks. The navigation message is up-dated at about 12-hourly intervals when the satellite passes near one of the four ground tracking stations. The up-dated message is a prediction of the satellite's orbit for the next 16 hours, and it is the deviations from this predicted orbit which limit the overall system accuracy when using the transmitted data. This problem can, however, be overcome by continuously monitoring the satellite's position in relation to known ground stations.

OBTAINING A POSITION FIX

As the satellite comes over the horizon, it is travelling rapidly towards the observer, emitting, as it does so, a series of signals at constant time intervals. Because the source of the signal is moving towards the observer, these signals arrive at the receiver at a rate faster than that at which they were transmitted. This frequency change, which is known as the Doppler shift, is directly related to the relative velocity between the satellite and observer. If this relative velocity is now integrated over a known time interval, then one can determine the change in distance between the satellite and observer in that time interval (the range change). It is the determination of this change which forms the basic calculation of the system, from which the receiver position can then be found.

Figure 1 shows the basics of the Doppler measurement technique. The navigation receiver is equipped with a stable reference oscillator from which the 400 MHz ground reference frequency f_G is derived. The receiver forms the difference frequency $(f_G - f_R)$, where f_R is the received frequency. Each Doppler measurement is then a count of the number of difference-frequency cycles occurring between the time marks received from the satellite. This count is a very sensitive measure of the change in distance between the satellite and receiver, as each count represents one wavelength change, which at 400 MHz is only 0.75m.

The equation defining the Doppler count of $(f_G - f_R)$ is the integral of this difference frequency over the time interval between receipt of the time marks from the satellite. So, taking period one of Fig. 1 for example:-

$$N_1 = \int_{t_1+R_1/C}^{t_2+R_2/C} (f_G - f_R) \; dt$$

Note that $(t_1 + R_1/C)$ is the time of receipt of the time mark transmitted at time t_1, after having travelled distance R_1 at velocity C.

To understand how this gives the slant range change it is helpful to expand the equation into two parts:

$$N_1 = \int_{t_1+R_1/C}^{t_2+R_2/C} f_G \, dt - \int_{t_1+R_1/C}^{t_2+R_2/C} f_R \; dt$$

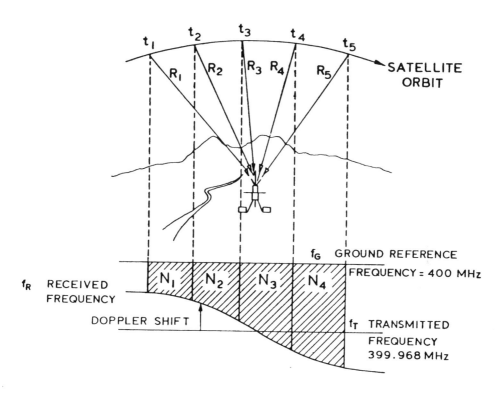

Fig. 1   Schematic for Position Fixing by Satellite Receiver (1)

The first part of the expression is straightforward to deal with, it is merely the integration of the constant frequency f_G. The second part however, contains the varying frequency f_R. This integral represents the number of cycles received between the time of receipt of the two timing marks. Clearly this quantity must equal the number of cycles transmitted between the transmission of the marks, hence we can now write:-

$$N_1 = \int_{t_1+R_1/C}^{t_2+R_2/C} f_G \, dt - \int_{t_1}^{t_2} f_T \, dt$$

where f_T is the transmitted frequency.

As the frequencies f_G and f_T are assumed constant during a satellite pass, the integral becomes trivial giving:-

$$N_1 = f_G \left[(t_2 - t_1) + \frac{1}{C}(R_2 - R_1) \right] - f_T (t_2 - t_1)$$

which, after rearranging, gives:

$$N_1 = (f_G - f_T)(t_2 - t_1) + \frac{f_G}{C}(R_2 - R_1)$$

f_G, f_T, t_1, t_2, C are known, N_1 is the Doppler count, hence the range change $R_2 - R_1$ can be determined. As mentioned earlier, the satellite also transmits details of the position in relation to time (the broadcast ephemeris), hence its location at t_1 and t_2 can be found. These details are fed into a computer, along with an initial guess at the receiver's position made by the operator. From the two sets of positions, the computer now calculates a set of theoretical range change values, which can then be compared with the measured values. If the assumed receiver position was correct, then the calculated values would agree with those measured. More usually however, one does not know the exact receiver position, in which case there will be a difference between the measured and calculated values. The computer uses this difference to calculate an improved guess at the receiver's position, and then calculates a new set of range change values, compares them, makes a new guess and so on, until the change in receiver position after an iteration falls within the accuracy limits appropriate to the situation, when the cycle is then stopped. The number of iterations required is actually quite small; with an initial position assumption in error by tens of kilometres, only two or three iterations are required to calculate the correct values. Even with the initial estimate in error by 200-300 km, the solution will still converge on the true values (1).

TRANSFORMATION OF THE RESULTS ONTO THE REGIONAL SURVEY CO-ORDINATE SYSTEM

The previous calculations give the receiver's position in relation to a number of known points on the various satellite tracks, the positions of which are described using a global co-ordinate system. This position must now be operated upon mathematically to transform its values into ones consistent with the co-ordinate system in the region in which the receiver is being operated.

Figure 2 gives the basics of the problem. The geoid is the gravitational equipotential surface approximately coinciding with mean sea level. It is an undulating irregular surface, and as such is unsuitable for geodetic work (elevations are, however, usually measured in relation to this surface). So a simpler surface is chosen, this is almost always an ellipsoid, the parameters of which are such that it is the "best fit" to the geoidal surface in that region (in the illustration it has been shown displaced downwards for clarity). The regional ellipsoids are, however, not suitable for use with a global satellite system, consequently a new ellipsoid was required, which would describe the shape of the whole Earth. The problem is, that the exact form of the globe can only be determined by careful measurements using the satellite system, hence the initial ellipsoid could only be defined arbitrarily. As the number of observations increases, so the initial ellipsoid is adjusted and re-adjusted to produce a "best fit". Continual revision of the satellite computing system would be both inconvenient and expensive, consequently these adjustments are given as corrections to be applied to the "standard" reference system. The broadcast ephemeris is given in terms of the ellipsoid known as NWL-10D, which is almost compatible with the NWL-9D system (2). The NWL-9D system is that used by the ground tracking stations, whose data are used for high accuracy work (see later). Both the NWL-9D and NWL-10D systems, due to the way in which they were defined,

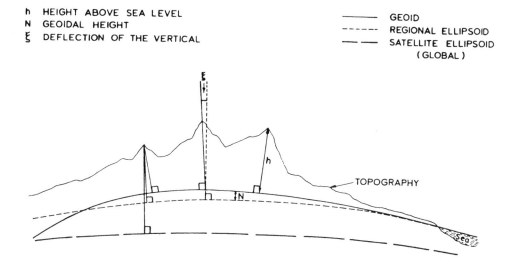

h HEIGHT ABOVE SEA LEVEL
N GEOIDAL HEIGHT
ξ DEFLECTION OF THE VERTICAL

—————— GEOID
- - - - - REGIONAL ELLIPSOID
— — — SATELLITE ELLIPSOID
(GLOBAL)

TOPOGRAPHY

Fig. 2   Showing the Relationship between the Three
Geodetic Surfaces involved in Satellite Survey,
and how Deflection of the Vertical arises.

cannot be connected directly to other classical geodetic systems. The corrections necessary to convert the NWL-9D system into a proper geodetic (CIO
pole and BIH zero-longitude based) reference system were given in 1975 as
+0.26" longitude rotation and -5.27 m reduction in the radius of the ellipsoid
(3), producing the new ellipsoid NWL-10F. Further observations showed that
this still was not satisfactory, and the latest (1978) values proposed are
+0.8" in longitude and -0.4 part/10^6 in scale, producing the system known
as DOPPLER 78 (4).   In the meantime, NWL-9D was changed to NSWC 9Z-2
with a different gravity field, but this change has not affected the user
(2).   It is clear that these minor adjustments will continue to be made as
the system accuracy and data available improve.

To summarize: The satellite's position is given in terms of a set of geocentric cartesian co-ordinates X, Y and Z, whose origin is at the centre
of the global satellite ellipsoid (NWL-10D for the broadcast ephemeris).
This has to be transformed into the NWL-9D system by a small correction
in the origin position and a small change in scale (in Europe the recommended values are ΔX = -3m, ΔZ = -11m and scale 1.3 parts/10^6 (2)).
NWL-9D then is adjusted to DOPPLER 78.   The co-ordinates are then in a
proper geodetic system and can be transformed into the required regional
ellipsoid by the appropriate origin translation, radius change and flattening
coefficient change.

Once in the regional reference system, the cartesian co-ordinates can be
converted to standard geodetic values of latitude, longitude and height above
ellipsoid using the "standard" equations below.

$$\tan \lambda = Y/X$$

$$\tan \phi = (Z + e^2 \nu \sin\phi) / (X^2 + Y^2)^{\frac{1}{2}}$$

$$N + h = \left[X/(\cos \phi \cos \lambda) \right] - \nu$$

where λ longitude
            ϕ latitude
            $(N + h)$ height of the point above the reference ellipsoid
            a semi-major axis of the ellipse
            e eccentricity of the ellipse
            ν radius of curvature of the prime vertical

$$\nu = \frac{a}{(1 - e^2 \sin^2 \phi)^{\frac{1}{2}}}$$

Note: a and e now of course refer to the regional ellipsoid values.

ACCURACY AND LIMITATIONS

Following the removal of errors due to unmodelled polar motion in 1973 (5), the errors for a single satellite pass have been reduced to:

1. Uncorrected propagation effects (ionospheric and tropospheric effects) 1-5m.

2. Instrumentation and measurement noise (local and satellite oscillator phase jitter, navigator's clock error) 3-6m

3. Uncertainties in the geopotential model used in generating the orbit 10-20m

4. Uncertainties in surveyor's altitude 10m

5. Incorrectly modelled surface forces (drag and radiation pressure acting on the satellites during the extrapolation interval) 10-25m

6. Ephemeris rounding error (last digit in the ephemeris is rounded) 5m

The root-sum-square of these errors lies in the range 18-35m.

Propagation effects: There are two sources of refraction error; the larger one is due to the ionosphere. The effect is to stretch the wavelength of the signal passing through, due to interactions between the wave and free electrons and ions. Work done during the first years of Transit's development showed that these errors could be corrected by combining the Doppler measurements made at two frequencies, hence the transmissions at both 150 and 400 MHz. The tropospheric error can only be removed by modelling its effects. Models of varying sophistication are available, usually the simpler models are sufficient, however, for high accuracy work full account of the tropospheric effects must be taken, with continuous measurements at the receiver location of temperature $\pm 1^\circ$C, pressure ± 2 mbar and humidity. The effect of neglecting these measurements manifests itself in the form of an error in the calculated receiver elevation (6).

Instrumentation: The clock error occurs because the power allocated, in the satellite, to transmitting the clock signal is very small;   only 62 milli-watts. As this signal has a period of nearly 10,000 μ sec it is not surprising that there are clock errors of the order 50 to 150 μsec (1). Since the Doppler frequency being counted has a period of 25-40 μsec, time recovery jitter can cause a count to begin several cycles early or late, where each count represents 0.75m.   The best technique for removing the effect of time recovery jitter is a precise clock that measures the time interval of each Doppler count, using the reference oscillator as standard.

Altitude: An error in the assumed altitude will affect the calculated longi-tude, if only one satellite pass is used to determine the position of the receiver; see Fig. 3.   If a second pass is incorporated in the calculations, where the angle of elevation of the second satellite track above the horizon is different from the first track, then the correct three-dimensional position can be determined.   Clearly, the more passes incorporated, the more accurate will be the final result.

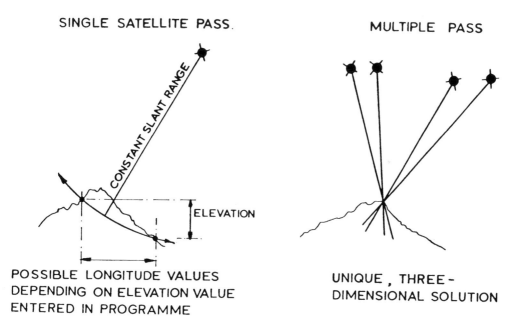

SINGLE SATELLITE PASS.

CONSTANT SLANT RANGE

ELEVATION

POSSIBLE LONGITUDE VALUES
DEPENDING ON ELEVATION VALUE
ENTERED IN PROGRAMME

MULTIPLE PASS

UNIQUE , THREE -
DIMENSIONAL SOLUTION

Fig. 3   Altitude Error and its Correction
using Multiple-Pass Techniques (1)

Uncertainties in geopotential model and surface force errors are continually being improved, but it is these uncertainties in position which will always limit the ultimate system accuracy.   As the broadcast ephemeris is only updated about every 12 hours, the errors clearly increase as one progresses towards the end of the extrapolation period.   The magnitude of these errors was acknowledged during the design stage of the satellites, with the satel-lite memory only capable of holding an ephemeris accurate to the nearest 10 metres.   On one occasion this broadcast ephemeris was found to be con-sistently in error for a week of observations, with the station co-ordinates displaced by 6m in both X and Y, indicating an inexplicable temporary shift in the reference system origin (7).

The solution to these problems is to track the satellites using ground stations of known co-ordinates. These observations then provide the data required for producing the precise ephemeris, which allows the user to determine his position using a single satellite pass, with a root-sum-squared error of about 6.3m (1). The main disadvantage is that this precise ephemeris can only be obtained after returning from the field. This is offset by the significant reduction in time required in the field; 5 to 10 hours for a +5m fix instead of about 48 hours when using the broadcast ephemeris. The bulk of this time is, in fact, made up of waiting for suitable passes, as a single pass typically lasts only 10 to 18 minutes. Due to a reduction in accuracy satellite passes with an elevation below about 20° and above 75° are usually edited, and occasionally more than one satellite is above the horizon with the possibility of the signal from one interfering with that from the other, thus making both redundant.

No significant breakthrough is expected in the Transit satellite system (in terms of accuracy). In 1979, using "orbit relaxation" programs such as that developed at Nottingham University (7) it was possible to determine the relative positions of points up to 300 km apart with accuracies of around 300mm. With the introduction of the new NOVA satellites, whose susceptibility to external forces such as atmospheric drag and radiation pressure will be much less than the present generation OSCAR satellites, orbit predictions will be improved. The present gravity model (WGS 72) is also under review, combined with a harmonization of the broadcast and precise ephemeris this could bring relative position accuracies down to about 100mm (8).

The main advance will be with the introduction of the Navstar Global Positioning System in 1986. This will consist of 18 satellites at an altitude of just under 20,000 km, spread evenly amongst three planar orbits. The navigation data will again be transmitted on two frequencies, 1575.42 MHz and 1227.6 MHz, to facilitate the ionospheric delay correction. The user will tune in to 4 of the 6-11 satellites visible at any one time, and compare the time of reception of the reference marks in the four messages with the clock in his receiver. By dividing the delay time between mark transmission and reception by the speed of light (corrected for ionospheric effects) the user can then calculate "pseudo-ranges" to the satellites and hence a 3-D position, using the fourth piece of information to calculate receiver clock error (the satellite clocks are stable to 1 part in 10^{12}); the true ranges and receiver position can then be calculated (9). Based on theoretical evaluations of the system in 1978, when the planned number of satellites was 24, an absolute positioning accuracy of about 1 metre was considered feasible, with the limiting factor being uncertainty in the satellite positions. By extending the measurement modes to long-baseline interferometry and Doppler, the predicted relative positioning accuracy will be better then 100mm for baselines up to 2,000 km in length (10).

ADVANTAGES OF THE TRANSIT SYSTEM AND ITS APPLICATIONS

It is a portable "all-weather" system. It does not have to be tied in to a local survey, nor do the stations have to be intervisible. The logistical requirements for its operation are known in advance, allowing improved planning and resource allocation. The independence of the individual stations facilitates a flexible approach to the actual execution of the work, so if, for example, there is an unexpected problem with the transport, then it does not matter if the last 'planned station is, in fact, occupied first.

In mountainous terrain it provides a rapid means of introducing the control necessary for survey by aero-triangulation or inertial survey systems, and in so doing provides considerable savings in manpower and time required.

Using the present system it is possible to establish control, to geodetic standards, in scale and orientation over lines in excess of 300km. With the introduction of the Navstar system, this length may be significantly reduced.

Affected only by the integrated effect of the gravitational field for the whole Earth, the satellite's orbits are effectively insensitive to small near-surface mass distribution variations. Such effects occur where mountain masses rise out of the continents, distorting the gravitational field and hence leading to irregularities in the geoid (see Fig. 2). If a theodolite is set up in such a region, then its axes will be aligned parallel and perpendicular to the geoidal surface at that point. As this surface is rarely tangential to the ellipsoid, then any determination of geodetic position by astronomical observations will be in error, due to the deviation from the ellipsoid (or from the vertical depending on which axis one refers to). A solution is to measure the strength of the gravitational field around a grid of points across the area, and then to calculate the deviation from these values. Far easier is to observe the station location using the satellite system, which is of course unaffected by the local deviation. By subtracting this position from that obtained by astronomical observations, one obtains directly the local deviation.

With the ultimate system accuracy it becomes possible to monitor global tectonic movements. A review in 1979 (11) stated that continental drift could be measured to an accuracy of 20 mm/year by continuous observation over a period of longer than 5 years. With the improved Transit system, or using the Navstar system, this interval would be reduced to 3-4 years. There is however, a system under development which, if all goes well, could allow monitoring on a 12-month basis, possibly even less. This is Very Long Baseline Interferometry (12). It is at present being tested on the San Andreas fault in America, and consists of two or more radio antennae monitoring extragalactic radio sources. These sources, such as quasars and Seyfert galaxies, although several hundreds of light-years in diameter, due to their extreme distance can be effectively regarded as point sources, and their signals on reaching the Earth as plane waves. Thus by monitoring the exact time of arrival or a particular signal, the difference in travel-time and hence relative position between the stations can be found. The main difficulties have been in achieving sufficiently accurate timing at any one station, in correlating the time information between the various stations and in modelling Earth rotation and atmospheric refraction. It is hoped however that the eventual relative positioning accuracy of this system will be of the order of a few tens of millimetres.

CONCLUSIONS

With the introduction of space technology into surveying, it is now possible to map any continent in the world to geodetic standards. Recent advances now also allow us to monitor the relative movements between these continents and to look more closely at the earth's crust in the regions where they meet.

REFERENCES

(1) Stansell, T.A. Positioning by Satellite. Electronic Surveying and Navigation, edited by S.H. Laurila, John Wiley & Sons 1976.

(2) Ashkenazi, V., Sykes, R.M., Gough, R.J. and Williams, J.W. First United Kingdom Doppler Campaign: Results and Interpretation. Phil. Trans. R. Soc. London A294, pp 253-259 (1980).

(3) Anderle, R.J. Error Model for Geodetic Positions Derived from Doppler Satellite Observations. NSWC/DL Tech. Rep. No. 3368, 1975.

(4) Anderle, R.J. Mean Earth Ellipsoid Based on Doppler Satellite Observations. Prepared for the Spring meeting of the American Geophysical Union, Florida, April, 1978.

(5) Piscane, V.L., Holland, B.B., and Black, H.D. Recent Improvements in the Navy Navigation Satellite System. J. Inst. Navigation Vol. 20, No. 3, pp. 224-229, 1973.

(6) Sykes, R.M. Translocation and Orbit Relaxation Techniques in Satellite Doppler Tracking. Ph.D. Thesis, 1979, Nottingham University, U.K.

(7) Ashkenazi, V. and Sykes, R.M. Doppler Translocation and Orbit Relaxation Techniques. Phil. Trans. R. Soc. London A294, pp. 357-364 (1980).

(8) Ashkenazi, V and Sykes, R.M. Absolute and Relative Positioning by Satellite-Doppler Techniques and Geophysical Applications. Sixth Annual European Geophysical Society Meeting, Vienna, September, 1979.

(9) Austin, J.A., G.P.S. - Global Positioning System. Military Electronics/Countermeasures. Hamilton Burr Pub. Co., Santa Clara, U.S.A. June, 1980.

(10) Anderle, R.J. Geodetic Applications of the Navstar Global Positioning System. Second International Symposium on Problems Related to the Redefinition of North American Geodetic Networks, Washington DC, April, 1978.

(11) Malyevac, C.A. and Anderle, R.J. Determination of Plate Tectonic Motion from Doppler Observations of Navy Navigation Satellites. Second International Geodetic Symposium on Satellite Doppler Positioning, Austin, Texas, 1979.

(12) McLintock, D.N. Very Long Baseline Interferometry and its Geodetic Applications. Ph.D. Thesis, May, 1980, Nottingham University, U.K.

Earthquakes

Tectonic studies in the Alpine-Himalayan belt

G. King and J. Jackson
Department of Geodesy and Geophysics,
Madingley Rise, Madingley Road,
Cambridge, CB3 0EZ.

ABSTRACT

Plate tectonics has successfully explained the behaviour of the major ocean basins. However, it has proved to be less successful in describing the processes of continental collision and its associated deformation. Seismic studies proved crucial in providing information on which plate tectonics was established. Recent developments in the study of seismic signals recorded on distant instruments and on portable seismographs temporarily installed in the Alpine-Himalayan belt are beginning to reveal the relation between extension, sedimentation, compression and major strike slip faulting such as the Karakoram fault.

RECENT STUDIES

Ten years have passed since the theory of plate tectonics was proposed. In this theory, the earth's surface is regarded as consisting of a small number of rigid spherical caps or plates which are in relative motion. The interiors of the plates are largely free from earthquakes but the intense deformation along their boundaries produces the major narrow seismic belts of the world. The plates are of the order of 100 km thick and the surface few kilometres consist either of light (low density) continental crust or heavy oceanic crust. Three types of plate boundary occur, new plate is created by the upwelling of molten material at the mid-ocean ridge crests and is destroyed where old dense oceanic crust is drawn back into the earth at ocean trenches. Where two plates slide past each other, transform faults occur.

The theory has proved remarkably successful in describing the behaviour of the large ocean basins. However, it has been recognised for some time that the presence of significant amounts of continental crust at a compressional boundary creates conditions where the simple plate description is inadequate. The difference of behaviour is illustrated in Fig. 1, see also reference (1). In the Atlantic basin the plate boundaries are clearly defined by narrow belts of earthquakes, but the seismicity becomes scattered when the deformation becomes involved with the continental crust of the Mediterranean region. It has long been known that continental rocks are more distorted than those of the oceans and it seems that much of this distortion

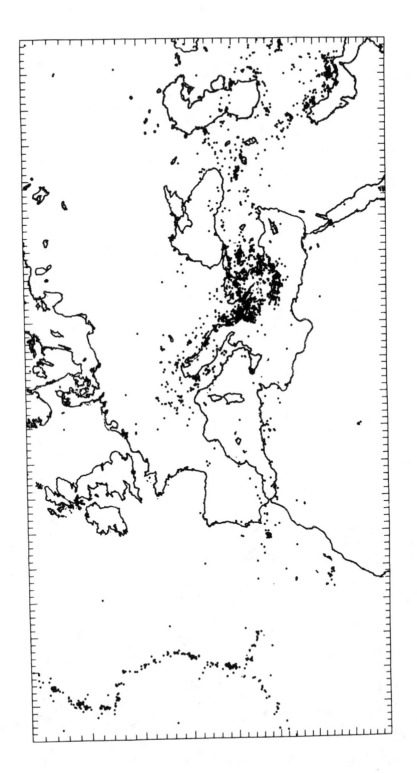

Figure 1 – Positions of Earthquakes between 1961 and 1974 between the Atlantic and Pakistan. Note the difference in distribution between the oceanic and continental regions.

is created in regions where two plates carrying continental material are in collision. The Alpine-Himalayan belt is the most spectacular contemporary example of such a collision.

In oceanic collision belts, seismic activity follows dipping zones to depths as great as 700 km tracing the locus of motion of dense oceanic crust as it is reabsorbed into the upper mantle of the earth. In regions of continental collision, earthquakes deeper than 100 km are rarely reported and even these reported depths can be in error.

The depth to which earthquakes extend is important for understanding how compression is accommodated when continents collide. It is presumed that the bouyancy of light continental rocks prevents their underthrusting into the mantle, but if this is true, where does the continental crust go, and how does it deform to get there?

Recently several authors (3) (4) have pointed out the apparent role of major strike-slip faults in the Alpine-Himalayan belt; which slide large wedges of continental material away from zones of collision, see Fig. 2, thereby avoiding the need for subduction or the excessive piling up and thickening of continental crust. These large strike-slip features are easily visible on satellite photographs and even on topographic maps. Seismic studies allow the current activity of these faults to be examined. Some, such as the Anatolian Fault of North Turkey (5) and the Altyn Tagh Fault of Tibet and China (4) are known to have motions as large as a few centimetres per year along at least part of their great length. Others, such as the Herat Fault of Afghanistan, or the Karakoram Fault of North Pakistan, show little seismic activity along much of their length. This may be because they currently accommodate no movement or because the movement is taking place without generating earthquakes detectable on distant seismometers.

In addition to the great strike slip faults there are also broad zones of extension and compression. It has been realised for many decades that compression and crustal thickening must play an important role in continental collision and mountain building, but the role of extension was not recognised (2). The mechanisms occurring in either process remain obscure except in a few places.

In the Aegean and Western Turkey, extension occurs (2). The surface geology shows many extensional (normal) faults some of which form major features on satellite photographs and some of which have moved by a metre or more during recent earthquakes, causing large areas of the land surface to subside. This process is continuing. When combined with other observations indicating that the Aegean is a sunken mountain range, recent authors conclude that the Aegean has sunk below sea level because the crust has been stretched by a factor of two and thinned from 40 km to around 20 km. The basin of the Aegean is now filling with sediment eroded from the surrounding mountains.

For some years, spurious locations of earthquakes deeper than 100 km were reported for the Zagros mountains of Iran. It was concluded that this represented some sort of subduction or overthickening of crust in this broad belt of compressional earthquakes. It now seems unlikely that earthquakes occur at depths much greater than 20 km. Two lines of enquiry have provided information to support this conclusion. The first uses temporary networks of portable seismic stations operated in several regions of the Zagros mountains, e.g. (6) (7). Distant stations can locate earthquakes to no better than 20 km laterally and often much worse vertically. Dense networks

218

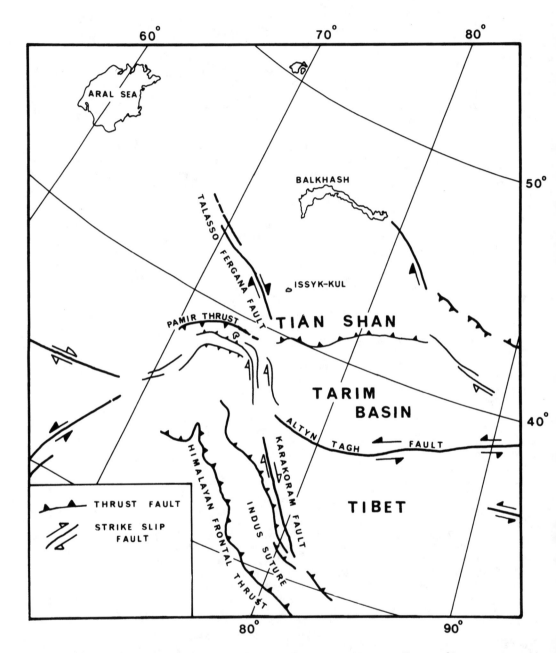

Figure 2 – Simplified map of recent tectonics in Asia. Heavy lines represent major active faults. Solid arrows indicate sense of motion, corroborated by fault plane solutions or surface faulting of earthquakes. Open arrows indicate sense inferred from analysis of photographs alone.

of local stations can achieve a resolution of 1 km. Although a temporary network is unlikely to catch a large earthquake, smaller earthquakes are symptomatic of the seismic behaviour of a region. The absence of any small shocks deeper than 20 km hinted strongly at the absence of activity below this depth.

A second method of examining earthquake depth uses computer modelling of the actual waveform of a seismogram recorded at a distant station. For some earthquakes, the waveshape is very sensitive to depth and this proved to be true for some Zagros earthquakes (8). It confirmed the results of the microearthquake studies using temporary networks.

In addition to earthquake location, the data recorded at both local and distant stations can define the angle and direction of slip motion on the fault causing the earthquake. Fault plane solutions confirm the tensional nature of Aegean seismicity and the compressional nature of that in the Zagros (8). However, a regional comparison of the angles of faults in various regions of the Mediterranean is revealing; see Fig. 3 of Jackson (this volume). Theory suggests that thrust faults should occur at shallower angles than normal faults. This is confirmed in many regions. The Hellenic Trench in the Ionian Sea and near Crete provides an example, and the dips of the thrust faults are even shallower than theory suggests. In the Zagros however, the thrust planes are very steep and show a remarkably similar distribution to those in the tensional Aegean region. This observation has led to the suggestion that normal (tensional) faults can be subsequently reactivated as reverse (compressional) faults if the regional stress system changes (8).

A careful seismic study around the Karakoram should attempt to identify what is happening. The following questions are critical:

(1) Is the Karakoram Fault still active?

(2) Are any other faults active?

(3) Do earthquakes occur at depths greater than 20 km? Some earthquakes are known to occur in Tibet at depths of 100 km (9).

(4) What is the thickness of the crust?

Hopefully some of these questions may be answered by operating our portable recorders in the Karakoram region during this forthcoming International Karakoram Project.

REFERENCES

(1) McKenzie, D.P. (1978) Plate tectonics and its relationship to the evolution of ideas in the geological sciences. Daedalus, Journal of Am. Acad. of Arts and Sciences 1, 97–124.

(2) McKenzie, D.P. (1978) Active tectonics of the Alpine-Himalayan Belt: The Aegean Sea and surrounding regions. Geophys. J.R. Astr. Soc. 55, 217–254.

(3) Tapponnier and Molnar (1976) Slip-line field theory and large scale continental tectonics. Nature 264, 319–324.

(4)   Molnar, P and Tapponnier, P (1975) Cenozoic Tectonics of Asia: effects of a continental collision.   Science 189, 419-426.

(5)   Brune, J.N.   (1968) Seismic moment, seismicity and rate of slip along major fault zones.   J. Geophys. Res. 73, 777.

(6)   Niazi, M., Asudeh, I., Ballard, G., Jackson, J., King, G., McKenzie, D. (1978)   The depth of seismicity on the Kermanshah region of the Zagros Mountains (Iran).   Earth and Planet Sci. Lett. 40, 270-274.

(7)   Savage, W.U., Alt, J.N., Mohajer-Ashari, A.   Microearthquake investigations of the 1972 Qir, Iran, earthquake zone and adjacent areas. Geol. Soc. Am. Abstr. 9 (1977), 496.

(8)   Jackson, J.A., (1980)   Reactivation of basement faults and crustal shortening in orogenic belts.   Nature, 283, 343-346.

(9)   Chen, W.P., Fitch, T.J., Nabelek, J.L. and Molnar, P.   (1980) An intermediate depth earthquake beneath Tibet.   Source characteristics of the event of September, 14, 1976.   J. Geophys. Res.

Earthquake activity and tectonics of the Himalaya and its surrounding regions

Teng Ji Wen and Lin Ban Zuo
Institute of Geophysics, Academia Sinica
People's Republic of China

ABSTRACT

Using teleseismic data from the Himalaya and surrounding regions plus observations from local seismic networks located on both sides of the Yarlung Zungbo River during 1975 – 1979, both large and small earthquake fault plane solutions are obtained, and the relationship between the seismicity of the plateau and plate tectonics is investigated. Shallow earthquakes are dominant on the Qinghai–Xizang (Tibet) Plateau, whilst intermediate earthquakes are common at the east and west extremities of the arc. In the central part of the arc intermediate depth shocks are rare. According to the fault plane solutions the directions of compressive stress are mainly north and northeast, i.e., basically perpendicular to the strike of the arcuate system.

The extremely thick crust and other geophysical data fields in this region suggest that the stress field is caused by the collision and compression of the Indian and the Eurasian plates.

INTRODUCTION

The Qinghai–Xizang (Tibet) Plateau is the highest in the world. The Himalayan mountain chain borders the southern fringe of the plateau and appears to be a southward projecting arc some 2400 km in length, about 300 km in width and with an average height of more than 6000 m. It is a young region of deformation where tectonic motions and earthquake activity are very pronounced. The Plateau has been regarded as the result of a collision between continental regions.

Since 1975, the Chinese Academia Sinica has organized several expeditions to the Qinghai–Xizang Plateau and a lot of field research has been carried out, including explosion seismology in lakes, magnetotelluric sounding, gravity and aeromagnetic surveys, paleomagnetism, geothermal activity, seismic surface wave observations and the installation of local networks of seismic stations (1).

This paper is based on the data from local networks of seismic stations installed in Zhamong and Dang Xiang during 1976 – 1977, as well as data collected from other sources.

GENERAL FEATURES OF THE SEISMIC ACTIVITY

The Himalayan mountain system is the youngest active region in the western part of our country and its seismic activity is higher than in the north. In Xizang region earthquakes are distributed over a belt several hundred kilometres wide (2).

Figure 1 shows that the shallow focus earthquakes are widespread and that intermediate focal earthquakes occur mostly at the eastern and western ends of the arcuate mountain system in the Hindukush and Burma mountains. However, there is also a scattered distribution of intermediate depth in the Himalayas.

Earthquakes with magnitude greater than 7 are shown in Fig. 2 (from reference (3)). Since 1833 many earthquakes with magnitude over 8 have occurred (Table 1).

No.	Year	Region	Magnitude
1	1833	Yunnan Chonqming	8.0
2	1897	India Assam	8.7
3	1902	Artush	8.5
4	1905	Kashmir	8.0
5	1905	Mongolia	8.3 x 2
6	1906	Malas	8.0
7	1911	Tajik	8.0
8	1912	Alma-Ata	8.4
9	1912	Burma	8.0
10	1920	Haiyuan	8.5
11	1927	Gulang	8.0
12	1931	Fuyun	8.0
13	1934	Boundary of Nepal-India	8.3
14	1950	Zayu	8.5
15	1951	Dangxung	8.0
16	1957	Mongolia	8.3

TABLE 1: EARTHQUAKES OF MAGNITUDE GREATER THAN 8.

The distribution of shallow earthquake foci in the northern Yarlung Zangbo River and the Dunaxung zone are shown in Fig. 3, a and b). In the northern Ganges plain (4) and Assam regions, the shallow earthquake zone dips northward (Fig. 3, c). A few intermediate focal earthquakes are centrally distributed within the transition zone between the Yarlung Zangbo River and the northern fringe of the Ganges plain. They are generally shallower in the south than in the north.

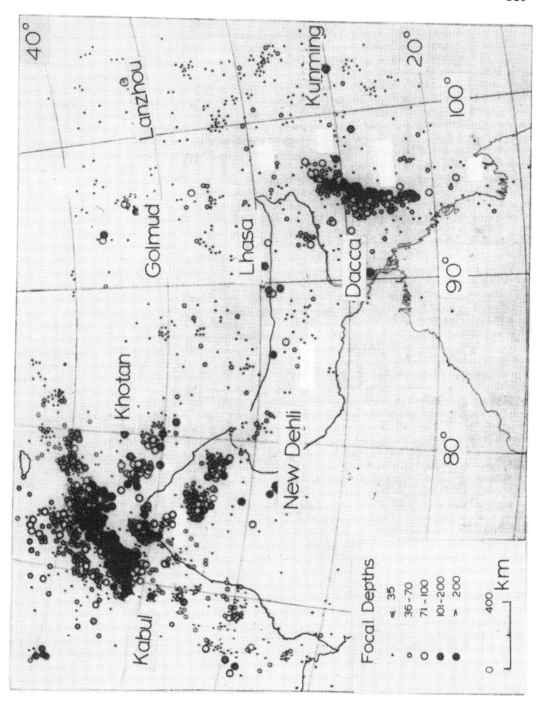

FIG. 1. THE DISTRIBUTION OF EARTHQUAKES OF VARIOUS DEPTHS.

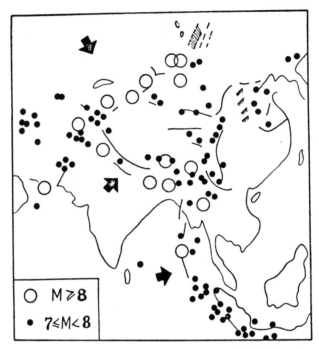

FIG. 2. DISTRIBUTION OF EARTHQUAKES IN TERMS OF MAGNITUDE.

SEISMIC ACTIVITY OF THE HINDUKUSH REGION

In the Hindukush region intermediate earthquakes occur to a maximum depth of 300 km, but are mainly distributed within a zone 200 km deep dipping northward.   However, in the Pamirs zone it inclines southward forming a "V" distribution (Fig. 4).   The "V" form distribution foci may be related to a bent   plate structure (Fig. 5) caused by the collision of the Indian subcontinent with the Asian continent.   The continuing north-south horizontal compressive force between the two continents causes earthquakes to take place frequently in this region.

SEISMIC ACTIVITY OF THE EASTERN PART OF THE ARC

The eastern part of the arc in northwest Burma is also an intermediate earthquake zone, the maximum depth being to 200 km[2].   It dips eastward at an angle of 40° - 45° (Fig. 6), and is presumably the result of an east-west rather than north-south compression.

The depth of seismicity and fault plane solutions in the above mentioned arcuate mountain system indicates that the shallow and intermediate earthquakes all involve thrusting, with a "V" shaped distribution in the Hindukush region, and an eastern dip in the eastern part of the arc.   In the central part of the Himalaya, in Kashmir, near Katmandu and nearby Gonhati, the focal depth of sporadic intermediate earthquakes is 70 - 96 km . Beneath Lhasa in Xizang there   are shocks of 82 - 180 km depth, i.e., the depths of the intermediate shocks in the south part of the Himalaya mountains are shallower than those of the north part (Table 2).   Whether

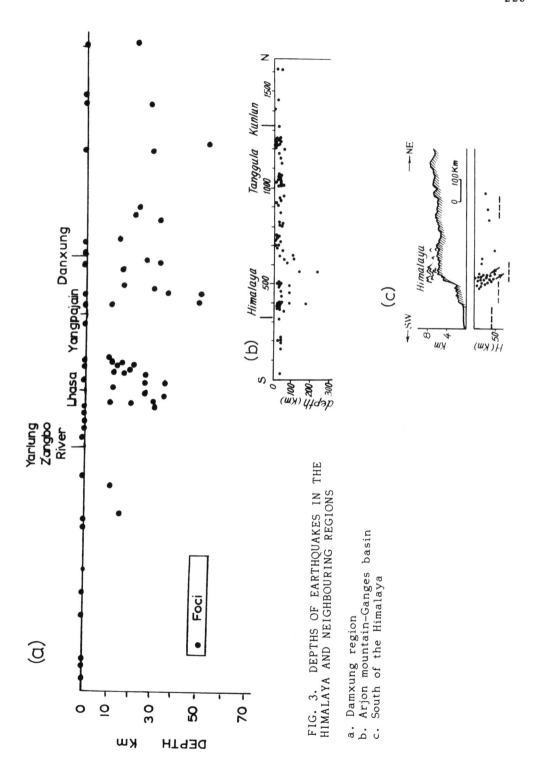

FIG. 3. DEPTHS OF EARTHQUAKES IN THE
HIMALAYA AND NEIGHBOURING REGIONS

a. Damxung region
b. Arjon mountain-Ganges basin
c. South of the Himalaya

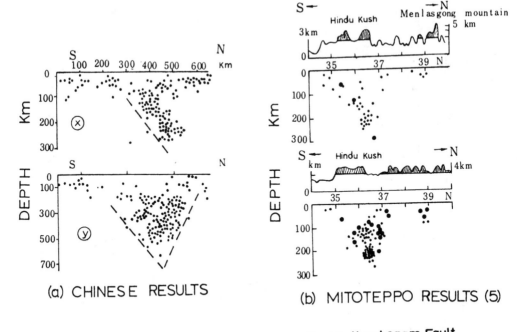

(a) CHINESE RESULTS

(b) MITOTEPPO RESULTS (5)

FIG. 4. THE DEPTH OF FOCI
OF EARTHQUAKES IN (x) THE
HINDUKUSH AND (y) THE PAMIRS
AREAS.

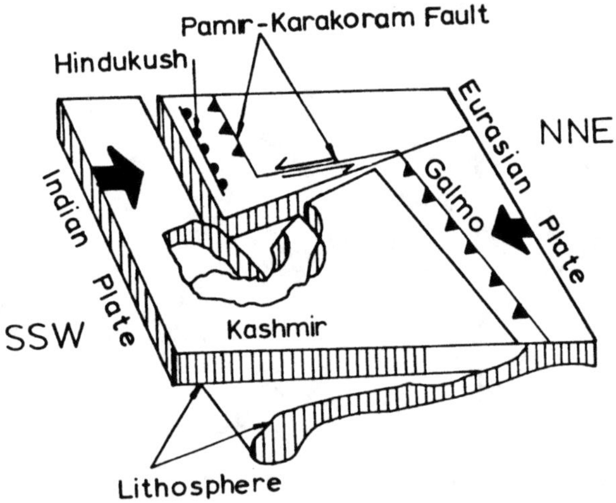

FIG. 5. A MODEL OF UNDER-
THRUSTING PLATES IN THE
HINDUKUSH AND PAMIRS.

a Benioff zone in the mantle (subduction zone) is present or not is still a question that requires further study and more data are needed.

FAULT PLANE SOLUTIONS IN THE TIBETAN REGION

Most shallow earthquakes that have occurred suggest that the belt is mainly undergoing horizontal compression. In the Qinghai-Xizang Plateau, the principal compression axis tends north and northeast (Figs. 7 and 8). The fault plane solutions in the southern Himalayan seismic zone show thrust faulting. Near Lhasa, normal faulting is found, corresponding to the normal faults of island arc-trench systems. Within the Plateau there is sinistral strike slip fault activity, probably parallel to the bend of the arc. Figure 9 shows that the focal mechanisms in the area vary

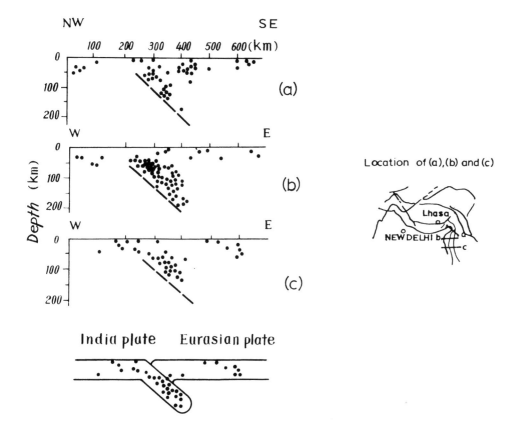

FIG. 6.   DEPTH OF FOCI OF EARTHQUAKES TOGETHER WITH SUBDUCTION
ZONE IN SECTIONS CROSSING THE ARCUATE MOUNTAIN SYSTEM OF INDIA
AND BURMA.

with depth (8).   The   relationship   between   the   principal   stress   axes
of shallow earthquakes and those of more than 70 km depth is very complex.
However, the general direction of compression is the same from Hindukush
to   Himalaya,   eastward   to   Burma.   Except   for   a   few   earthquakes,   the
compressive stress axes   tend mostly north to northeast.   However, locally
the   stress   field   is   complex,   e.g.,   in   the   Zhamu   area   and   its   vicinity,
31 earthquakes and 67 first motions, recorded by a local network of stations,
do not give a simple composite fault plane solution (Fig. 10 a). But, if
divided   into   two   groups   sensible   composite   solutions   are   obtained,   with
different directions of the P axes. (Fig. 10, b and c).

SUMMARY OF GEOPHYSICAL DATA

In   this   region   the   total   intensity   of   the   geomagnetic   field   shows an
east-west pattern.   Yarlung Zangbo River is a high positive aeromagnetic
anomaly belt (150 – 450).   Palaeomagnetic data show   that the north
and   south   sides   of   the   Yarlung   Zangbo   River belong to the Eurasian   plate

Region	Date	Location	Focal Depth (km)
North part of) Yarlung Zangbo) river)))	21 st May, 1935 21 st Jan., 1941 14 th Sept., 1976	Xizang Gyanze Xizang Conanan Dawang Xizang Lhasa vicinity	140 180
North fringe of) Ganges plain and) south part of) Himalaya mountain) system)	7 th June, 1962 27 th Feb., 1970 19 th June, 1975	Kashmir Nepal Boundary of China-Nepal	88 96

TABLE 2:   FOCAL DEPTHS NORTH AND SOUTH OF THE HIMALAYA.

and the Indian plate respectively.

Bouquer gravity anomaly is about −200 mgal to −500 mgal, while in the region between the northern fringe of the Ganges plain and the Yarlung Zangbo River there is a positive isostatic anomaly region. Mount Jolmo Lungma region is +120 mgal, and here the crust has not reached isostatic equilibrium.

To the south of the Yarlung Zangbo River and its north fringe is a geothermal zone where hydrothermal explosion, geysering, and boiling springs occur. Although this is typical geothermal activity in a recent volcanic active zone, there is currently no volcanic activity.

The crust thickness of the Plateau zone is extremely thick and to the north of the Yarlung Zangbo River it is 70 – 73 km (10), whereas south of the river it is 45 – 68 km. The Yarlung Zangbo River is a great deep fault belt. A low velocity laver in the subcrust was found, its thickness being about 10 km and its velocity 5.64 km/s (Fig. 11). To the north part of the crust thickness changes gently and further to the north, in the Qaidam Basin, the crust thickness is 52 km (10). South of the Ganges plain it is 40 km or so (12), and, therefore, the Xizang Plateau is a tremendously thick crustal region (Fig. 12), which is not symmetrical in the north and south. Within the collision and compression belt the Himalaya mountains were lifted up to 8848 m, and the Ganges plain subsided about 1000 m. It is thus a region where vertical motion is most pronounced. The tremendously thick crust material has resulted in strike slip motion and caused the dominant easterly direction of material motion (Fig. 13).

CONCLUSIONS

In the Qinghai-Xizang Plateau the crust is tremendously thick, seismic activity is high and distributed over a wide area, while in the Himalaya zone earthquakes are distributed along the arcuate mountain system.

Intermediate depth earthquakes are chiefly centred in the eastern and western ends of the arcuate mountain system. In the central part

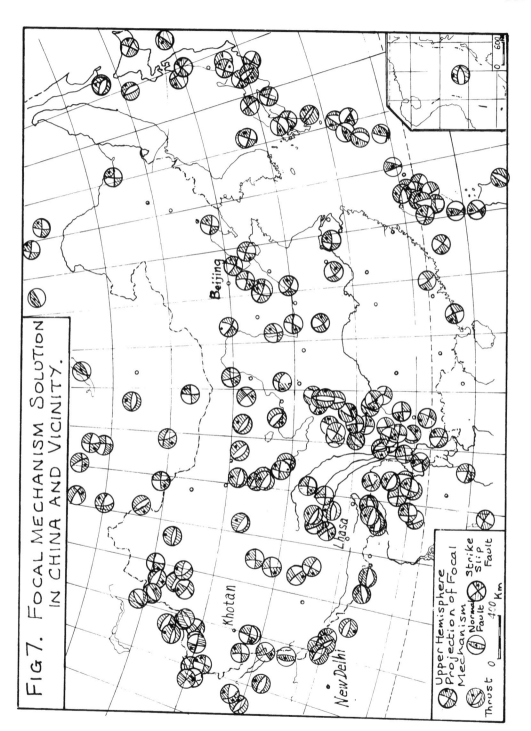

FIG 7. FOCAL MECHANISM SOLUTION IN CHINA AND VICINITY.

230

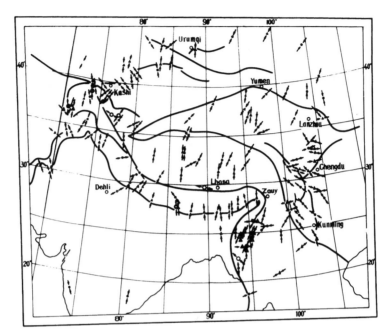

FIG. 8.   HORIZONTAL PROJECTIONS OF PRINCIPAL COMPRESSIVE AXES OF
SHALLOW EARTHQUAKES IN QINGHAI-XIZANG PLATEAU.

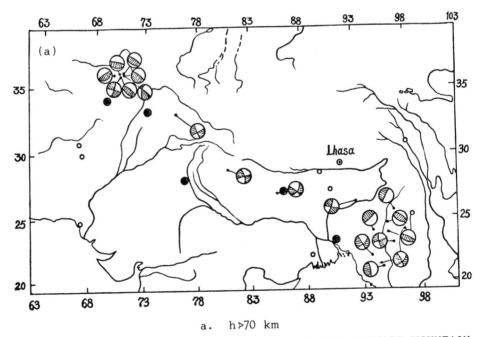

a.   h⊳70 km

FIG. 9.   WAVE FAULT PLANE SOLUTION OF HIMALAYA ARCUATE MOUNTAIN
SYSTEM AND ITS VICINITY. (Part a) (CONT'D)

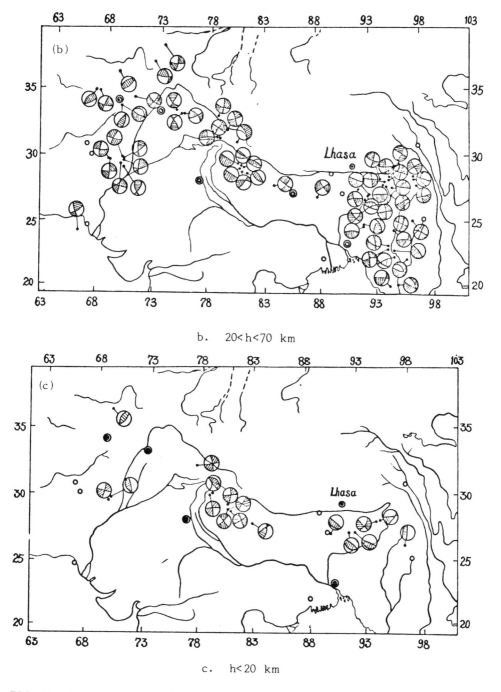

b.   20< h<70 km

c.   h< 20 km

FIG. 9.   WAVE FAULT PLANE SOLUTION OF HIMALAYA ARCUATE MOUNTAIN
SYSTEM AND ITS VICINITY.   (Parts b and c).

232

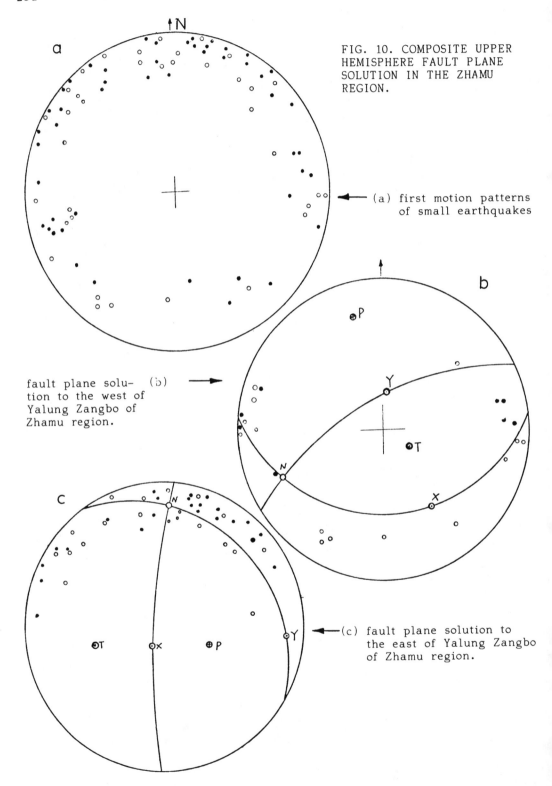

FIG. 10. COMPOSITE UPPER
HEMISPHERE FAULT PLANE
SOLUTION IN THE ZHAMU
REGION.

(a) first motion patterns
of small earthquakes

fault plane solu- (b)
tion to the west of
Yalung Zangbo of
Zhamu region.

(c) fault plane solution to
the east of Yalung Zangbo
of Zhamu region.

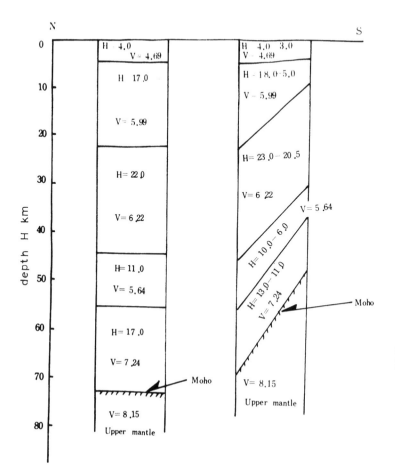

N   S

depth H km

FIG. 11. COLUMNAR SECTION OF THE QINGHAI-XIZANG PLATEAU (TIBETAN PLATEAU): (A) NORTH OF YARLUNG ZUNGBO RIVER; (B) SOUTH OF YARLUNG ZANGBO RIVER.

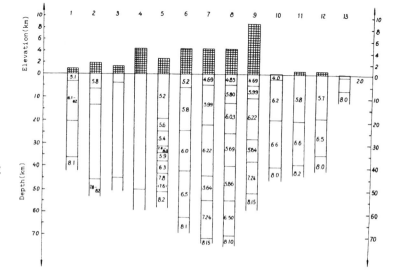

FIG. 12. CRUSTAL STRUCTURE OF THE QINGHAI-XIZANG PLATEAU (TIBETAN PLATEAU) AND NEIGHBOURING REGIONS.

234

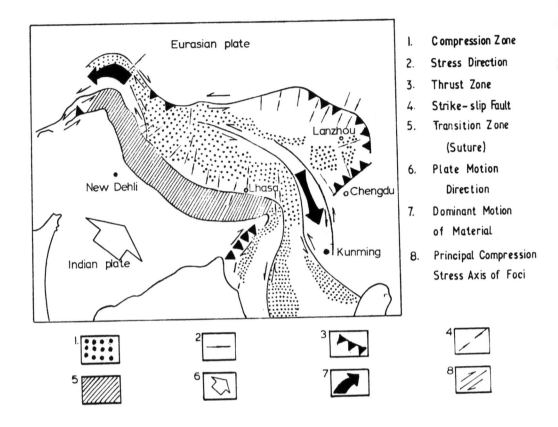

FIG. 13. SIMPLIFIED MAP OF THE TRANSITION BELT
BETWEEN INDIAN PLATE AND EURASIAN PLATE.

they are more sporadic, being shallower in the south than in the north. The fault plane solutions demonstrate underthrusting. The Yarlung Zangbo river is a deep fracture zone dipping slightly southward.

REFERENCES AND BIBLIOGRAPHY

1) TENQ JI-WEN, (1980). 'Characteristics of geophysical fields and continental plate tectonics of the Qinghai-Xizang plateau and its neighbouring regions', Acta Geophysica Sinica, Vol. 23, No. 3.

2) HUAN WEN-LIN ET AL, (1980). 'The distribution of earthquake foci and plate tectonics on the Qinghai-Xizang plateau', Acta Geophysica Sinica, Vol. 23, No. 3.

3) LUO ZHUOLI, (1979). 'On the Action of the Northward Push and. Collision of the Indian Plate and the Characteristic Features of the Recent Tectonic Stress Field and Seismicity in the Himalayan Arc

4)   YANG BING-PING ET AL, (1980). 'Some characteristics of small earth-quakes in the Zhamong and Damxung Districts of the Xizang Plateau', Acta Geophysica Sinica, Vol. 23, No. 3.

5)   MITOTEPPO, (1971). 'Seismicity Around the Himalaya Region', Science 41(4).

6)   JEMENITSKYA, R. M., (1975). 'Crust and Mantle of the Earth', Moscow 'Mineral Wealth'.

7)   YAN JIA-GUAN ET AL, (1980). 'Recent tectonics of the Qinghai-Xizang Plateau', Acta Geophysica Sinica, Vol. 23, No. 4.

8)   YEH HUNG ET AL, (1975). 'The analysis of the recent tectonic stress of the Himalaya Mountains arc and its vicinities', Scientia Geologica Sinica, No. 1.

9)   ZHANG LI-MIN ET AL, (1980). 'The average stress fields of Zhamong and Damxung regions in the Xizang Plateau', Acta Geophysica Sinica, Vol. 23, No. 3.

10) TENG JI-WEN ET AL, (1980). 'Explosion seismic study for velocity distribution and structure of the crust from Damxung to Yadong of Xizang Plateau', Acta Geophysica Sinica, Vol. 24, No. 1.

11) TENG JI-WEN ET AL, (1974). 'Deep reflected waves and the structure of the earth crust of the Eastern part of Chaidam Basin', Acta Geophysica Sinica, Vol. 17, No. 2.

12) ARGRA, S. K., (1973). 'Contribution of DSS to evaluation of crustal structure near Ganrihidanur', Geophys. Res. Bull. 13, No. 1-2.

Basement fault reactivation
in young fold mountain belts

J.A. Jackson
Department of Geodesy & Geophysics
Madingley Rise, Madingley Road,
Cambridge CB3 0EZ

ABSTRACT

During the early evolution of some fold mountain belts, old listric normal faults in a stretched and thinned basement beneath the sedimentary column may become reactivated as thrust faults. The reversal of motion on these faults allows considerable shortening to occur without subduction or excessive thickening of continental crust. This reactivation hypothesis is supported by the present day seismicity of the Zagros collision zone and the structure of sedimentary basins.

INTRODUCTION

Although plate tectonics has been very successful in describing the overall deformation and kinematics of the large ocean basins, no comparably simple description accounts for the general behaviour of the continents. Recent studies of continental deformation (1-3) have drawn attention to some of the notable differences between oceanic and continental behaviour, and have emphasized the dramatic role of major strike-slip faults in the Alpine-Himalayan Belt, which seem to slide large wedges of continental material away from zones of collision, thereby avoiding the excessive thickening of continental crust. Such large strike-slip features are easily visible on satellite photographs, and their spectacular nature, together with various models proposed to explain their large-scale significance (2-5), have to some extent diverted attention from the large areas of the Alpine-Himalayan belt which are nonetheless undergoing active shortening or extension.

In particular, little progress seems to have been made towards solving the classical space problem in orogenic belts: if the upper crust is shortened several tens of km by folding and thrusting, what happens to the basement? Very often the sediments involved in foreland folding and thrusting are separated from their basement by a decoupling horizon, often of evaporites or shale, which provides a surface of décollement separating structures above from those below. Various authors have discussed subduction as a mechanism for removing continental crust (6, 7) and concluded that its buoyancy should hinder its subduction and lead to crustal thickening instead (7, 8).

SEDIMENTARY BASINS

Helwig (9) pointed out in 1976 how this space problem in shortened orogenic belts is greatly diminished if the basement underneath the folded belt is thin to start with. Subsequent shortening then thickens it, but not to an abnormal degree. Since then McKenzie (10) showed that the subsidence of continental sedimentary basins can be quantitatively described by a simple model in which the lithosphere is rapidly stretched and thinned, sinks quickly to maintain isostatic equilibrium, and then sinks more slowly as it cools in a manner similar to that described by the successful oceanic models (11). The initial stretching of the basement at shallow depths is thought to occur by listric normal faulting and thereafter stretching ceases and subsidence continues without further faulting. This model is supported by the crustal thickness and heat flow of the Aegean, which is still being stretched by normal faulting, and appears to account well for the subsidence history of the North Sea (12) and many of the intra-Carpathian basins (13) all of which show an initial normal faulting episode followed by subsidence with no further faulting. Royden et al (14) have shown that the gross features of the same model will account for the stretching and subsidence preceding the break-up of a continent to form an Atlantic-type margin. It thus seems probable that thick sedimentary sequences generally form on a thin extended basement which was stretched by normal faulting. Evidence for basement thinning is best seen in seismic refraction profiles and has been demonstrated in the North Sea (15, 16) and Pannonian Basin (13). The Aegean is in an earlier stage of development, with high heat flow (17), thin crust (18, 19) and active normal faulting (20). There is thus every reason to think that Helwig's (9) suggestion has some validity and that the basement under the thick piles of sediments that make up fold mountain belts was thin before the onset of compression. The rest of this discussion concerns the early stages in the development of a fold mountain belt following continental collision.

It is proposed that thick sediments were initially deposited on a subsiding basement which had been stretched by normal faulting (Fig. 1). The faulting was very probably of a low angle listric nature, spread over a region several tens of km wide as observed in the Great Basin (22). As stretching proceeded, ultimately an Atlantic-type margin was formed with thick sediments on the continental shelf overlying a thin basement. In the early stages of continental collision, the basement takes up the shortening by reversing the sense of motion on the pre-existing normal faults (Fig. 1), which are now used as thrusts (reverse faults). The basement tends to return to its original thickness while the overlying sedimentary column, having been deposited on an extended basement, is forced to take up the shortening by folding. Thus in the early stages of collision the space problem is avoided without thrusting continental basement into the mantle or thickening it beyond its original (unstretched) state.

SEISMICITY OF THE ZAGROS MOUNTAINS, IRAN

The Zagros mountains of Iran provide an example of a young fold mountain belt currently shortening as a result of the collision between Arabia and Iran. The structure of the Zagros is very simple superficially, with a thick conformable sequence of Paleozoic-Mesozoic-Tertiary shelf deposits warped into gentle folds in a single process in the latest Tertiary (23, 24, 25). Long linear fold axes and seismic activity that is predominantly thrusting of the same orientation (1, 26) (Fig. 2) both indicate a general NE-SW shortening. Although strike-slip faulting is present on the northeast

238

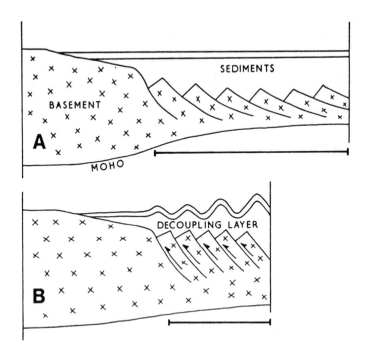

Fig. 1   Cartoons to illustrate the reactivation of basement faults
         see reference (21).

A.   Sedimentary basin formed by deposition on a basement which has been
     stretched and thinned by listric normal faulting.

B.   When the basin contracts motion is reversed on the normal faults which
     are now used as thrusts.   The sedimentary column takes up the shorten-
     ing by folding.

front of the Zagros, especially in the north (27, 28), it is not seen in the
fault plane solutions of the folded belt (Fig. 2) and the relative motion
between Arabia and Iran is normal to the strike of the belt.

The Folded Belt (23) of the Zagros is seismically very active with earthquakes spread over an area about 300 km wide and with a distinct north-
eastern boundary roughly coincident with the Zagros Thrust Line (or Main
Zagros Reverse Fault,   Fig. 2 and (28)), which in turn marks the edge
of the Zagros sedimentary trough.   These earthquakes are probably all
shallower than 40 km (see (29) for discussion) and there is no reliable
evidence that they increase in depth towards the northeast margin of the
Zagros.   Consequently there is no evidence for seismic shortening on a
single shallow-dipping plane, as is commonly assumed for areas of oceanic
underthrusting.   Fault plane solutions in the Zagros consistently show
thrusting with comparatively high angle (40°–50°) fault planes; (23) and
Figs. 2 and 3.   This is in marked contrast to the shallow angle (0°–10°)
thrusting which occurs in areas where oceanic underthrusting of continents

is, or was, taking place, such as the Hellenic Arc, Makran and Eastern Himalaya (30). The seismic activity of the Zagros Fold Belt reflects high angle reverse faulting on a large number of faults distributed across the whole width of the belt.

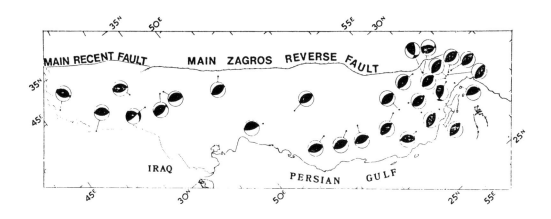

Fig. 2   Fault plane solutions from the Zagros from (26), to illustrate the high angle thrusting in the Folded Belt southwest of the Main Zagros Reverse Fault (called the Zagros Thrust Line by Stocklin (25) and others).

Black quadrants are compressional.

Although microearthquake surveys in the Zagros have shown that small earthquakes do occur within the sedimentary column (31, 32, 33) it has been suggested that the largest shocks, which reach m_b 6.0-6.3, have focal depths of about 15 km, i.e. beneath the sedimentary cover and in the crystalline basement. Berberian and others (28, 34) reached this conclusion from a study of damage distributions and because of the lack of surface faulting in the Zagros after major earthquakes. It is known that at the base of the sedimentay column is the thick infracambrian Hormuz Salt Formation, and that additional salt is present in the Jurassic and especially in the Tertiary Gach Saran Formation (24). Any basement faulting is unlikely to propagate to the surface through these ductile horizons. It is also unlikely that shocks the size of the 1972 Ghir earthquake (m_b=6.0), with a source dimension of about 30 km (35, 36), could be contained within 6 km of sediment thickness without showing surface faulting. The most convincing evidence of basement faulting comes from detailed examination of the long period teleseismic waveforms from Zagros earthquakes. Shocks in the magnitude range m_b 5.5-6.0 are common in the Zagros, and their far-field long period radiation is usually very simple. Studies of the waveforms (37), demonstrate that in three areas of the Zagros, the Dezful Embayment, Ghir and Khurgu (Fig. 2), earthquakes occurred at depths of 12-15 km. In each case, this is below the probable thickness of the sediments (38, 39, 40). There is therefore the strong probability that the basement beneath the Zagros Fold Belt is deforming by high angle (40°-50°) reverse faulting.

Fig. 3.

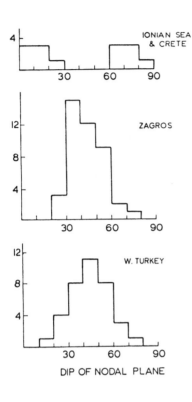

DIP OF NODAL PLANE

Histogram to illustrate the high angle (40°-50°) nature of the fault planes in the Zagros compared to those in the Ionian Sea and Crete, which are both part of the Hellenic arc where active subduction (intermediate depth earthquakes and andesitic volcanism) is taking place and thrusting is on shallow dipping (0°-10°) fault planes. The Zagros histogram is similar to that of western Turkey, where active extension by normal faulting is occurring. Dips are taken from the fault plane solutions in (1), (20) and (26) and are mostly well controlled in the Zagros, even though the strikes may not be.

EVOLUTION OF THE ZAGROS

There is much evidence that, during the Mesozoic and Tertiary, what is now the folded belt of the Zagros was a subsiding continental margin (24, 25, 41, 42). Since the basement was almost certainly cut by normal faults which caused its extension prior to the onset of subsidence and sediment deposition in the Permo-Triassic, it is likely that the present day reverse faulting is happening on the old normal fault surfaces which have been reactivated. This would explain the high (40°-50°) rather than low (0°-10°) angle nature of the fault dips, and the similarity of the dips of fault planes in western Turkey (currently extending by normal faulting) with those in the Zagros (Fig. 3). Reactivation of old faults is known to be a common and important phenomenon (45) and the reactivation of old normal basement faults as thrusts has been reported from Chile (46). The possibility of reactivation in the Zagros was alluded to by Falcon (23) in 1969, and Stocklin (25) proposed that the Zagros Thrust Line itself was a reactivation of an old Precambrian rift boundary which controlled the distribution of the Hormuz salt. The N.W. trend of the Zagros is found in the Precambrian of the N.E. Arabian shield (25). The same trend is evident in the sediment isopachs of the northern Zagros (Lurestan and Kuhzistan) throughout the Mesozoic (47, 48).

An objection to this suggested basement involvement is the evidence from seismic relection lines in old fold mountain belts, such as the eastern Rockies (44, 49), which shows apparently undeformed Precambrian shield dipping gently beneath the foreland folds. However, it is probable that in these older orogenic belts the amount of shortening involved is considerably greater (about 160 km, or 50%, in the Canadian Rockies (44, 50)) than that which has so far taken place in the Zagros Fold Belt (about 20-50 km, or 20%, (23)) which is in a much younger state of development. The contraction mechanism proposed here will presumably work until the reverse motion on old normal faults restores the basement at least to its original thickness. Thereafter the basement will either become thicker than its pre-stretched state, or be thrust into the mantle. It is likely that the sediments of the folded belt of the Zagros are still almost above the original basement on which they were deposited. As shortening continues, the folded cover will start to migrate southwest over the undeformed Arabian shield, separated from its basement by the décollment at the level of the Hormuz salt. The foreland folds of the Rockies, Appalachians and Jura are no longer above their original basement which was left behind in the internal zones. Those orogenic belts which are known to have thick crust are all older and in more advanced states of development than the Zagros and have involved more shortening. It is thus an interesting prediction of the scheme proposed here that, in the initial stages of continental collision, the basement under the orogenic belt should be thinner, not thicker, than that of its neighbouring craton. At present no adequate refraction data from the Zagros is available to test this.

In conclusion, the origin of a young folded belt proposed here accounts well for the present day seismicity and deformation of the Zagros. Such an origin involves the reactivation as thrusts of listric normal faults in a stretched and attenuated basement on which thick sediments of the Atlantic-margin type were deposited, and does not necessitate either abnormal thickening or subduction of continental crust in the early stages of continental collision. How long this process can continue is not clear. It is unlikely to account for all the features of older orogenic belts which have involved considerably greater shortening. However, even if operative in only the early stages of collision, it greatly reduces the space problem encountered in palinspastic reconstructions of orogenic belts (43).

REFERENCES

(1) McKenzie, D.P. Active tectonics of the Mediterranean region. Geophys. J. Roy. astr. Soc. 30, 109-185 (1972).

(2) Tapponnier, P. and Molnar, P. Active faulting and tectonics in China. J. Geophys. Res. 82, 2905-2930 (1977).

(3) Molnar, P. and Tapponnier, P. Cenozoic tectonics of Asia: effects of a continental collision. Science, 189, 419-426 (1975).

(4) Tapponnier, P. and Molnar, P. Slip-line field theory and large scale continental tectonics. Nature, 264, 319-324 (1976).

(5) Tapponnier P. and Molnar, P. Active faulting and Cenozoic tectonics of the Tien Shan, Mongolia and Baykal regions. J. Geophys. Res. 84, 3425-3459 (1979).

(6) Bally, A. Orogenic Images. in Thrust and Nappe Tectonics Conference. Geol. Soc. Lond. Spec. Pub. No. 9 (1980)

(7) Molnar, P. and Gray, D. Subduction of continental lithosphere: some constraints and uncertainties. Geology, 7, 58-62 (1979).

(8) McKenzie, D.P. Speculations on the causes and consequences of plate motions. Geophys. J. Roy. astr. Soc. 18, 1–32 (1969).

(9) Helwig, J. Shortening of continental crust in orogenic belts and plate tectonics. Nature, 290, 768–770 (1976).

(10) McKenzie, D.P. Some remarks on the development of sedimentary basins. Earth & Planet. Sci. Lett. 40, 25–32 (1978).

(11) Parsons, B. and Sclater, J.G. An analysis of the variations of ocean floor bathymetry and heat flow with age. J. Geophys. Res. 82, 803–807 (1977).

(12) Christie, P. and Sclater, J.G. An extensional origin for the Buchan and Witchground Graben in the North Sea. Nature, 23, 729–732, (1980)

(13) Sclater, J.G., Royden, L., Howrath, F., Burchfiel, C., Senken, S., and Stegner, L. The role of continental stretching in the Neogene subsidence of the Pannonian Basin. Earth & Planet. Sci. Lett. (1980).

(14) Royden, L., Sclater, J.G. and von Herzen, R.P. Continental margin subsidence and heat flow: important parameters in formation of hydro-carbons. Am. Assoc. Petrol. Geol. Bull. 64, 173–187, (1980)

(15) Zeigler, P. Geology and hydrocarbon provinces of the North Sea. Geo. Journal 1.1, 7–32 (1977).

(16) Christie, P. The crust and upper mantle beneath the North Sea basin. Ph.D. Thesis, University of Cambridge, (1979).

(17) Jongsma, D. Heat flow in the Aegean Sea. Geophys. J. Roy. astr. Soc. 37, 337–346 (1974).

(18) Makris, J. and Vees, R. Crustal structure of the central Aegean Sea and the islands of Evvia and Crete, Greece, obtained by seismic refraction experiments. J. Geophys. 42, 329–341, (1977).

(19) Makris, J. Crustal structure of the Aegean Sea and Hellenides from Geophysical Surveys. J. Geophys. 41, 441–443 (1975).

(20) McKenzie, D.P. Active tectonics of the Alpine–Himalayan Belt: the Aegean Sea and surrounding regions. Geophys. J. Roy. astr. Soc. 55, 217–274 (1978).

(21) Jackson, J.A. Reactivation of basement faults and crustal shortening in orogenic belts. Nature, 283, 343–346 (1980).

(22) Proffett, J.M. Cenozoic geology of the Yerington district, Nevada, and implications for the nature of Basin and Range faulting. Geol. Soc. Am. Bull. 88, 247–266 (1977).

(23) Falcon, N.L. Problems of the relationship between surface structure and deep displacements illustrated by the Zagros Range. in Time and Place in Orogeny. Geol. Soc. Lond. Spec. Pub. (1969).

(24) Stocklin, J. Structural history and tectonics of Iran: a review. Am. Assoc. Petrol. Geol. Bull. 52, 1229–1258.

(25) Stocklin, J. Possible ancient continental margins in Iran.
      in Burk, C.A. and Drake, C. (Eds.) The geology of continental
      margins. Springer, New York.

(26) McKenzie, D.P. and Jackson, J.A. Active Tectonics of Iran and E.
      Turkey. in preparation.

(27) Tchalenko, J. and Braud, J. Seismicity and structure of the Zagros
      (Iran): the Main Recent Fault between 33 and 35°N. Phil. Trans.
      Roy. Soc. Lond. 277A, 1-25 (1974).

(28) Berberian, M. Geol. Surv. Iran Reports 39 and 40 (1976, 1977).

(29) Jackson, J.A. Errors in focal depth determination and the depth of
      seismicity in Iran and Turkey. Geophys. J. Roy. astr. Soc., 61,
      285-301 (1980).

(30) Jackson, J.A., McKenzie, D.P. and Fitch, T.J. Active thrusting and
      the evolution of the Zagros fold belt. in Thrust and Nappe Tectonics.
      Geol. Soc. Lond. Spec. Pub. No. 9 (1980).

(31) Savage, W.U., Alt, J.N. and Mohajer-Ashaji, A. Microearthquake
      investigations of the 1972 Qir, Iran, earthquake zone and adjacent
      areas. Geol. Soc. Am. Abstr. 9, 496, (1977).

(32) Von Dollen, F.J., Alt, J.N., Tocher, D. and Nowroozi, A. Seismological
      and geological investigations near Bandar Abbas, Iran. Geol. Soc. Am.
      Abstr. 9, 521 (1977).

(33) Atomic Energy Authority of Iran. Bulletin of the Seismographic
      Network, Bushehr region (1978).

(34) Berberian, M. and Papastamatiou, D. Khurgu (North Bandar Abbas,
      Iran) earthquake of March 21, 1977: a preliminary field report and a
      seisomotectonic discussion. Bull. Seism. Soc. Am. 68, 411-428 (1978).

(35) Jackson, J.A. and Fitch, T.J. Seismotectonic implications of relocated
      aftershock sequences in Iran and Turkey. Geophys. J. Roy. astr. Soc.
      57, 209-229 (1979).

(36) Dewey, J.W. and Grantz, A. The Ghir earthquake of April 10, 1972 in
      the Zagros mountains of Southern Iran: seismotectonic aspects and some
      results of a field reconnaissance. Bull. Seism. Soc. Am. 63, 2071-2090
      (1973).

(37) Jackson, J.A. and Fitch, T.J. Basement faulting and the focal depths
      of the larger earthquakes in the Zagros mountains (Iran). Geophys. J.
      Roy. astr. Soc. (1981).

(38) Morris, P. Basement structure as suggested by aeromagnetic surveys in
      S.W. Iran. Oil Service Co. Iran, report (1977).

(39) James, G.A. and Wynd, J.G. Stratigraphic nomenclature of Iranian Oil
      Consortium Agreement Area. Am. Assoc. Petrol. Geol. Bull. 49 2182-2245
      (1965).

(40) Comby, O., Lambert, Cl. and Coajon, A. An approach to structural
      studies in the Zagros Fold Belt in the EGOCO agreement area. 2nd
      symposium of geology of Iran, Teheran, March, 1977.

244

(41) Haynes, S.J. and McQuillan, H. Evolution of the Zagros suture zone, southern Iran. Geol. Soc. Am. Bull. 85, 739–744, (1974).

(42) Stoneley, R. On the origin of ophiolite complexes in the southern Tethys region. Tectonophysics, 25, 303–322, (1976).

(43) Laubscher, H.P. The large scale kinematics of the western Alps and northern Apennines and palinspastic implications. Am. J. Sci. 271, 193–226 (1971).

(44) Bally, A.W., Gordy, P.L. and Stewart, G.A. Structure, seismic data, and orogenic evolution of the southern Canadian Rocky Mountains. Can. J. Earth Sci. 3, 713–723 (1966).

(45) Sykes, L. Intraplate seismicity, reactivation of pre-existing zones of weakness, alkaline magmatism and other tectonism post-dating continental fragmentation. Rev. Geophys. Space. Phys. 16, 621–688 (1978).

(46) Winslow, M. A new mechanism for basement shortening under a foreland fold belt in the southernmost Andean foothills, S. America. in Thrust and Nappe Tectonics, Geol. Soc. Lond. Spec. Pub. No. 9 (1980).

(47) Setudehnia, A. The Mesozoic sequence in southwest Iran and adjacent areas. J. Petrol. Geol. 1, 3–42, (1978).

(48) Szabo, F. and Keradpir, A. Permian and Triassic stratigraphy, Zagros basin, southwest Iran. J. Petrol. Geol. 1, 58–82 (1978).

(49) Royse, F., Warner, M.A. and Reese, D.L. Thrust Belt structural geometry and related stratigraphic problems – Wyoming – Idaho – Northern Utah. in Bolyard, D.W. (ed) Deep Drilling Frontiers of the Central Rocky Mountains. Rocky. Mtn. Ass. Geol. (1975).

(50) Price, R.A. and Mountjoy, E.W. Geological structure of Canadian Rocky Mountains between Bow and Athabasca rivers: a progress report. Geol. Assoc. Can. Spec. Pap. 6, 7–25 (1970).

Housing and natural hazards

Experimental studies of the effect of earthquakes on small adobe and masonry buildings

D.F.T. Nash
(Department of Civil Engineering, University of Bristol)

R.J. Spence
(Department of Architecture, University of Cambridge)

ABSTRACT

The majority of earthquake deaths occur as a result of the collapse of small dwellings of masonry or adobe, supporting massive roofs. The widespread introduction of improved methods of building construction is hampered by a lack of understanding of the behaviour of such buildings under earthquake forces. The complexity of the problem makes analytical study very difficult, and few experimental studies have so far been carried out. The paper discusses the experimental methods available and considers the problems of modelling in the laboratory small buildings made of traditional low-strength materials. Experimental programmes in the authors' departments which are in their early stages are briefly described.

INTRODUCTION

Although accurate statistics are difficult to obtain, the number of fatalities caused by earthquakes this century has been estimated to be around 1.5 million. Most of this loss of life has resulted from roof collapse in unreinforced masonry and adobe low-cost housing. Self-built housing of this type comprises around 95% of the housing stock of such nations as Iran, Turkey and Chile.

The materials and building methods employed in such housing vary greatly around the world, but certain characteristics are common to many areas. Of these the two which particularly result in collapse during earthquakes are:

i)   the use of unreinforced brittle building materials,

ii) the prevalence of heavy roofs.

In the light of this it is perhaps surprising that little research has been carried out into the seismic behaviour of such buildings. There must be many reasons for this including:

i)   the fact that individually each house represents a very small financial investment which gives it a low priority in preventative measures against earthquakes,

ii)   a common belief that no solution is possible without resort to modern materials and building methods,

iii)   a tendency amongst researchers to avoid subjects which do not lend themselves to theoretical analysis,

iv)   an assumption that the problem is only a short-term one which will be overcome rapidly by the general development of third-world countries.

This last assumption is now being questioned, and it now seems certain that the great majority of people living in unreinforced masonry and adobe structures will continue to do so for the foreseeable future.

There is therefore a great need to investigate the seismic performance and behaviour of such traditional buildings in order to find effective improvements for existing and new housing. Such improvements must lie within the range of existing technologies and should be socially and economically appropriate.

This paper examines the range of experimental methods available to investigate the performance of low-cost masonry and adobe structures.

THE NEED FOR EXPERIMENTAL STUDIES

It was once said that "In research, theory is important, but experimental studies are more important, and the analysis of phenomena caused by nature is best". This certainly applies to the study of the seismic protection of unreinforced masonry and adobe low-cost housing.

Useful conclusions about the performance of existing building types in earthquakes can be reached from a study of damage following major earthquakes. A number of such studies have been made, of which an outstanding one is that by Razani (1). He presents design criteria for such buildings as follows:

i)   the roof and floors must remain monolithic and connected,

ii)   a vertical load system must survive,

iii)   a lateral load system must survive.

Razani also makes a number of recommendations as to how these criteria might be achieved.

Unlike many other engineering problems, the economic investment in an individual dwelling of this type bears no relation to the complexity of the analysis required to understand its behaviour. The design criteria above imply the limitation of damage rather than its avoidance. Since the building materials are extremely brittle, the behaviour of such buildings must be considered after cracking has occurred, a situation which is not readily amenable to analytical treatment.

Thus to advance our understanding of the seismic behaviour of such buildings, we have to rely on experimental studies. Such studies fall into two categories

i)   observation of the performance of existing and modified building
       types during earthquakes,

ii)   dynamic tests on buildings, model buildings and building elements.

So far there are not any reports of such buildings being instrumented.
It is probably unrealistic to envisage a significant programme of instrumentation given that the return period of destructive earthquakes is
relatively long.

Dynamic tests on buildings or model buildings offer the most ready
means of assessing the effectiveness of proposed modifications to the
design. Penzien (2) has stated that "the main objective of conducting
dynamic tests on structures is to improve and verify mathematical
modelling which is intended to realistically represent prototype behaviour
under seismic conditions". While this may well apply to dynamic tests
on "engineered" structures, the primary objective of such tests on low-
cost housing at this stage is surely to assess empirically the effectiveness
of introducing simple modifications to a design.

EXPERIMENTAL METHODS

Experimental methods for the investigation of seismic building
performance fall into two categories – destructive and non-destructive.
Such tests can be carried out either on full-scale buildings, building
elements or model buildings.

Non-destructive tests are confined to studying building behaviour
in the elastic range. Random vibrations, impulses and steady-state
harmonic vibrations can be used (3) to measure natural modes,
frequencies, damping coefficients and other dynamic properties. These
low-amplitude values may then be compared with values derived from
the theoretical models which are used to predict a building's performance
under earthquake loading. While such work is primarily concentrated
on 'engineered' structures such as arch dams which are designed to
remain elastic under earthquake loading, the same approach may be
adopted for low-cost housing to investigate the conditions under which
cracking is initiated. Recently some field studies were carried out by
Ohta et al (4) in Turkey following the Caldiran earthquake in November
1976. Vibration tests were carried out on several adobe and masonry
buildings to measure the natural periods of vibration and their results
are reproduced in Table 1.

	Natural Periods	
	T_1 (short span)	T_2 (long span)
Adobe	0.16 sec	0.135 sec
Stone	0.155 sec	0.13 sec
Block	0.18 - 0.2 sec	0.165 sec
R.C.	0.13 - 0.2 sec	0.24 - 0.27 sec

after Sakai in (4).

Table 1.   Field Vibration Tests on Small Buildings

Such work could usefully be extended and supplemented by tests on models.

Tests in which the building is strained beyond the elastic limit are essential to understanding its behaviour during an earthquake. Two approaches are commonly adopted:

i)   Static or slow-cycle tests.   Full-scale buildings may be tested on a tilting table (5, 6) or building elements may be tested in a load frame (7, 8 and others).   Such tests have the advantage that the boundary conditions are known throughout the test and strains may be observed.   This is particularly useful for examining the conditions under which cracking is initiated and investigating the restraining influence of reinforcement.   It does not permit any examination of dynamic interaction between different parts of the building.

ii) Dynamic tests on a shaking table.   Such tests may be on full-scale buildings (9) but for reasons of economy are more likely to be made on models (10).   The base motion may be a recorded earthquake or sinusoidal vibration.   Providing the problems of scaling can be overcome, the response of a particular building may be observed directly and the effects of simple modifications investigated.

Recently we have carried out some preliminary model tests at the Universities of Bristol and Cambridge (11, 12).   Working at a scale of 1 : 10 two types of model were developed:

i)   Models to study elastic response.   Tests were made using single point excitation by electro-magnetic vibrators to measure natural frequencies and mode shapes.   These indicated a very complex response combining lateral vibration of the walls with shear and torsion.   The lowest natural frequencies indicated for full-scale buildings were around 15 Hz, rather higher than expected.

ii) Models to investigate failure.   Tests made on a harmonic shaking table were able to reproduce building failures in the laboratory, although there were considerable problems with the model materials.

In addition tests were made of a model single wall on an electro-magnetic shaking table.   Walls made from unmortared model bricks proved extremely stable even when subjected to out of plane earthquake vibration. This suggests that Coulomb friction could usefully form the basis of an improved design.

It is hoped to extend this work as modelling materials are developed.

In reviewing previous work in this field it has become clear that there has been very little work (10) in which the links between engineering properties in the field, structural arrangement, behaviour of building elements under load and dynamic behaviour of buildings have been made. Such a coordinated experimental programme is essential if real progress is to be achieved.

REQUIREMENTS FOR MODEL TESTS

Small scale model tests have been used for many decades by structural engineers, but until recently their use was mainly confined to static, elastic behaviour of structures. The use of simple inexpensive models of low-cost housing to study their dynamic behaviour appears attractive at first sight. However the similitude requirements for dynamic modelling present considerable problems.

The basic modelling theory for dynamic structural models is well known and has been presented by several authors. Nazarov (13) gives a comprehensive survey of the general theory but it is useful to consider particular cases here. A primary requirement when modelling structures beyond the elastic region is geometric similarity; it is necessary to ensure that the strains in the model are equal to the strains at full-scale. Two common approaches when designing a model are as follows:

i) No change of acceleration. This has the advantage that tests may be carried out under natural gravity. However a model material is required whose strengths and dynamic moduli are scaled down and a compressed time scale is used.

ii) No change of material. While this has obvious advantages, for complete similarity tests must be carried out with increased acceleration and enhanced gravity in a centrifuge. A compromise solution is generally attempted, by the use of additional weights hung from the structure. Weights may be placed on the structure but this alters the inertia forces. Alternatively the density may be increased without changing the other material properties.

The correct scale factors for various properties for these cases are given in Table 2.

	General	Case I (acceleration unaltered)	Case II (material unaltered)
length	α	α	α
strain	γ	1	1
stress	β	$\alpha\delta$	1
density	δ	δ	1
time	$\alpha\sqrt{\dfrac{\delta\gamma}{\beta}}$	$\sqrt{\alpha}$	α
elastic moduli	β/γ	$\alpha\delta$	1
displacement	$\alpha\gamma$	α	α
velocity	$\sqrt{\dfrac{\gamma\beta}{\delta}}$	$\sqrt{\alpha}$	1
acceleration	$\beta/\alpha\delta$	1	$1/\alpha$
coulomb friction	1	1	1
damping	1	1	1

Table 2. Scale Factors of Complete Model Similarity

MATERIALS FOR MODEL TESTS

There is a dearth of information on materials which have been found suitable for dynamic testing of model masonry or adobe structures. For the testing of concrete structures it is normal to use micro-concrete, and Bosjancic (14) suggests that similar materials can be used for modelling masonry. Unfortunately he gives no details of the modelling materials, but he discusses the use of plastic-coated wires to model reinforcement correctly. Sabnis and White (15) also discuss the modelling of concrete structures.

The only extensive model testing of adobe structures of which we are aware has been carried out at the University of Mexico (10). Models at a scale of 1 : 2.5 have been tested on a shaking table, and it appears that reduced strength adobe was used. Conditions of complete model similarity were not achieved but useful results were obtained when comparing different types of strengthening systems.

Modelling Materials				
Material	σ_c (N/mm^2)	σ_t (N/mm^2)	E kN/mm^2	ρ kg/m^3
sand + 10% wax		1.46		
sand + 5% wax		0.82		
sand + 2% wax	5.36	0.32		1500
sand + 1% wax		0.07		
sand + 20% kaolinite	2.50	0.24		1870
sand + 10% kaolinite		0.17		
sand + plaster (3 : 1)	17.0		0.2	
sand + plaster (6 : 1)	1.5			

Prototype Materials			
mud mortar	4.75	0.03	0.42
adobe	1.0	0.3	

Table 3. Modelling Materials and Prototype Materials

In our recent model testing at Bristol and Cambridge Universities (11, 12) two modelling materials were used:

i)   Models to study elastic response.   These were constructed mono-lithically out of a mortar of Kaffir D plaster and sand. This is a brittle material which reaches its full strength rapidly, but was too strong for tests to destruction.

ii) Models to investigate failure.   In an attempt to reproduce building failures in the laboratory models constructed from a sand–wax mortar were tested on a shaking table.   This material is weak and brittle, but its strength was still similar to that of mud mortar when it should have been reduced by the scale of the model (10 times).   Such a reduction in strength would have made it very difficult to handle.

Details of some of the mechanical properties of these materials are given in Table 3, but it is clear that much more work is required to develop suitable materials for scale modelling.   Probably the scale of the model should be increased to at least 1 in 5.   This would reduce the problems with the model material but the increase in scale is limited by the size of the shaking tables.   In his paper on model materials, Bosjancic (14) recognises this, and calls for urgent cooperation between laboratories interested in development of model techniques.

CONCLUSIONS

Clearly there is an urgent need for a coordinated programme of experi-mental studies of the seismic behaviour of unreinforced masonry and adobe low-cost housing.   A link needs to be established between the engineering properties of building materials in the field, the behaviour of structural elements under known conditions of loading, and the dynamic behaviour of buildings or models tested.   In this way the effect of simple modifications may be established which could then be incorporated in an empirical design guide.   However for such recommendations to be of any use, they must take full account of the existing technical skills and social and economic conditions in the situation for which they are intended.

REFERENCES

1)   RAZANI, R.   Seismic protection of unreinforced masonry and adobe low-cost housing in less developed countries: Policy issues and design criteria.   Int. Conf. on Disasters and Small Dwellings.   Oxford 1978.

2)   PENZIEN, J.   Theme report on dynamic tests on structures.   Proc. 6th WCEE, New Delhi, India, 1977.   Vol. III, p 2687.

3)   WARD, H. S.   Experimental techniques and results for dynamic tests on structures and soils.   Proc. 6th WCEE, New Delhi, India, 1977. Vol. III, p 2810.

4) OHTA, Y. ed. Engineering seismological studies on the 24th November
    1976 Caldiran earthquake in Turkey. Faculty of Engineering,
    Hokkaido University, Japan. March 1980.

5) RAZANI, R. Investigation of lateral resistance of masonry and adobe
    structures by means of a tilting table. Proc. 6th WCEE, New Delhi,
    India, 1977. Vol. II, p 2130.

6) MUNSKI, K. D. and KEIGHTLEY, W. O. Tilting platform for measuring
    earthquake resistance of small buildings. Proc. 6th WCEE, New Delhi,
    India, 1977. Vol. III, p 2915.

7) CHINWAH, J. G. On the shear resistance of brick walls against
    seismic loads. Proc. 3rd European Symposium on EE, Sofia, Bulgaria,
    1977, p 847.

8) KRISHNA, J., CHANDRA, B. Strengthening of brick buildings against
    earthquake forces. Proc. 3rd WCEE, New Zealand, 1965. Vol. III,
    p 324.

9) CLOUGH, R. W., MAYES, R. L. and GULKAN. Shaking table study
    of single-storey masonry houses. EERC, University of Berkeley,
    California, 1979.

10) HERNANDEZ, O., MELI, R. and PADILLA, M. Estudios experimentales
    sobre la seguridad ante sismos de las casas de adobe. Instituta
    de Ingenieria, University of Mexico, 1979.

11) COBURN, A. W. Modelling the effects of earthquakes on ·adobe
    buildings. Dissertation for Diploma of Architecture, Pt 1, University
    of Cambridge. June 1980.

12) MARSHALL, C. P. and TREVELYAN, J. Investigation into the behaviour
    of low-cost housing during earthquakes. Undergraduate Project,
    Department of Civil Engineering, University of Bristol, 1980.

13) NAZAROV, A. G. Theoretical foundation of the investigation of earth-
    quake resistance of structures by models. Proc. 3rd WCEE, New
    Zealand 1965, Vol. II, p 812.

14) BOSJANCIC, J. Model materials suitable for dynamic tests of models
    in the plastic range. Proc. 6th WCEE, New Delhi, India, 1977.
    Vol. III, p 2745.

15) SABNIS, G. M. and WHITE, R. N. Small scale models of concrete structure
    subjected to dynamic loads. Proc. 6th WCEE, New Delhi, India 1977,
    Vol. III, p 2766.

Traditional housing in seismic areas

R.J.S. Spence
A.W. Coburn
Department of Architecture, University of Cambridge

ABSTRACT

A survey of the literature concerning the earthquake-resistance of traditional buildings in developing countries is presented. Certain types have regularly been found to perform well, notably those in which a braced timber frame has been used. It is argued that a study of such traditional building types may be of some relevance in devising future strategies for modification of rural low-cost buildings, but that the earthquake hazard should be seen as only one of many factors which have influenced traditional forms of construction.

INTRODUCTION

It is well known that the vast majority of casualties caused by earthquakes occur in developing countries, and result from the collapse of low-cost dwellings built by traditional methods in rural areas. Earthquake engineering, which has been so successful in enabling large buildings to be built in modern materials with a high degree of earthquake resistance, has so far had little impact on the methods of construction used for such low-cost buildings. This is not because these buildings have been totally ignored by engineers, as has sometimes been suggested. In most countries which are regularly visited by earthquakes, some research into the traditional construction methods has been carried out, and codes of practice or construction manuals have been published which describe how these methods could be modified to render them resistant, or at least less vulnerable, to earthquakes. But too often the modifications proposed turn out to require the use of substantial amounts of scarce and expensive materials (such as cement and steel), or skilled workmanship, and thus to be beyond the reach of the vast majority of people who must build with virtually no cash outlay and using only their own labour and traditional skills. (1)

Yet, in many seismic areas, it has been reported that there are certain particular building types which traditionally incorporate earthquake-resistant design principles or construction features, or which have been found to perform particularly well in earthquakes. This is an important observation, because it suggests that there are alternative techniques of building earthquake-resistant dwellings which require neither modern materials nor skilled craftsmanship, and opens the possibility that these techniques might be adapted for use elsewhere, should suitable materials be available. And it is important for another reason: if these traditional building types can be shown to be a conscious response to the earthquake hazard, this may provide an insight into the way that earthquake awareness is developed and passed on in a society, and how this awareness, in turn, leads to modifications in the form and construction of housing. Such insights would be very helpful in attempting to devise suitable strategies for future modifications of low-cost housing in seismic areas.

This paper is a preliminary attempt to collect and examine the evidence of earthquake-resistant design in traditional housing in all parts of the world. The study is not based on field work; the information presented is taken entirely from published papers and reports. The main sources have been the Proceedings of the six world conferences on earthquake engineering, and the reports of the UNESCO reconnaissance missions to the centres of recent earthquakes. Figure 1 shows the geographical location of the various building types discussed, superimposed on the seismicity map of the world.

SEISMIC RESPONSE OF TRADITIONAL BUILDINGS

By far the most useful data are those given in surveys of the performance of traditional dwellings in earthquakes, classified according to types of construction; but unfortunately few such surveys have been carried out. Arioglu and Anadol (2, 3) have reported limited surveys conducted after four earthquakes in Turkey between 1969 and 1975, those of Gediz, Burdur, Bingol and Lice, all with magnitudes of 6.3 or greater, and with intensities reaching VIII or IX in the epicentral region. In all these earthquakes, masonry buildings whether of stone or adobe, performed very badly; but in the Gediz earthquake region, 75% of the rural dwellings were found to be of a two-storey braced timber frame system (locally called 'Baghdadi') which was highly earthquake resistant; 50% of such buildings were undamaged, and only 9% were irreparably damaged in the earthquake. The seismic histories of the regions concerned are also reported; Gediz has had no earthquake of magnitude greater than 6.0 during the present century, whereas Burdur (7.1 in 1914), Bingol (7.0 in 1954) and Lice (7.0 in 1971) have all suffered large earthquakes in living memory. Thus there is, on this evidence, no correlation between recent seismic history and earthquake resistant techniques.

Joaquin (4) reports the results of damage surveys following 10 earthquakes in Chile between 1918 and 1966 having intensities of VII to IX; some 20,000 buildings were included in these surveys, mainly in towns and cities. Nevertheless the results are of considerable interest, because of the range of building types included. The relationship between damage and intensity has been plotted for five different types of construction, and these relationships are summarised in Figs 2 and 3. The best performance was achieved by modern reinforced brick masonry buildings designed to withstand lateral loads. But again, timber framed buildings of traditional construction performed better than unreinforced brick masonry, and adobe construction – used by the majority in the most densely populated part of the country – performed very poorly. It is also reported that subsequent to the 1939 Chilean earthquake, the prohibition of adobe buildings was considered, but could not be effected because of the absence

255

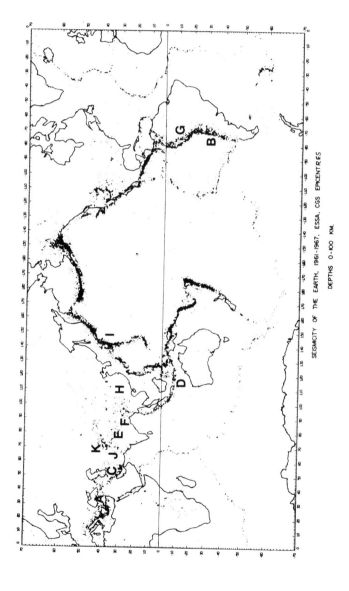

SEISMICITY OF THE EARTH, 1961-1967, ESSA, CGS EPICENTRES

DEPTHS 0-100 KM.

FIG. 1. Seismicity map of the world. Dots indicate the distribution of seismicity in the mid 20th century; see (22). Letters indicate approximate locations of traditional building types discussed.

256

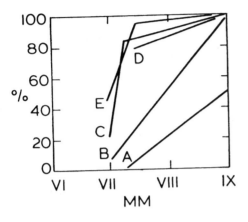

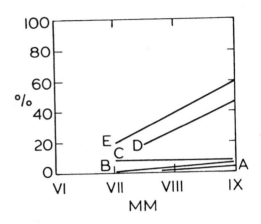

FIG. 2
Proportion of damaged buildings

FIG. 3
Proportion of totally destroyed
buildings

Approximate distribution of damage to buildings classified according to
construction techniques in Chile, based on surveys of ten earthquakes
1918 – 1966, after Joaquin (4).

Key A = Brick masonry, reinforced
    B = Wooden frame, without fill
    C = Wooden frame with heavy fill
    D = Brick masonry, unreinforced
    E = Adobe.

of an economic alternative.

During the last 15 years, UNESCO reconnaissance missions have visited
the centres of a number of large earthquakes shortly after the event,
and have documented the damage. Although the damage surveys have
not been on a statistical basis, the reports note where particular buildings
performed well or badly. Four such reports have been concerned with
earthquakes in Iran, at Dasht-e-Bayaz (1968), Karnaveh (1970), Ghir (1972)
and Gisk (1977) (5, 6, 7, 8). In rural Iran, absence of timber has led
to the general use of vaults and domes for roofing, and many lives were
lost through the collapse of such roofs. However, significantly better
performance was noted where the domes were quasi-spherical rather than
cylindrical, or where tie-rods or tie-beams had been incorporated at the
springings of vaults or domes to eliminate horizontal thrusts on the walls.
Another interesting observation was that tall masonry structures such as
chimneys, wind-towers or minarets were frequently undamaged. The lightweight portable structures of the nomads performed well; they constituted
practically the only houses in the Zagros mountains which survived the
Ghir earthquake, and were used for emergency housing after the Karnaveh
earthquake.

A manual on earthquake-resistant construction from Bali, Indonesia

(9), attributes a high degree of earthquake resistance to traditional buildings because of their regular plan form, lightweight roof supported on a ring beam, braced timber columns and strong heavy foundations (Fig. 4).

In India, Arya and Chandra (10) report that similarly aseismic forms of construction were used in the earthquake-prone regions of Kashmir and Assam. In Kashmir braced timber stud-wall construction with brick nogging called Dhajji Diwari was used, while in more humid Assam, the framework was of bamboo poles with a light cladding of Ikra reed.

Evans (11) has reported that the Peru earthquake of 1970 showed 'quincha' construction to be much superior to adobe. A quincha wall consists of vertical timber poles about 1 m apart, with bottom bracing and a horizontal cap pole. Several equally spaced horizontals complete the basic frame. Vertical canes are closely woven between the poles and the assembly plastered with mud.

Both China and Japan have a history of devastating earthquakes, and in both countries the traditional architecture has always been based on the timber frame, even where stone and brick were known and available. Needham (12) has suggested that a knowledge of the superior earthquake-resistance of timber may have been a contributory reason for its use, but certainly other cultural factors were involved. The traditional detailing at the head of the column (where complex bracket clusters supported deeply overhanging eaves) and at its foot (where it rests on a stone pad) are not directly derived from the need for earthquake resistance, but it has been pointed out (13, 14) that the large deformations and distortions which these details can tolerate without failure are a great asset, contributing both to flexibility and to damping.

PAKISTAN

In the North and West of Pakistan there are regions of high seismic activity, and numerous large earthquakes have occurred during this century (15). In the treeless region around Quetta, where mud is commonly used for walling, the earthquakes of 1931 and 1935 caused considerable damage and loss of life, and the only buildings reported as resisting the shaking were those built in modern materials, usually incorporating some form of steel reinforcement (16, 17). In North West Frontier Province, the walls of the houses are built of stone, and some important differences in performance were noted by the UNESCO reconnaissance mission following the Pattan earthquake of 1974. (18)

> There are two types of construction (a) with timber-supported roofs and non-bearing rubble-fill walls, in which the heavy roofs are carried independently by wooden unbraced columns, and (b) with similar roofs supported by bearing rubble-fill walls.

> In both cases it is customary to use very thick walls coursed at intervals of 0.4 to 0.6 metres with horizontal baulks of timber at front and back tied together through the thickness of the wall by timber ties. Roofs for both types consist of closely-spaced timber rafters covered with rushes and layers of tamped loessic clay in thickness sufficient to carry heavy domestic animals such as cows. The use of the roof of one house as the yard of the next house above is unavoidable,

258

"From our legend, according to our ancestors, earth is supposed to be Bedawang Nala (Turtle-god worshipped in Bali). This earth could tremble at any time..."

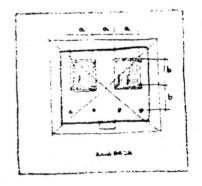

"Because obviously it is proportionally divided into 'head' 'body' and 'feet'.
(a) Head: lightweight, low, strong and solid.
(b) Body: Strong, rigid.
(c) Feet: Heavy, strong and firm."

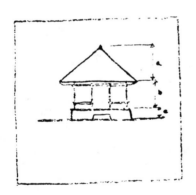

"...The Balinese traditional buildings resist earthquakes well. It is similar to a man, strong and resistant."

"The plan of the building is simple and practical. Each section is divided according to its duty."

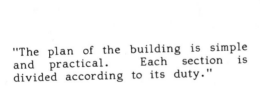

FIG. 4. Extracts from Balinese manual on earthquake-resistant construction (9)

"A lightweight roof, rigid and neat, divides the weight evenly on the ringbeam which will carry the force to the columns."

"The chosen timber columns, which rest on foundations support the roof. If the columns are tied together by a "bamboo-bed" (1), a kind of structural frame, or by (2) stout bracing, they will be capable of movement during the earthquake. The brick walls work just like clothing. Strong, heavy and stable foundations would shift the centre of gravity downwards so that the whole building is more stable..."

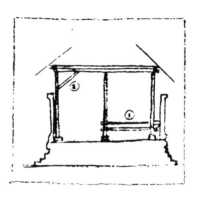

"...A great earthquake could tear the brickwall clothes, but the construction of the building must stand still, in order to protect the residents."

FIG. 4. (cont) Extracts from Balinese manual on earthquake-resistant construction (9)

as almost all settlements form terraces up the steep valley sides.

Dwellings of the bearing type construction suffered severely due to collapse of bearing walls. Other dwellings with independent wooden columns withstood the shaking much better, although one or more of the non-bearing walls collapsed. Rubble-fill walls, consisting of angular rocks, withstood the shaking better than those built of river-worn rounded or semi-rounded boulders (Fig. 5).

FIG. 5. Detail of method of construction used for houses near Pattan, NWF Province Pakistan: timber coursing and cross-ties, roof independently supported. (18)

A special type of structure in the area is that of the "Shingris" or tribal forts (Fig. 6), with massive 0.8 to 1.2 m thick walls, coursed with timber as described above, which rise to heights of 15 m above ground. These towers, as well as other structures of similar constructions, suffered no damage.

A similar though somewhat different house-type is found in the Chitral valley and has been described by Illi (19) and Hassum-ul-Mulk and Staley (20). Its walls are built in the manner described above, but the roof is supported on large timber beams which extend from wall to wall, but are supported by two or more internal timber columns. These columns functionally demarcate the internal space, which is otherwise undivided. Above the centre of this space, between the four main pillars, the roof is raised by an arrangement of timber beams laid diagonally between the main beams; more such diagonal beams are laid above, forming eventually a hole about 0.45 m square open to the sky, which allows smoke to escape and light to enter (Fig. 7). Referring to the wall construction, Hassam-ul-Mulk and Staley observe that "the Chitralis claim that this combination of wood and stone is particularly resistant to earthquake shocks"; and that "when there is an earthquake the women run to the sher-o-tun (main pillar) and cling to it as the safest place in the house." Illi, however, maintains that at the slightest tremor everyone would immediately leave the house. In either case, it is clear that there is a conscious awareness of the earthquake risk (small tremors are frequent in this area). In this case, however, we do not know how such houses would actually perform in a severe earthquake.

FIG. 6. Tribal fort (Shingri) of the Hazara-Swat area, NWF Province, Pakistan. Dry masonry walls coursed with timber and tied through the thickness of the wall. (18)

FIG. 7. Sketch of the interior of Chitrali house. Note heavy timber beams supported by internal columns, and roof opening over central space. After Hassum-ul-Mulk and Staley. (20)

ASSESSMENT AND CONCLUSIONS

The evidence presented does certainly show that there are, in some seismic zones, traditional building types which have been found to perform particularly well. In most cases, these types are based on braced timber frame construction, although timber frame construction does not always perform well; under certain circumstances vault and dome construction has performed well in earthquakes, and tall masonry or timber structures are often surprisingly resistant.

However, these essentially engineering observations, useful as they are, tell us little about the way such building techniques developed. It is possible to imagine earthquakes as an evolutionary mechanism in the development of the house form. In any such event, only the fittest, the best built structures will survive, and these will form a model for the rebuilding of the whole community. The resistant structures might be those which almost accidently have incorporated earthquake-resistant techniques, or may have been introduced from elsewhere, presenting a new model for the community. The knowledge of the earthquake-resistant building techniques might then become part of the tradition, and be handed down unconsciously from generation to generation; or in bureaucratic societies, the successful techniques could become incorporated into the regulations governing what may be built. What is possibly a very early example of this latter process has been observed in the rebuilding of the ancient city of Taxila (18); excavations show that around 25BC the town was hit by a destructive earthquake, which led to basic changes in the architecture of the rebuilt city. The height of the damaged buildings was reduced from four to two storeys, and the lower storey was left half-buried beneath the fallen debris, constituting a sort of basement. Rebuilt houses were more strongly constructed and dug down into the ground to make similar basements. There are many similar examples in modern times, where reinforced concrete frames, tied jack-arches or reinforced brickwork have become mandatory in the rebuilding following earthquakes: but such a process is more likely to affect urban than rural building.

There are also examples to the contrary: the frequency of earthquakes is often such that a new generation of buildings has replaced the old between earthquakes. People forget the event and mistakes are often repeated. Buildings are sometimes altered and adapted in ways which weaken them significantly, and religious beliefs may prohibit the notion of building in such a way as to plan for, or evade such an act of God.

In any case, the role of earthquakes in the development of building form should not be overemphasised.

Catastrophic as they are, destructive earthquakes are very rare events in any one community and as major hazards are likely to be much less significant than other hazards such as floods, high winds, or rock falls; and the significance of natural hazards has to be taken along with the constraints of climate and building materials in determining built form. Further, as Rapoport and others have argued (21), these physical constraints must be seen in the light of the whole socio-cultural context, and with an understanding of the religious world-view of the people, if the development of building form is to be properly understood.

The conclusion of this brief study is, therefore, that there are very little data available on which to assess the part which earthquakes have played in the development of traditional building types. More detailed studies are needed, particularly in the highly active seismic zones, where the risk of recurrence of an earthquake within the life of one building or its occupants is high. Such studies, as well as examining in detail the structure and form of construction of local house-types, should look at the other likely determinants of house form, at the available building materials, the climate, and other natural hazards; should attempt to assess the geographical distribution of different house types and their relationship with ethnic and cultural groupings, and with the economic status of their occupants; should examine the relationship of local building types to the seismic history of the area, and attempt to assess the level of earthquake awareness of the people; and should try to understand the overarching effect on all this of their religious beliefs.

This is one of the objects of the multidisciplinary housing study which is to take place in the Karakoram region of Pakistan in the next few months, forming a part of the R.G.S. expedition to the area.

REFERENCES

1) RAZANI, R., "Seismic protection of unreinforced masonry and adobe low-cost housing in less developed countries: policy issues and design criteria". International Conference on Disasters and Small Dwellings, Oxford Polytechnic, April 19 - 21, 1978.

2) ARIOGLU, E and ANADOL, K., "The structural performance of rural dwellings during recent destructive earthquakes in Turkey (1969 - 1972)". Fifth World Conference on Earthquake Engineering. Rome 1973, Vol. 1, pp 529 - 538.

3) ARIOGLU, E. and ANADOL, K., "Response of rural dwellings to destructive earthquakes in Turkey (1973 - 1975)". Sixth World Conference on Earthquake Engineering, New Delhi, 1977, pp 249 - 254.

4) JOAQUIN, M. E., "Seismic behaviour and design of small buildings in Chile". Fourth World Conference on Earthquake Engineering, Santiago 1969, B-6.

5) AMBRASEYS, N. N., ANDERSON, G., BUBNOV, S., CRAMPIN, S., SHAHIDI, M., TASSIAS, T. P., TCHALENKO, J. S. Dasht-e-Bayaz Earthquake of 31 August, 1968, UNESCO 1214/BMS.RD/SCE, Paris, May 1969.

6) AMBRASEYS, N. N., MOINFAR, A., TCHALENKO, J. S. The Karnaveh Earthquake of 30 July, 1970. 2380/RMO.RD/SCE, Paris, April 1971. UNESCO.

7) AMBRASEYS, N. N., MOINFAR, A., TCHALENKO, J. S. Chir Earthquake of 10 April, 1972. UNESCO 2789/RMO.RD/SCE, Paris, October 1972.

8) AMBRASEYS, N. N., ARSOVSKI, M., MOINFAR, A. A. The Gisk Earthquake of 19 December, 1977, and the Seismicity of the Kuhbanan Fault-Zone, UNESCO, FMR/SC/GEO/79/192, Paris 1979.

9) LINUH, Balinese Earthquake Manual, B.I.C., Pusat Infirmasi Teknik Pembangunan, Bali, Indonesia.

10) ARYA, A. S. and CHANDRA,. B., "Earthquake-resistant construction and disaster prevention". Souvenir Volume, Sixth World Conference on Earthquake Engineering, New Delhi, January 1977.

11) EVANS, F. W., "Earthquake engineering for the smaller dwelling". Fifth World Conference on Earthquake Engineering, Rome, June 1973.

12) NEEDHAM, J. Science and Civilisation in China, Vol. 4 Part 3, Civil Engineering and Nautics, 91, Cambridge 1971.

13) ALEX, W. Japanese Architecture, Prentice Hall, 1963.

14) TANABASHI, R., "Earthquake resistance of traditional Japanese wooden structures". Second World Conference on Earthquake Engineering, Tokyo, 1960, pp 151 - 163.

15) WASTI, S. T. and AHMAD, S. N., "Improving the earthquake resistance of rural dwellings in Pakistan". Cento seminar on recent advances in earthquake hazard minimisation, Tehran, November 1976, (Cento Science Report No. 27), pp 215 - 222.

16) WEST, W. D., "The Baluchistan earthquakes of August 25th and 27th, 1931". Memoirs of the Geological Survey of India, LXVII, 1934.

17) WEST, W. D., "Preliminary geological report on the Baluchistan earthquake of May 31st, 1935". Records of the Geological Survey of India, Vol. LXIX, 1935 - 36, pp 203 - 240.

18) AMBRASEYS, N., LENSEN, G. and MOINFAR, A. The Pattern Earthquake of 28 December, 1974. UNESCO FMR/SC/GEO/75/134, Paris, 1975.

19) ILLI, D., "Archetypal forms of construction in the Hindu Kush", to be published.

20) HASSUM-UL-MULK, S. and STALEY, J. Houses in Chitral: traditional design and function. Folklore, Vol. 79, 1968, pp 92 - 110.

21) RAPOPORT, A. House form and culture, Prentice Hall, 1969.

22) DOWRICK, D. J. Earthquake-resistant design, Wiley, 1977.

House types and structures in Chitral District

Israr-ud-Din

Dept. of Geography, Peshawar University

ABSTRACT

This paper records the distribution of house types in the Chitral Valley. It then analyses the relationship between the house types and such factors as tribal distribution, the physical environment, availability of building materials, the hazards faced by the different groups and the varying cultures they have developed.

INTRODUCTION; THE REGION

Chitral district is situated in the northern-most zone of Pakistan between longtitudes $71°51E$ and $74°E$, and latitudes $35°N$ and $39°N$. It is surrounded by Afghanistan on the north and west, the Northern Area of Pakistan on the east, Swat and Dir districts of Pakistan on the south; see Fig. 1. It covers an area of 5,727 square miles.

Physiography

The region is extremely rugged and mountainous with deep, narrow and tortuous valleys through which run the river Chitral and its many tributaries; see Fig. 2. It is enclosed by the vast system of the Hindu Kush and Karakoram and their various offshoots which are 15,000 to 25,000 feet high. Here are more than 40 peaks over 20,000 feet high which finally culminate into one towering mass forming Tirich Mir (25,263 feet). The region is separated from the rest of the country by the Hindu Raj range, a branch of the Karakoram traversing NE to SW, the only contact being through passes over 10,000 feet. The area is drained by the River Chitral and its many tributaries, and human settlements and the establishment of villages is possible only along the rivers, on terraces, alluvial fans or on abandoned river courses. The river rises in the Baroghil area from the Chiantar glacier and flows for about 220 miles through the area and enters Afghanistan at a village called Arandu which is the lowest point in th region with an elevation of 3,577 feet above sea level. On its way the river collects many other rivers and streams

---

Editors Note: To convert distances to the metric system, note that 1 mile = 1760 yds = 5280 ft = 1.609 km.

266

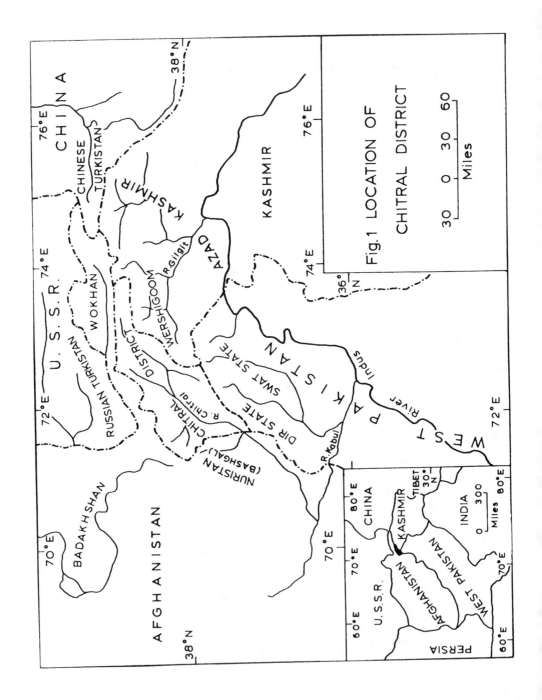

Fig.1 LOCATION OF CHITRAL DISTRICT

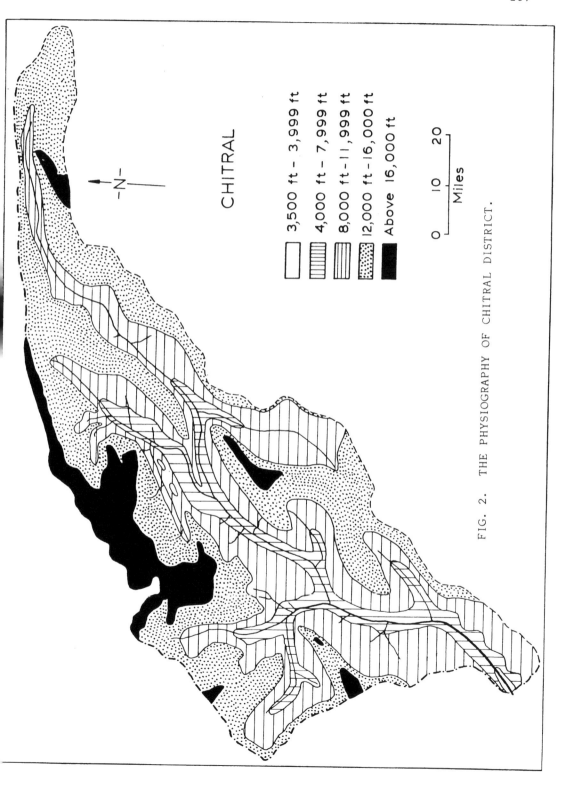

CHITRAL

3,500 ft – 3,999 ft
4,000 ft – 7,999 ft
8,000 ft – 11,999 ft
12,000 ft – 16,000 ft
Above 16,000 ft

0   10   20
Miles

FIG. 2.   THE PHYSIOGRAPHY OF CHITRAL DISTRICT.

which are an important source of life in the area.

The Chitral River valley has an average width of 3/4 mile. Sometimes, however, it opens to about 3 or 4 miles wide as in Baroghil, Buni, Chitral Town and Drosh, and at other times it narrows to a defile of less than 200 yards, e.g., Darband, Mashalik, Karbitari etc. The side river valleys are even narrower and there the average cannot be more than 1/4 mile.

The most important characteristic feature of the area is the presence of the alluvial fans which are found throughout the region and on which are almost all the settlements and villages. The origin of these alluvial fans can be attributed to the temperature and humidity extremes of the area which create the principal weathering agents. Avalanches rushing down the hill slopes in spring or winter, hill torrents and streams which rise abnormally when the snow melts in summer, and floods which are caused by the summer torrential rains, have since time immemorial brought with them millions of tons of material to be deposited at various places in the valley floor and some to be partly drained into the rivers. A succession of alluvial fans have developed at the mouths of streams all over the district.

Climate

The Chitral District experiences extreme and dry climates and the whole area is continental. In summer it ranges from very hot in the lowlands to warm in the uplands and cool in the higher elevations. The highest temperature recorded in Drosh is 110°F for the month of July. Even in the highlands the temperature rises appreciably high. In winter most of the valleys are in the path of cold winds and blizzards which sweep through from the north and it becomes bitterly cold, though it is less severe in the lowlands than the uplands. The lowest temperature recorded at Drosh is 10°F in the month of January. The area gets 10 to 40 inches of rainfall from the southwest to the northeast. This is why all the forests are found in the southern parts of the district. The rains mainly come in winter and spring, e.g., between December and April and which amount to 67% of the total annual rainfall. These rains often come in torrents and result in widespread floods that cause great damage to the area.

Forests

Forests, which are mostly determined by the distribution of rainfall, are found in the south and southwestern parts of the region. These forests occur between 3,000 feet and 10,000 feet above sea level and comprise deodar (Cedrus deodora), Spruce, Fir and Chir (Pinus longifolia). Oak is also found but scattered and mainly on the southern aspects of the hills. Birch and Juniper are found in abundance. The hills in the rest of Chitral are almost barren. However, in the valley bottoms occasionally one comes across thick wooded patches of birch and willowbrakes and small aromatic and xerophytic shrubs. Poplar, walnut and plane trees are found in most parts of the region below 8,000 feet and they are planted wherever soil is suitable and water is available and so they go side by side with the settlements. The plantations are all confined to the valley floors except where hill terraces and gentle slopes with suitable soil and sufficient water are available. Fruit trees, e.g., apples, pears, peaches, mulberries, apricots and grapes are also planted in the area. No trees whatsoever are found above the altitude of 12,000 feet.

Settlement distribution and siting

In Chitral we find settlements spreading from 3,727 feet elevation at Arandu, the lowest point in the region, to the 12,000 feet contour line at Baroghil. As mentioned earlier, most settlements are found on the alluvial fans, or on certain river terraces, wherever soil fertility coincides with easily available water. Villages are also located in the beds of abandoned river courses where similar conditions obtain. Besides the vast tracts of uninhabited areas due to adverse physical factors, there are many such habitable stretches scattered in the region which are at present not settled because of precarious conditions of water supply. Settlements are generally sited on the raised side of alluvial fans which contain mostly infertile and stony lands. Thus the fertile lands are spared for cultivation. Hill torrents and streams are also important determining factors, several being avoided because of their being prone to flooding. The banks of deeper and less dangerous streams are, on the other hand, favoured sites for settlements.

There are also certain socio-economic factors in settlement foundations and locations which result in the establishment of certain individual hamlets and dwellings amidst the cultivated lands. From place name evidence and ties of kinship, it is evident that many such settlements are due to the increase in population on older sites. Another reason is the feudal system that prevailed in the region till the 1950's. The then ruler of the area had supreme power and had full authority to seize any land and grant at will. In this way his favourites were given tracts of village land, where they settled and surrounded themselves with a number of agricultural labourers or tenants and their relatives. The inheritance system prevalent in the region also plays an important role in this respect. Because of this system the holdings of villagers are scattered in fragmented pieces of different sizes. Many who inherit land in different parts prefer to settle near their holdings.

In the Mulikhow area, fertile terraces and the gentle sloping nature of the hills have made it possible to locate settlements there. Settlement growth, however, started on the valley floor but the increase in population has impelled the settlers upward alongside the streams arising from springs. The avoidance of landslides and mudflows, which are common in the area, is another factor in this movement. In many villages this pattern also results from seasonal trans-humance, several families having houses at different levels. The Kalash tribesmen build their houses on hillsides to gain space for cultivatable areas. Defence was probably as important a factor in the past because, until about one hundred years ago, there was a constant threat of attacks on the Kalash valleys from the valleys of Kafiristan (now Nuristan in Afghanistan) by the Red Kafirs.

The Wokhi of the Baroghil area live in scattered dwellings and site their houses with a view to shelter from the cold northern winds of winter.

In Hairan Kot, the only wholly Pathan area in Chitral Town, houses are built along the slope in such a way that every one has its door almost on its neighbour's roof. One reason for this may be the prevailing traditions in the district of Dir and Swat, from which the tribe originates, where most of the houses are built in the same fashion for defence purposes.

Distribution of Population

The total population of the area is 160,000 and the density is about

30 persons per square mile; see Fig. 3. This is low because of the vast glacier bound valleys and barren mountains, which are uninhabited. The distribution of population follows the lines of streams and rivers and is concentrated on the alluvial fans where water can be easily obtained or on the gentler slopes or hill terraces which have fertile stretches and where water is available. If we consider the density of individual villages then a different picture would present itself for Chitral proper enjoys a density of about 1,000 persons per square mile and in the rest of the villages it would vary from 11 persons per square mile in Baroghil to 755 persons per square mile in Jughoor. The average density for all the villages is 433 persons per square mile.

HOUSE TYPES AND STRUCTURES

Broadly speaking, the houses in Chitral are divided into two types, e.g., houses with lantern-type ceilings and the flat roof houses; see Fig. 4. The main characteristic of the lantern-type ceiling houses is the presence of a bulge created in the centre of the roof by building up beams of the roof cross-wise to form an octagon shape. The central structure is carried by four strong wooden pillars erected in the middle of the house. An orifice or 'koomal' varying in diameter from place to place, is left open in the bulge, just above the hearth, to serve as a chimney and also for the purposes of windows and ventilator. The second group have completely flat roofs without any bulge in the middle. This type includes further sub-types, e.g., the 'Shalma' type and the 'Angeeti' type. The two are different from each other in the sense that the 'Angeeti' Khattan has no smoke hole in the middle unlike the 'Shalma' type, but has a chimney placed on one of the walls, mostly on the back wall. The fireplace in 'Angeeti' type houses is also made in the wall, not in the centre of the floor and modern types of windows and ventilators are also provided in the walls.

Houses with lantern-type ceilings

This group of houses is found predominantly in all parts of the district with different variations and includes the Khowar Khatan or Baipash type, the Bashgali type, the Kalash type and the Wokhi type. It would be worthwhile to examine the different types of houses which have resulted from the socio-economic, traditional and physical conditions of the area.

1. Khowar Khatan or Baipash type

This type has its origins among the Khow tribe which forms the majority of the population throughout the district. The type is also found in the valleys of Gizar, Yaseen and Ishkuman valleys in the Northern Area of Pakistan. The reason being that the Khow population spreads as far as Gupis in Gilgit Agency (Northern Area) and thus, through traditional and historical contacts with the people belonging to other language groups, has made an impact on the house types of the area.

It is believed that this type was introduced into the region a long time ago from different parts of Central Asia from where the different waves of the Khow immigrated in the course of history. The earliest of these people are thought to be of Aryan stock who invaded and occupied these valleys about three thousand years ago coming through the northern passes of the Hindu Kush. At one time, it is believed, the whole of the area from Nuristan (Afghanistan) to Astor (Gilgit

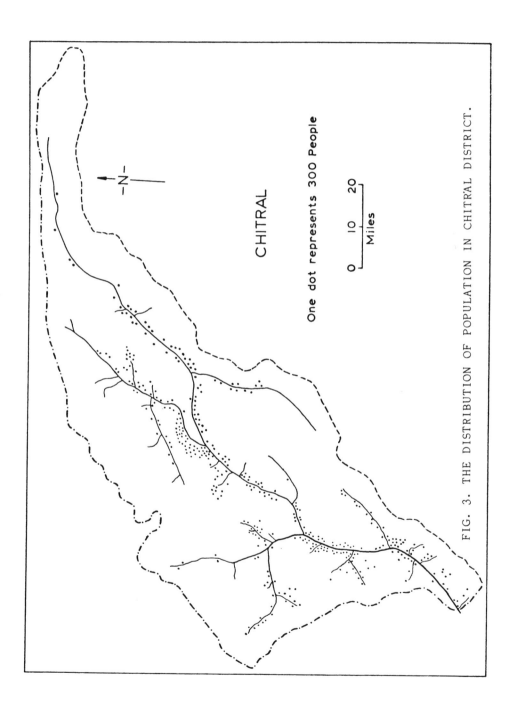

CHITRAL

One dot represents 300 People

0   10   20
Miles

FIG. 3. THE DISTRIBUTION OF POPULATION IN CHITRAL DISTRICT.

272

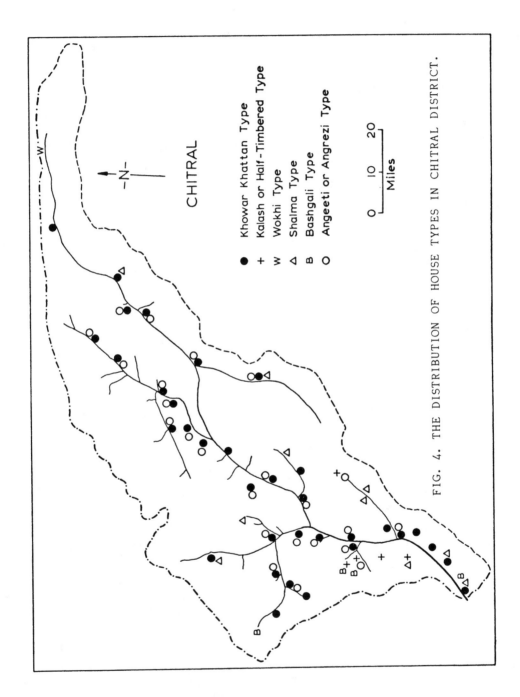

FIG. 4. THE DISTRIBUTION OF HOUSE TYPES IN CHITRAL DISTRICT.

Agency) in the Northern Area was occupied by one homogeneous race. But subsequently, as Biddulph suggests (1) the area was split into two parts, 'by a wedge of Khow invasion, representing members of different but related tribes coming from the north'. It is thus concluded that the Khow came and settled in these valleys later than the first wave of Aryans, belonged to the same race but had adopted certain traits of the Ghalcha speaking people of the north through long contacts with them in the area before crossing over into Chitral. In later periods until the modern times, people have been coming into the area from the surrounding regions of Badakhshan, Wakhan, Russian and Chinese Turkistan, Gilgit Agency, Dir and Swat districts of Pakistan and parts of Afghanistan in the form of refugees, invaders or as followers of the ruling class who were at first adventitious and sporadic squatters but they made themselves at home subduing, dispossessing and oppressing the existing primitive stock.

These later immigrants despite their different origins and backgrounds, absorbed themselves into the original Khow by living in the same villages, intermarriage and by adopting their language, Khowar, and their customs and other ways of life. In later years Islam, which they all had embraced, played an important role as a unifying factor. Thus these people, having ethnologically and historically different backgrounds and consisting of hundreds of clans and families are one people today and all called the Khow; see Fig. 5.

The Khowar 'Khatan', which is the traditional house of the tribe, seems to have been introduced to the area from the time they have been living in these valleys. The typical "Khowar Khatan" is always square and one storeyed. Attached to it is a verandah, in the case of the warmer areas, or 'dahlenz' (closed corridor) in the colder areas. It is believed that the verandah is a recent addition. Its height is normally twelve feet and the floor area is 25 x 20 square feet. The floor area of the house would vary according to the number of family members and also their economic condition and the height is mainly determined by the physical and climatic conditions of the area; the colder areas having houses of height about six or seven feet. The house has no windows, nor ventilators. The smoke hole is in the roof which is mostly about fourteen inches square and serves the purposes of letting in light and letting out smoke. The door is about 6 feet x 4 feet in height in the lower and warmer areas, and about 4 feet x 2.5 feet in the colder parts; see Figs. 6 and 7.

The whole roof structure is mainly supported by four large wooden posts in the middle and 8 to 10 small posts free standing against the walls. The walls give partial support to the roof and also provide the weather protective envelope to the activities carried on inside the house. The four central posts carry beams of about 8 x 8 inches, or of varying thickness according to the nature of the wood, over which the beams form a diamond-shaped ceiling, stepped up in four (or in certain cases, five) levels of about six inches each, culminating in the carved sides of the square smoke-hole. The parts outside the four central pillars are made straight with the help of joints and planks of different thickness. Grass or brushwood is laid on the roof timbers to preserve the wood and then is plastered thickly with straw mixed mortar, rising to a mound around the smoke hole. Thus from the outside the roof of the house is flat on all sides and bulging in the middle. The roof also slightly overhangs the walls and flat stones are placed

274

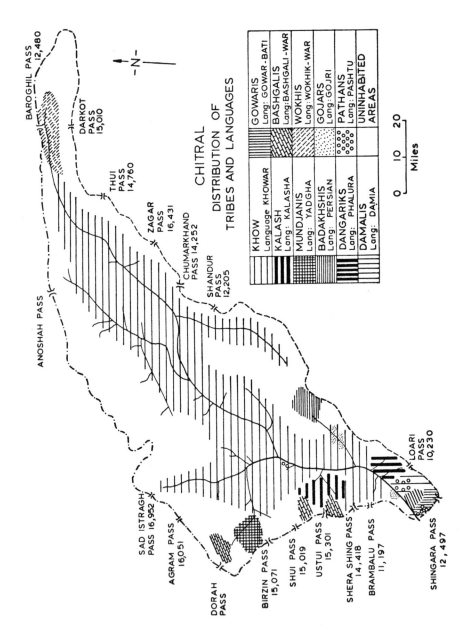

CHITRAL
DISTRIBUTION OF
TRIBES AND LANGUAGES

KHOW Language KHOWAR	GOWARIS Lang: GOWAR-BATI
KALASH Lang: KALASHA	BASHGALIS Lang: BASHGALI-WAR
MUNDJANIS Lang: YADGHA	WOKHIS Lang: WOKHIK-WAR
BADAKHSHIS Lang: PERSIAN	GOJARS Lang: GOJRI
DANGARIKS Lang: PHALURA	PATHANS Lang: PASHTU
DAMALIS Lang: DAMIA	UNINHABITED AREAS

Miles
0   10   20

BAROGHIL PASS 12,480
DARKOT PASS 15,010
THUI PASS 14,760
ZAGAR PASS 16,431
CHUMARKHAND PASS 14,252
SHANDUR PASS 12,205
ANOSHAH PASS
SAD ISTRAGH PASS 16,952
AGRAM PASS 16,051
DORAH PASS
BIRZIN PASS 15,071
SHUI PASS 15,019
USTUI PASS 15,301
SHERA SHING PASS 14,418
BRAMBALU PASS 11,197
SHINGARA PASS 12,497
LOARI PASS 10,230

FIG. 5.   THE DISTRIBUTION OF TRIBES IN CHITRAL DISTRICT.

FIG. 6.   OCTAGON SHAPED CEILING OF KHOWAR KHATAN.

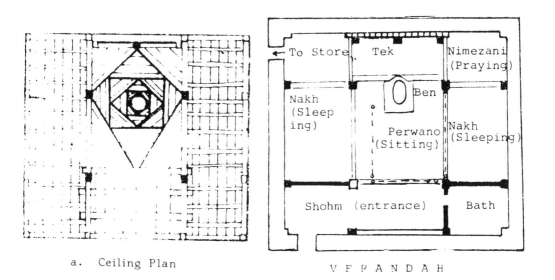

a.   Ceiling Plan

b.   Floor plan

FIG. 7.   SECTIONS OF A KHOWAR KHATAN OR BAIPASH.

at the edge of the roof to protect them from melting snow or rain. The roof plastering is done in such a way that the rain water easily drains towards gutters, which are fixed at the edge of the roof. Wood shortage is an acute problem in most parts of the area which forbids the use of wood for building the roof. Roofs, which are therefore built with mud plaster, are made mostly flat to avoid the danger of being washed away by the rain or melted snow if it is made steep. The roofs are also used for drying maize and storing fodder. In some cases they are also used as threshing floors. This is probably the most satisfactory explanation for the building of a flat roof even in areas with abundant wood. The wood used in the construction of a house in Chitral varies according to its availability. In upper areas poplar is generally used as there is no other wood locally available. In the rest of the region sufficient pine is available for construction purposes; see Fig. 8.

Materials used in walling of the houses include slate stones, cobble, sun baked bricks, earth, etc. Many different types of stones, e.g., slate, schist, crystalline limestone, granite, gneiss, etc., are found in abundance in various parts of the district, and walls are constructed mostly in local stones. The people generally use irregular masonry for the construction of the walls which will be composed of undressed field stones, quarry faced stones, or straight undressed field stones, in each case obtained from the ground close by the building under construction. The stones are, however, placed in such a skilful manner that the outer side of the wall often looks plane and smooth. The interior of the walls are normally packed with small loose pieces of waste stones or mud.

There are areas where, for various reasons, people use sunbaked bricks with stone foundation. In certain cases, depending on the availability of suitable stone, only the upper half of the wall is made of brick and the lower from stone. While making the walls, the jointing of the materials is always the great concern of the builders so that stability against the roof-load as well as weather conditions can be achieved. In Chitral the walls of the 'Baipash' are not necessarily load bearing. With the available materials it would not have been possible to make them very strong to hold the heavy roof load. Therefore the walls are built only to give partial support to the roof, as mentioned earlier. But the area experiences frequent earthquakes, though mostly not of severe intensity, causing considerable damage to the walls by creating cracks in them or by causing total collapse. Moreover, rains sometimes greatly damage the walls. All of these problems are seriously considered while constructing the walls. The walls are 2.5 to 3 feet thick and are lined by horizontal logs at heights of about 3 feet. In areas of abundant wood such logs are placed one foot apart. Straw mixed mortar is used for cementing the materials together. The outer side of the walls are also thickly plastered with the same material and finished with clay distempering.

The floor plan of the house is also worth considering. The whole floor is divided into the following portions, each having separate functions and uses; see Fig. 7(b).

(i) The Shohm: This is the entrance space where shoes are placed before entering other portions of the house. In the case of poor people, certain parts of this portion are also used for storing agricultural and other implements and for keeping newly

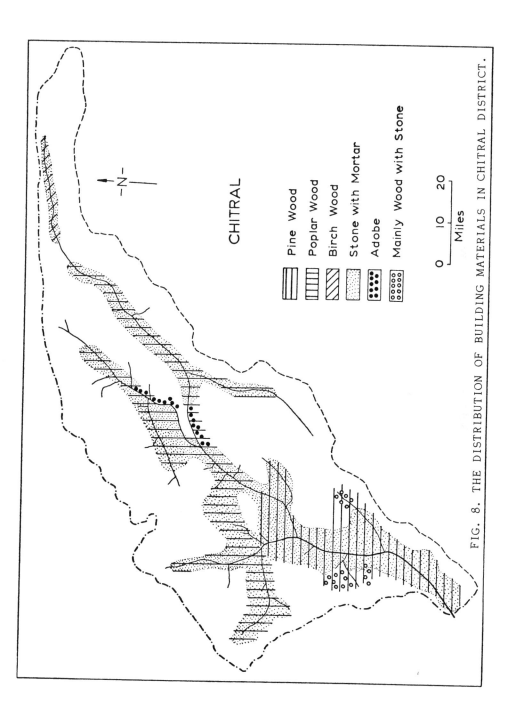

FIG. 8. THE DISTRIBUTION OF BUILDING MATERIALS IN CHITRAL DISTRICT.

born calves or kids.

(ii)   Shung: This is one corner of the 'Shohm' but raised for storage of wood and for keeping chickens. In certain cases this corner is walled on all sides and turned into a bathroom to be used by the womenfolk.

(iii)   'Perwano' and 'Ben': The portion on three sides of the hearth is separated from the 'Shohm' by wooden boards called 'Taktabandi' and used for sitting purposes. This part is further divided into two parts, i.e., the portion towards the Shohm, called the Perwano, is used by younger male members of the family while the other portion, called the Ben, is used mostly by the father or elderly person in the house.

(iv)   'Tek': This is the back portion of the house raised, from end to end, about a step from the rest of the area. This portion is used by the female members of the house. Along the wall on the back of the Tek are cupboards or shelves to keep utensils etc. On one side of the Tek is the praying place, mainly for the women, while on the other side is an entrance through a small door into an attached store room.

(v)   The 'Nakh': On two sides of the Perwano and Ben and along the side walls are the Nakh which are separated by 'Taktabandi'. These portions are used for sleeping. In the colder areas the 'Nakh' are filled with grass and straw over which rugs, quilts, etc., are placed to be used as beds. In warmer parts cots are placed in them to be used for beds.

2. The Kalash type

This type belongs to the Kalash tribe who live in the valleys of Bamburat, Birir, and Rumbar. The tribe is composed of about 3,000 souls who are non-Muslims and have different ways of life, customs and traditions. They have, strangely enough, succeeded in maintaining their old beliefs and culture in the face of increasing influence of Islam and centuries of domination by alien people. It is believed that the Kalash immigrated to the different valleys of Chitral sometime in the 10th or 11th century A.D. from certain parts of the Nuristan Valley in Afghanistan. During the 10th century of the present era, Sabugtagin and his son, Mahmud of Gazni, the then Kings of Kabul, were waging war against the infidel tribes in the region of Jalalabad and Lughman. These tribes, who are now inhabitants of Nuristan (Afghanistan), could not stand the attacks and so had to retreat, pushing back the tribes inhabiting the upper valleys between Lughman and Lower Bashgal. The Kalash who were then living in this area could not face the attacks of the retreating tribes and so in their turn invaded the lower parts of Chitral which they occupied as far as the villages of Baranis or Rashun (about 30 miles north of Chitral Town). They remained rulers of this part of Chitral for about three hundred years when in 1320 A.D. they were defeated and subjugated by the Khow who had by then accepted Islam. With the passage of time, the Kalash who remained in the main valleys with the Muslim Khow, were greatly influenced by them and so accepted their religion and adopted their customs and language. On the other hand, the Kalash who lived in the valleys mentioned above and those who had retreated there later were, due to their seclusion but mainly because of the tolerant character of their neighbours, able to continue their old practices.

The typical house of the Kalash is double-storeyed and mainly made of wood; see Fig. 9. The main room which is on the upper storey

FIG. 9. A KALASH HOUSE IN BUMBORAT VALLEY.

is built on the same pattern as the Khowar Khatan. A balcony is attached to the house or the upper storey made by laying wooden planks. Roof adjustment and plastering of the roof are also made in the same way as in the former case. The walls are timber based in which horizontal logs are used with rubble stones embedded in mud mortar. The placement of the horizontal logs varies from less than 1 foot to 3 feet depending on the nature of the site of the house. If a house is built on a slope of the hill, then the horizontal logs are placed at smaller intervals and if the site is on a comparatively stable place then the interval is up to 3 feet. The foundations of the walls are however, built with stones only. The horizontal logs are jointed together at the corners with great skill. The walls are plastered with mortar with straw from inside, and sometimes from outside. The area has abundant pine, spruce, and fir, and therefore there is no problem of availability of wood materials for construction purposes.

The height of the whole house from the ground is about fifteeen feet, out of which the upper storey, which is the living part, is ten feet and is reached from the ground by a ladder. The floor plan is almost the same as in the case of the Khowar Khatan.

3. Bashgali type

The Bashgali tribe, or Shiekhan as they are called by others, live in different parts of Chitral district such as Gobor in the southern part, Rumbur and Bumborat in the northwestern part and Langur-bat in the south. Their total number is approximately 4000. The Bashgali are the descendents of those immigrants who were formerly non-muslims and lived in the Nuristan valleys of Afghanistan. They have been called Red Kafirs by foreign writers to distinguish them from the Kalash who are termed Black Kafirs, and the area was named Kafiristan. Their immigration to their present abode took place in the last decade of the 19th century when the conversion of their community was enforced by the Amir of Afghanistan. In consequence, most of them were converted to Islam but these people took refuge in Chitral and settled in the areas mentioned. They practised their old religion until the 1920's at which time all of them voluntarily embraced Islam.

The Bashgali houses – see Figs. 10 and 11 – are very much like the ones built in Nuristan valley today as the environmental conditions of the new area (except Gobor) where the tribe has now settled, are the same. The houses are mostly situated on the steep mountainside in order to spare the fertile and flat lands in the valley bottoms for cultivation and also to attain a good view of the valley. The Bashgali houses are often two-storeyed, sometimes even three. The upper floor of the house comprises the main room called the 'ama' the roof of which is supported by four decorated wooden cedar pillars around the fireplace. The roof is built up similarly to the lantern-type ceiling of 'Khowar Khatan' and the Kalash type. The entrance to the main room is the only entrance to the house. The lower floor, which serves as a storeroom, is entered through an opening in the floor in a corner of the main room. This opening is usually closed with a trap door.

The typical feature of a Bashgali house is that the two beams under the roof run at slight angles to the entrance wall and traverse the roofed verandah at the same level and together with the uppermost beam in the lateral walls. Here they are supported by a row of pillars with very fine decorations and by the ornamental front structure of the house.

The wall of the main room or the 'ama' is built of horizontal logs kept in place by vertical poles, which are a little shorter than the depth from floor to ceiling in the 'ama' and which in turn are supported on both sides of the wall by an upper and lower wooden clamp (containing two holes) that have been inserted horizontally through the wall so that their two ends project from the wall. The walls may consist entirely of horizontal wooden logs or timbers but usually in place of every second log there is a layer of stones and mud.

It is characteristic of house building of the Bashgali tribe that up to five houses can be built together simultaneously giving room for several households within the same family. The upper storey of the house always contains a closed verandah with the main house or 'ama'. If the owner cannot afford to build the decorated verandah immediately the house may stand unfinished, the beams of the roof projecting into the open air for years.

In the Gobor area where there is a problem of wood shortage, the Bashgali cannot afford to build such ambitious wooden houses as in other valleys.

FIG. 10.   A BASHGALI HOUSE IN BUMBORAT VALLEY.

FIG. 11.   ANOTHER BASHGALI HOUSE IN BUMBORAT VALLEY.

There the houses are mostly one-storeyed made of stone and mortar and birch or poplar wood, which is brought from other villages nearby. The distinguishing feature of these houses is that they have square or, less often, round wicker receptacles covered with mud, called 'chakki' or 'guzzuli' used to store grain and are fixed on roof tops. These storage receptacles resemble a jumble of chimney pots on the top of the house. This shows how, in changed circumstances, grain stores have been relocated to the top of the roofs instead of on the ground floor.

4. Wokhi house type

This type is found in the Baroghil area which is inhabited by the Wokhi tribe who are a recent migrant tribe to the area from Wakhan in Afghanistan. The area is about 12,000 feet above sea level and is situated in the neighbourhood of huge glaciers and is covered with snow for more than seven months of the year. The houses though are a lantern-ceiling type but are only seven or eight feet high; see Fig. 12. The roofs have chimney holes of about nine inches in diameter which is the only way to let in light. During rain or snow or during the night the hole is closed by a lid being placed over it. The door is built only 30 inches high. The building materials are mainly stone for walling and birch wood for roofing. The roof covering is done with stone slabs; grass and manure-mixed mortar is used for plastering. The walls are also covered with manure-mixed plaster and every precaution is taken to leave no gaps anywhere. The entrance to the house is made through many closed corridors, numbering from two to six, depending on how much one can afford.

A few variations regarding the floor plan of Wokhi houses are worth mentioning;

   (i)   The Wokhi house has five central pillars, two on one side and three on the other for carrying the beams of the central octagon over them.

   (ii)   One corner of the 'Shohm', from where entrance is made into the house, is turned into a closed corridor by erecting walls on all sides up to the roof. The way in is left between the two central pillars on the side of the entrance.

   (iii)   The 'Tek' is raised high, compared to other house types, and is about two to three feet. The fireplace is made like a deep oven in the 'Tek' but underneath the smoke hole towards the front.

   (iv)   The side portions including the 'Shohm' are furnished with woollen rugs and other materials and used both for sitting and for sleeping.

   (v)   At the corners of the Shohm and Nakh, big wooden boxes are placed for storing grain and other belongings.

FLAT ROOF HOUSES

Though the lantern-ceiling type of houses mentioned above have flat roofs around the central bulge, the flat roof houses have no central bulge

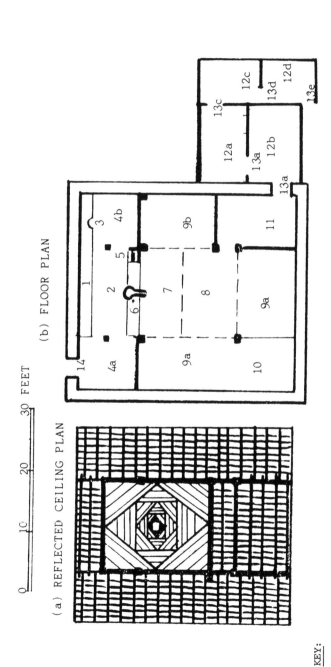

(a) REFLECTED CEILING PLAN

(b) FLOOR PLAN

0   10   20   30 FEET

KEY:

1. BRICKCHIKISH: RAISED SITE ALONG THE BACK WALL FOR
    PUTTING UTENSILS ETC.
2. CHIKISH: ABOUT 3 FEET RAISED PLATFORM USED BY WOMEN-
    FOLK FOR SITTING
3. DIL-DUNG: SMALL FIRE PLACE
4a. BALAND CHIKISH (LEFT): USED FOR ENTRANCE TO STORE.
    ALSO USED FOR PUTTING WOODEN BOX FOR GRANARY
4b. BALAND CHIKISH (RIGHT): USED FOR PUTTING UTENSILS
    AND SOMETIMES FOR COOKING
5. STAIRS
6. TANOOR: DEEP OVEN FOR COOKING AND HEATING

7.   YORCH BALA: FOR PUTTING WOOD
8.   YORCH PAIN: ENTRANCE
9a.   LUP RAISH)
9b.   PUT RAISH)   FOR SITTING AND SLEEPING
9c.   PAST RAISH)
10.   BALAND RAISH: 3 FEET HIGH PLATFORM FOR PUTTING
        BOX FOR GRAINS ETC
11.   KUNJ: THRESHHOLD
12.   DAHLENZ: CLOSED CORRIDOR
13, 14.   DOORS

FIG. 12.   TWO SECTIONS OF A WOKHI HOUSE.

at all. In this case the roof is completely flat from end to end. This type of house has further subtypes, e.g., the 'Shalma' type and the 'Angeeti' or 'Angrazi' type.

1. 'Shalma' type

This type is popular among the tribes called Gojur and Dangarik who inhabit different parts of the district. The Gojur are a semi-nomadic tribe and so, in spite of inhabiting some of the well-wooded valleys, live in wretched huts no better than those of their animals. Their houses, which are called 'doogoors' or 'bothies', are built by piling up flat stones to make the walls which are without any regular height or breadth. The roof, which is flat, is built by placing dried grass and pieces of wood on one or two beams and a few branches and then covered with mud. The chimney hole in the roof is also left open.

The Dangariks who live in Ashrat valley, also make houses with a flat roof. Their houses are of a better standard than those of the Gojur. The Dangariks migrated from Tangir valley in the Northern Area about seven generations ago and first settled in the Ashrat valley and later spread to the surrounding valleys. Originally, they are related to the Shina people living in the Gilgit Agency region in the Northern Area where the flat roof type of houses are common.

The area is well-wooded with pine and other important species of construction wood. Therefore a lot of wood is used in lining the walls and also in roofing. Smoke holes are left open in the roof and no windows or ventilators are provided in any other part of the house. This house covers an area of about 20 x 25 feet.

The roof structure is erected on a number of beams carried by many pillars standing in about three rows, two of them along the walls and one in the middle. Thus the walls, which are about two feet thick are not fully load bearing and give only partial support to the roof. Joists and wide boards are placed over the beams for covering the roof. Grass or brushwood is laid on the roof timbers to preserve the wood, and this is further covered with about nine inches of clay. Rafters project slightly outside the top of the walls and a rough parapet is formed of flat stones at the edge of the clay roof. The roof, though mainly flat, slopes slightly to the back of the house where a wooden spout carries away the rain water. The walling is done with rough stones set in mud mortar stabilised by horizontal logs at a height of two feet. The walls are then plastered, outside and in, with clay mixed with chopped straw.

2. 'Angeeti' or 'Angrazi' type

The characteristic feature of this house is that it has a flat roof without a bulge or smoke hole. The chimney is placed on one of the walls in which the fireplace is made. Windows and ventilators are also provided. This type is called 'Angrazi' or English because this was introduced in the area during the British period. The adoption of this type is the result of the people's greater contact with other parts of the world, a contact which has greatly increased during the past 50 years or so. This house type is now often combined with the Khowar Khatan or other types of houses and is frequently used as a guest room.

Recently certain well-to-do families have been using corrugated steel
sheets for roofing such houses. Thus sloping roofs are being introduced
by placing the steel sheets over criss-cross wooden frames connected
to the walls. Cement and lime has also been commonly used for the
construction of these houses.

In these houses the pillars, if at all needed, are erected along the
walls, standing freely to support the roof. Wood lining of the walls
is also done. In certain cases, when cement is available, only wall
plates are used to replace most of the pillars along the walls. No
central pillars are used as in other types of house and the floor plan
is also simple without any partitions.

THE HOUSE PLAN

Houses are generally divided into two parts; 1) 'Dur' or human section,
and 2) 'Shal Mudi' or animal section; see Fig. 13. These two sections
are often attached to each other though in some cases because of certain
local factors to be described, the 'Shal' or goathouses are built separately.

The 'Dur' or dwelling section in well-to-do families is divided into
two parts called 'andran' or interior and 'beri' or exterior. The former
is used by the family and the latter by guests. The two parts are enclosed
by high walls which provide separate courtyards or havailis for each.
Thus not only is privacy preserved, but security from theft is also assured.
These 'havaili' are also made into small gardens.

The number of rooms depends on the social status and total number of
family members. The well-to-do families have about six rooms for family
use, two guest rooms and four rooms for servants. An ordinary 'dur'
consists of two rooms, except for the Gojur, Wokhi and Kalash tribes who
have only one room and a store. The Khow are generally fond of attaching
a small garden to their houses, so even with the ordinary 'dur' one often
finds small gardens attached. The house is entered through a verandah,
but in the colder areas a 'dahlenz' or closed corridor is built instead.
The Wokhi type of house mentioned earlier is always attached to a
'dahlenz' because of the high altitude at which they are situated. The
Kalash and the Bashgali use only a verandah or a balcony. The Gojur
who live mainly in the warmer southern parts of the country use neither
a verandah nor a dahlenz.

The 'mudis' or byres are, except in the Kalash valleys where the whole
animal section is separated, always attached to the human section of the
dwelling. But the 'shal' or goathouse in most parts of lower Chitral
is built separately. A 'shal' or goathouse normally includes a shed,
two rooms for goats, one barn and a hut for the shepherd. A byre has
one shed, three rooms (separately for oxen, cows and calves) and a barn.
The number of rooms increases according to climate, and the economic
conditions of the people. For instance, in the colder area where a long
spell of cold weather and snow necessitates much storage accommodation,
we find more than one barn. Those who possess large flocks have to
provide more rooms.

In view of what is stated above, we can distinguish two principal
types of houseplans in the region. These are single houses and multiple
dwellings. The former have human and animal accommodation on the same
site attached to one another and the latter have either the 'shal' or the

286

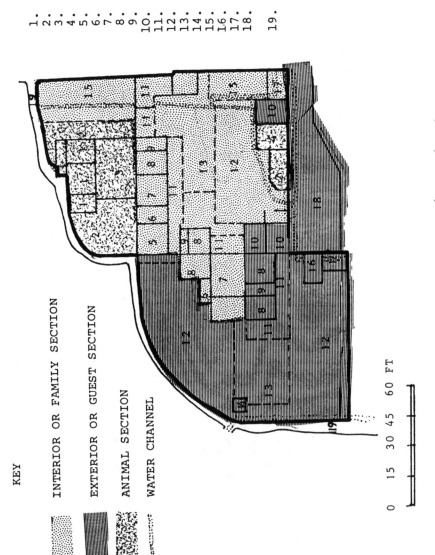

KEY

INTERIOR OR FAMILY SECTION

EXTERIOR OR GUEST SECTION

ANIMAL SECTION

WATER CHANNEL

1. BYRE
2. CATTLE SHED
3. BARN
4. STABLE
5. KITCHEN
6. STOPE
7. BAIPASH
8. 'ANGEETI' OR 'ANGRAZI'
9. BATH
10. 'SHALMA' (for servants)
11. VERANDAH
12. ORCHARD
13. YARD/LAWN
14. WATER TANK
15. KITCHEN GARDEN
16. GARAGE
17. TOILET
18. OPEN SPACE TOWARD
    MAIN ENTRANCE
19. STREET CONNECTING
    WITH VILLAGE HOUSES.

0   15   30   45   60 FT

NOTE: The goat house is at a
         separate site on the hill-
         side, about ½ mile away.

FIG. 13.   PLAN OF A KHOW HOUSE.

287

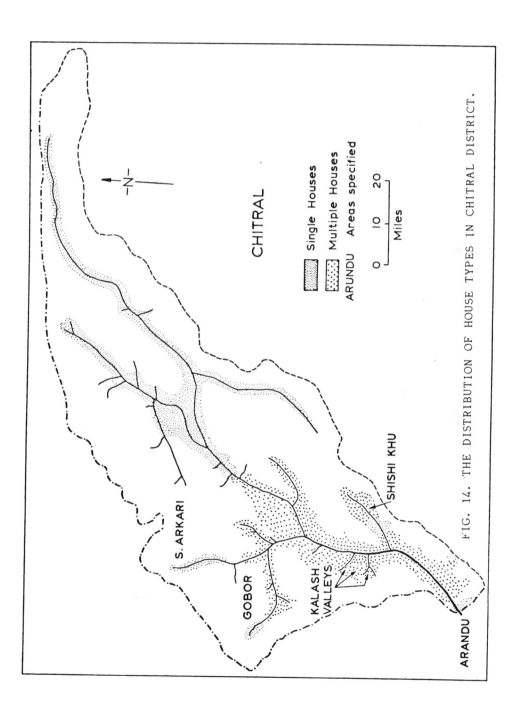

FIG. 14. THE DISTRIBUTION OF HOUSE TYPES IN CHITRAL DISTRICT.

whole animal section on separate sites.

Distribution of Single House Type

This type of house is found in Upper Chitral. But the villages of Gobor and Siah Arkari in Lotkuh tehsil (Fig. 14) and Arandu villages in Drosh tehsil, have also the same type of houses. In Upper Chitral mainly sheep are reared and stock are accommodated in an attached animal section. Because of the long severe winters, animals are fed indoors and it becomes necessary to keep them as near the house as possible so that they can be looked after properly.

The houses of Gobor and Siah Arkari villages are without separate accommodation for stock because of the long winters and for security reasons, particularly as they lie close to the border. In Shish Kuh the pastoralist belongs to the Gojur tribe mainly. In most cases, they live with the animals under the same roof. In Arandu village on the border of Afghanistan there arises the problem of theft, both from inside the village and from across the border. A number of Gojur also live there possessing most of the livestock. Thus the social and security factors coincide here.

The Multiple Type

This type prevails in Lotkuh tehsil (except Gobor and Siah Arkari villages), the Kalash valleys, Urtsoon, Damil, Ashrat and Beori valleys and certain villages between Kogoozi and Drosh where they keep goats. In all these areas only goathouses are built separate from the rest of the house. Sheep and cattle are kept, however, attached to the 'dur'. In the Kalash valleys of Birir, Bumborat and Rumboor, the women are forbidden to look after flocks, especially the goats, so the animals are kept away from women 'lest they (the animals) would become impure and die'. The same custom prevails amongst the Bashgali tribe living in the same valleys.

Many people are superstitious and are afraid of losing their goats if they kept them amongst the 'people' because of 'evil-eye'. Others, more realistically, keep them away from fields because they damage the crops.

CONCLUSIONS

This study indicates that the house types, structures and plans in the Chitral District are the result of man's interaction with nature as reflected by the differences in relief, altitude, slopes, scantiness of cultivatable areas, agricultural cum pastoral nature of the economy, availability of different types of building materials in different parts, and diversity of the people ethnically. The people are quite alive to the different hazards which they are often faced with, e.g., landslides, rockfalls, landslips, flash floods, avalanches, heavy winter snowfall, torrential summer rains and earthquakes. They have tried through centuries of trials and errors to adjust themselves to such circumstances. With the exposure to the outside world during the last 50 years or so new trends in house types are also being introduced in the region which most of the population is slowly and steadily adopting in addition to their old traditional houses.

REFERENCES

1) BIDDULPH, J., (1880). Tribes of the Hindu Kush, Calcutta (Reprint by Indus Publications, Karachi, 1977).

BIBLIOGRAPHY

BIDDULPH, J: Dialects of the Hindu Kush, Journal of R.A.S., London, Vol. XVII, 1885, pp 133 - 134.

EDELBERG, L.: The Nuristan House, Culture of the Hindu Kush (Selected Papers from The Hindu Kush Cultural Conference held at Moesgard 1970). pp 120 - 123.

GRIERSON, Sir George A.: Linguistic Survey of India, Vol. VIII, Part II, Calcutta, 1919.

ISRAR-UD-DIN: A Social Geography of Chitral State; M.Sc. Thesis, University of London, 1965.

ISRAR-UD-DIN: Settlement Patterns and House Types in Chitral State: Pakistan Geographical Review, Vol. 21, No. 2, July 1966, pp 21 - 38.

ISRAR-UD-DIN: The People of Chitral - A Survey of Their Ethnic Diversity, PGR, Vl 24, No. 1, January 1969, pp 45 - 57.

ISRAR-UD-DIN: Settlement Patterns in Peshawar and Malakand Divisions, Board of Economic Enquiry, University of Peshawar, Pakistan, No. 76, 1972.

POTT, J: 'Houses in Chitral', Architectural Association Journal, London, Vol. 80, No. 89, March 1965, pp 246 - 248.

The vulnerability and reduction of damage risk in small houses subject to natural hazards

I. Davis

Dept. of Architecture, Oxford Polytechnic

ABSTRACT

The determinants of house forms and settlement patterns are described with particular reference to their vulnerability to natural hazards. Three responses to make low-income houses resistant to extreme forces are identified: the introduction of bye-laws, a total change in house form and the modification of traditional housing. Each is considered in the light of its practical relevance and past performance. Finally, a basic approach to housing modifications is identified which itemises the wide range of factors which have to be considered.

SETTLEMENTS AND THEIR VULNERABILITY TO NATURAL HAZARDS

In the 10-year period between 1965 and 1975 over 3.5 million lives were lost through natural disasters (1). Within this period over half a million deaths were reported in the Chittagong cyclone and sea surge of East Pakistan (now Bangladesh) of 1970. Then in 1976 242,000 were killed and 164,000 injured in the Tangshan earthquake in China. Both of these disasters caused exceptional devastation, but they may be fore-tastes of the scale of disasters that are increasingly possible due to the marked increase in vulnerability of settlements (2). See Fig. 1.

There is no evidence of any increase in seismic activity, or an increase in the incidence of high winds – hurricanes and cyclones. However, it is now established that the type of extensive flooding that has occurred in the autumn of 1978 in India was partially caused by deforestation on the southern slopes of the Himalayan Mountains of Nepal. This is an example where the agent of disaster (or the hazard) is being adversely affected by the ecological imbalance of the deforestation which is a symptom of poverty. It is now known that the majority of people who die in natural disasters do so as a direct result of the vulnerability of their settlement location (and possibly the constructional form of their houses) to the forces that are exerted in earthquakes, high winds or flooding. It is also known that the victims of disasters are the poorest sections of the population within the poorest countries of the world. There is in effect direct correlation between poverty, vulnerability and casualties. (3)

GLOBAL TRENDS

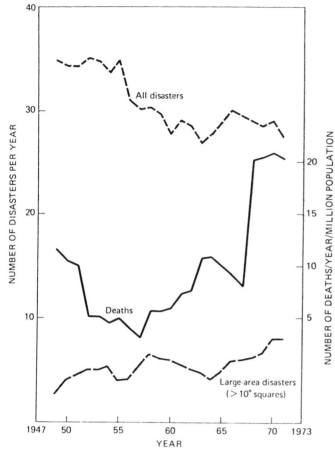

FIG. 1 – MAJOR NATURAL
DISASTERS, 1947 – 1973

A five year moving aver-
age of major internation-
ally reported disasters
(excluding drought). For
this period the average
of all reported natural
disasters appear to decl-
ine and then remain con-
stant, although both death
rates and large-area dis-
asters are increasing.
(from Ref. 2 Page 3).

House forms and settlement patterns are responsive to a variety of closely inter-related elements, some of which can be considered as internal factors, others as external. The internal elements relate to the basic needs of the occupant of each house. The owner's economic level will influence or determine the location, size and form of construction of his home, and his family needs will embrace such factors as size of unit, as well as their living patterns. At a deeper level the form of the house and its relationship to other houses may have cosmological significance relating to the religion or philosophy of its occupants. External factors include a response to local climate; available building materials and con- struction skills; the nature of the site; availability of water supply, and the local land tenure system.

All of these factors relating to individual families have to be compounded in any consideration of an overall settlement. Here the relationships of one house to another are dictated by patterns of status, kinship or patrimony. (4)

The various elements identified above are the everyday affairs which collectively determine the form of each house and the settlement pattern. In effect the houses and their relationship to each other are a manifestation of the priorities and values of a given community. So if a hazard is a regular event, occurring, for example, within a two year period, it is likely that houses and settlement configurations will have evolved to cope with the risk. However, when there is a long-return period of a particular hazard the occupants may either be unaware of the risks they face or alternatively they may have decided that it is not worth bothering to cope with it. In the latter condition a mental equation is performed: "is the risk of earthquake (or other hazard) great enough to forego all the advantages of site or form of house construction?" (In the appendix of this paper I have included a series of charts which relate four types of settlement to possible reasons for their siting and house form, and to the range of possible hazards they face).

The analysis of hazards and their threat to human life is normally referred to as 'risk assessment' or 'vulnerability mapping'. This new science is still in its infancy where analytical tools are still being developed to understand and quantify the levels of risk facing communities. Much of the focus of this work in past years has related to protection of investment in the form of buildings and infrastructure – (roads, services, railway tracks, bridges, etc) within the industrialized world. There has been minimal priority of attention towards the risks facing the houses in the poorest sections of communities within the poorest countries, and it is no coincidence that these countries are frequently subject to extreme hazards. Professor Razani from the University of Shiraz in Iran has written:

> "Review of the technical papers published in earthquake engineer-
> ing journals and in the proceedings of the past regional or world
> conferences on earthquake engineering points out that while collapse
> of unreinforced masonry and adobe low-cost housing (UMALCH) has caused
> more than 80% of the earthquake fatalities during this century through-
> out the world, only less than 5% of the published papers deal with this
> type of building. The percentage of the total research funds allocated
> worldwide to the study of the seismic behaviour and design of these
> types of buildings (excluding reinforced masonry research) is much less
> than 5%." (5)

From the limited body of knowledge that has accumulated within the past two decades certain matters have come to light. Analysis of medical records indicates that the vast majority of casualties in earthquakes occur when poorly or inappropriately built houses collapse. A key determinant in establishing the vulnerability of a settlement is that of climate. For example in dry, arid climates (such as Central Turkey) high density building is an advantage since buildings protect each other from extremes of heat or cold. However, these settlements are vulnerable to earthquakes in that there is inadequate space between houses for buildings to collapse without crushing people escaping from their homes in response to initial tremors.

The most dangerous situations are dry, arid, seismic zones where there is a shortage of timber (such as parts of Central Turkey, Iran and Central Asia). In these situations the vulnerability of the settlements results from both the siting of the buildings as well as their construction. Where unreinforced stone or adobe is used for walls and vaulted roofs, there are obvious climate advantages but serious dangers in the event

FIG. 2 - HOUSING EDUCATION CLASS IN PROGRESS IN POAQUIL, GUATEMALA, FOR LOCAL CARPENTERS AND HOUSE BUILDERS.

They are sitting on the corner posts of the house they will proceed to build.   (Photo Oxfam)

of earthquakes.

The least dangerous climate zones are warm, humid areas (such as parts of Nicaragua and Guatemala). In these locations timber, bamboo and thatch will normally be available and can form the basis of safe lightweight building techniques.

In most societies there is an essential difference between rural and urban house building. Rural houses are low in investment whilst high in maintenance. In general, rural housing is more vulnerable to earthquakes since it is likely to be built without regard to seismic criteria, whilst higher investment housing is more likely to be safe.

A detailed analysis of disaster types and casualties indicates that cyclones result in most deaths and injuries, and the primary cause of death is drowning not specifically the collapse of buildings. However, in the case of earthquakes (particularly in the dry, arid context identified above where spanning materials are in short supply) virtually all deaths and injuries are caused by the collapse of buildings or external walls, or from fires ignited after the earthquake.

Where the hazard is flooding or high winds the vulnerability of settlements is of a different order to that of earthquakes due to the more frequent return periods, and far more precise information being available on the area at risk. For example, the cyclone and sea surge that occurred in Andhra Pradesh, India, in 1977 was an extreme event but cyclones normally occur in this region about four times each year (although the incidence of a sea surge is a rarer phenomenon). In this region it is significant that houses in the most vulnerable location, in the coastal regions, have made significant adaptations to cope with high winds. Houses, mainly occupied by fishermen, are frequently round in their plan form, with steeply pitched thatch roofs. This is an ideal configuration in such a context. However, despite such adaptation in construction the siting remains very dangerous as a result of other priorities such as the obvious need of fishermen to be close to the sea. In similar manner, rural families in Turkey frequently choose to live on steep slopes. This location is highly undesirable in a very active seismic zone, it is also subject to other more frequent hazards such as rockfalls, landslides, erosion and in certain high altitude locations to avalanches. However, outweighing all of these risks, there are the more immediate considerations such as the need:

(i)   to obtain climate protection by living on the lee-side of a slope providing shelter from cold northerly winds:

(ii)   for good drainage, and for good water supplies often from the spring line on a hillside;

(iii)   to use flat land for farming and to use the less productive slope for living;

(iv)   to provide a good view over the farmers land – particularly to observe his animals to note anyone attempting to steal his livestock.

The equation I have referred to above on hazards and benefits will vary greatly depending on the economic level of the people concerned. However, the reality is that any study of disasters is by definition a

FIG. 3 — MODIFIED HOUSE BEING BUILT IN GUATEMALA

The major changes from the local traditional house are the lightweight corrugated iron sheet roof in lieu of the heavy tiles (shown in the foreground) and the timber framing into which adobe blocks are being inserted. (Photo Oxfam)

study of the poor, 95% of disaster related deaths occur among the two-thirds of the world's population that occupy developing countries. (2)
As a general rule the poorer the community, the more they will be pre-occupied with immediate short-term survival needs, with little sense of priority for unknown and unexpected hazards. It is this awareness that is the dominant issue in considering this subject.

PROVIDING SAFE HOUSES AND SETTLEMENTS

Recognising the realities I have referred to above, disaster prone communities have frequently taken evasive action to protect themselves or their possessions (including their land, animals or buildings). Normally such measures <u>follow</u> a disaster rather than <u>anticipate</u> the event. In some rare instances the response may include the abandonment of a particularly dangerous setting for a safer location, or in the case of flood risk the establishment of flood control devices (locks, dykes, bunds and emergency overspill routes for flood waters). In areas subject to high winds communities have often attempted to build shelter breaks with belts of trees or high mounds of earth. Then there are examples where buildings have been modified, after disasters, from their previous constructional form.

There are three ways that societies have attempted to improve the resistance of housing to disaster threats:-

(a) To introduce bye-laws governing the construction or siting of dwellings;

(b) To radically change the house form to an engineered structure, possibly pre-fabricated, away from the locality; and

(c) To modify the existing traditional construction or siting to make it resistant to disaster forces.

Many countries faced with vast problems of reconstruction have rapidly enacted bye-laws, or land-use controls to prevent families rebuilding their homes in an unsafe manner or location. Such bye-laws are based on several pre-suppositions which may not be correct, such as the assumption that people can read bye-laws; secondly, government officials drafting bye-laws assume an inspectorate capable of verifying whether house builders are conforming to the laws; and thirdly, they assume that such laws are legally enforceable. It is also apparent that many bye-laws prohibit certain materials which may be the entire local building materials of a given region.

These comments are not intended to dismiss bye-laws as being unimportant, since they are an essential provision in any hazard-prone environment. However, they are limited in their application - being primarily appropriate for middle class housing as built by contractors. They are also appropriate and essential for all public buildings (schools, dispensaries, mosques, etc.) (6). Their relevance to the owner builder, particularly when his house is being built for a minimal sum, is less apparent. In such contexts a radically different approach is necessary.

In a paper presented to the Oxford Conference 'Disasters and the Small Dwelling' Ken Westgate spoke of the problem of introducing land-use controls in such contexts:

"It is possible to envisage land-use planning operating on an

FIG. 4 – MODEL CYCLONE RESISTANT HOUSE IN ANDHRA PRADESH
DURING CONSTRUCTION.

Anti-cyclone features include the Centre Anchor post, metal corner straps
(see FIG. 6) to fit frame members to corner posts and beams. Posts are
creosoted and wrapped in polythene at ground level to retard termite attack. Diagonal cross bracing has still to be fitted. (Photo Oxfam)

entirely different level. Firstly, acceptance of the fact that low-income spontaneous settlements in urban areas of the Third World exist is necessary, as is the fact that these settlements are likely to continue their growth and consolidation. Secondly, viewing low-income settlements as viable communities with an evolving social cohesion renders it possible to consider micro-level, grassroot land-use planning to create a less vulnerable environment. Utilising, as far as possible, indigenous pot-ential in all relevant fields, such land-use planning may consider the siting of future dwellings, the creation of strong central community buildings, the dissemination of building practices and the provision of basic services at low-cost. Sensible and viable land-use planning coupled with adequate controls need not be the privilege of a wealthy consumer society." (7)

The second approach (to introduce a radically different type of house incorporating a structure engineered to withstand extreme hazards) has rarely been successful. The major weaknesses in this concept relate to the social change that such housing inevitably demands as well as the generally weak performance of such houses in economic or climatic terms. Stuart Lewis has itemised the main difficulties encountered with the Prefabricated Post Disaster Housing Programme that the Turkish Government have operated for many years. He refers to the factors that are at the root of the rejection of the housing by the affected rural population:-

1. The prefabs were conceived by urban dwellers in Ankara, unable to identify with needs of users;

2. they have proved to be difficult to repair, modify or extend.;

3. they were designed for flat sites - traditional villages are often on a south facing slope;

4. regional variations in traditional buildings resulting from place and means of livelihood are not possible with standard units;

5. they have poor protection from climatic changes (in traditional construction the building mass acts as a heat sink - modifying day/night temperature);

6. they do not incorporate any provision for animal husbandry (tradi-tional buildings usually relate to winter housing for animals and storage of fodder). (8)

There is a very fundamental issue at stake here, namely the need for governments to recognise that they cannot afford to provide shelter for their poorer citizens. The International Council of Scientific Unions (ICSU) have sponsored a study which includes the statement:-

"It must be accepted that most shelter will have to be self-provided, such self-provision takes place initially in a poor way but it is maintained by continuous efforts at improvement... Instead of rejecting this approach to shelter provision, governments should accept as a starting point this willingness of people to help themselves." (9)

The third approach is to build on this self-help principle with the added dimension of modifying the housing. There are two issues to consider here, 'what to do, in order to make the house safe' and 'how to implement such a programme' given the local priorities referred to at the outset of this paper. The first is a strictly technical issue, the second a social,

FIG. 5 - COMPLETED ARTIC HOUSE

This photograph indicates two anti-cyclone features: the almost square plan, since this is a better plan form than an elongated rectangle. Another feature is the raised mound of earth on which the house stands. This is intended to keep the house clear of the localised flooding that occurs in the torrential rainfall of cyclones.

cultural, even economic concern.

During the past decade some very important studies have been completed which consider the resistance of small dwellings to high winds (10) and to earthquakes (5, 11). In all of these studies the authors have chosen typical traditional houses for their analysis. However, the form and construction of houses are infinitely variable and for this reason specific studies have to be made on a region by region basis, gradually narrowing down to variations in houses village by village. This type of analysis is envisaged in the Housing and Hazards Study of the Karakoram Project.

Documented experiences of such house modification programmes are still limited since they remain comparatively rare events. In past decades there have been occasions when individuals have attempted to modify their house form or construction but the knowledge of the technique used, or the performance of such structures in subsequent disasters is virtually non-existent. Looking at historical records there are examples from classical or pre-classical history. Nicholas Ambraseys from Imperial College has written:-

> "After destructive earthquakes, towns were often rebuilt on an extensive plan with marked changes in building techniques such as unusual types of foundations, consisting of a grid of wooden beams on which the structures are built, the introduction of timber-bracing of houses and the abandonment of ordinary reinforced brickwork. It is often assumed that these changes are due to techniques brought into a region by new settlers, or by invaders. This is not always the case." (12)

One particularly remarkable example of local adaptation to earthquake risk occurred on the Greek island of Lefkas in the Adriatic Sea. The frequency of earthquakes is such that the local inhabitants have evolved indigenous practices to reduce the destructive effects of earthquakes. The main technique used being the development of a timber frame with masonry inserted between the wooden members. (13)

It is not the purpose of this paper to describe such aseismic techniques in any detail. However, I want to observe that the resistance of buildings to earthquake or high wind forces has much to do with the concept of an integrated framed structure, to hold the building together, and secondly the ability of the building to absorb some of the seismic or wind force energy in its structure. In the fourth European Symposium on Earthquake Engineering held at Imperial College in 1972, A. A. Moinfar of the Technical Research and Standards Bureau of Tehran pointed out that most low-income houses are built not engineered:-

> "It is at this level that good workmanship and careful choice - (where there is choice) - of materials count for much more than any resistant design. It is possible to reinforce brick and stone buildings with steel, and such buildings show greatly increased resistance to collapse from earthquakes, but if the roof is not properly connected to the walls, then such defects will far outweigh any defensive measures." (14)

The first major programme of house modification working totally within a local building tradition occurred in Guatemala following the 1976 earthquake. This programme was directed by Fred Cuny (the Director of an American based firm of disaster consultants called INTERTECT) who

FIG. 6 - DETAIL OF CORNER POST OF ARTIC HOUSE.

The metal strap is to provide additional strength, this is to prevent the lifting force that is encountered in high winds. The bamboo diagonal braces are indicated.

developed a highly significant reconstruction programme. It differed from the programmes of the other agencies in that it did not attempt to build lots of houses. Cuny and Oxfam's Field Director, Reggie Norton, saw little point in duplicating what people could well undertake themselves; they saw the role of an external agency as being strictly a support activity. Since the Indian families in the devastated highlands of Guatemala possessed highly developed building skills, they argued that it was pointless to build large quantities of houses. What they (as an expatriate agency) could do was to provide key materials at prices which people could afford, and secondly provide expertise to assist families in building safe houses. This was to give advice on both the safe siting of buildings, and ways of making adobe houses resistant to a future earthquake. The Oxfam/World Neighbours programme consisted of building model houses which incorporated safe constructional techniques in various towns and villages. The houses were then used for extension programmes where classes were held to train people to build safe houses. (15, 16, 17); see also Figs. 2 and 3.

A further development of this programme occurred in Andhra Pradesh, India, following the cyclone of November 1977. Through the active involvement of Intertect an organisation was formed called ARTIC (Appropriate Reconstruction Training and Information Centre). One aspect of this group's work was to develop ways of strengthening houses to make them cyclone resistant, whilst making minimal changes to the basic traditional village house. Oxfam and the Salvation Army were instrumental in establishing this work and as Peter Winchester has described:-

> "Experience from previous disasters had taught these agencies that innovations which used materials foreign to an area, or that incorporated unfamiliar technology, were of little lasting value and often interfered with local economic structures. Therefore, the aim of the Centre was to work through village communities and support local industries and trades, only introducing new ideas which, after explanation and demonstration, were approved by the villagers themselves. It was hoped that in this way village communities would be strengthened and made more self-reliant, while improving their health standards and the life of their houses." (8); see Figs. 4, 5 and 6.

A wide range of visual aids were developed by ARTIC including simple manuals and posters for use at a village level; see Fig. 7.

Initial evaluations of the Guatemalan and Andhra Pradesh housing programmes have been made. Results from Guatemala are rather more encouraging than from India. (16) In some Guatemalan villages there has been a wide and continual acceptance of the new housing system long after their teachers have left, whilst in Andhra Pradesh there is little evidence that local carpenters and house builders have incorporated the structural improvements such as cross-bracing into their building traditions. On a recent visit two years after the cyclone it was apparent that in one village, house owners had removed the diagonal bracing members – perhaps to sell or use as firewood. In other instances the innovations were only partially incorporated in the first instance. The director of this particular project has answered my query about this failure:-

> "The fact that the new designs have not been incorporated is obviously partly our fault because we had not communicated entirely

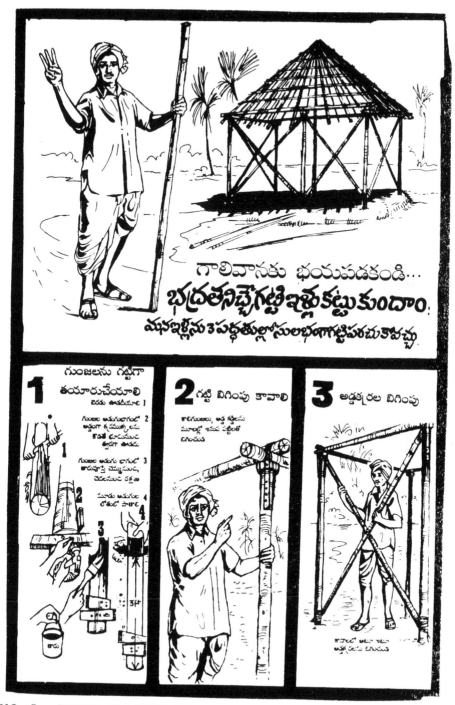

FIG. 7   POSTER PRODUCED BY ARTIC FOR USE IN ANDHRA PRADESH AS
VISUAL AID FOR SAFE HOUSE CONSTRUCTION.

the benefits of the new ideas, but partly the people's who have
resisted the change. Perhaps they have felt that the thick walls that
they employ compensate for the cross bracing, and so have dismissed
the idea."

This candid assessment underlines the key role of education in this
process. The subject of transferring ideas is at the very centre of this
approach to housing and there are many examples of failure in the
education process. In the Oxford Conference on 'Disasters and the Small
Dwelling' of 1978. Julius Holt and John Bowers both presented papers
which identified the particular problems of rural education, particularly
when there are cross-cultural problems between the teacher and his
audience. (18, 19) The concepts identified by A. Fuglesang are essential
reading material for any officials contemplating modification programmes
which seek to transfer new ideas to a village level. (20)

DEFINING A BASIC APPROACH TO MODIFICATION

Both the house modification programmes in Guatemala and India occurred
in the aftermath of disasters. Perhaps the most ambitious project of this
nature that has occurred before a disaster is the study for the Peruvian
Civil Defence Organisation on 'Indigenous building techniques and their
potential for improvement to better withstand earthquakes' (21).

During this project, which was undertaken by a team from Carnegie-
Mellon University with INTERTECT, a very detailed methodology was devised
for the analysis of vulnerable settlements. This was documented by Cuny
as a 'scenario for housing improvement' (22). He identified a logical
process of analysis in the following steps: 1) identify high-risk areas;
2) identify areas with concentrations of vulnerable structures; 3) determine
housing demand; 4) determine receptivity to new ideas; 5) conduct a socio-
logical profile of the community; 6) select a community/site; 7) study the
normal building process; 8) develop training aids; and finally, 9) conduct
a pilot project.

The project proceeded to follow these steps and pilot projects were
undertaken and subsequently evaluated. Several important studies emerged
from this project which have helped to advance our understanding of this
subject (23, 24, 25).

In the Karakoram Project: Housing and Hazards Study a multi-
disciplinary international group has been formed comprising a structural
engineer; an archaeologist; an architect/ethnologist; an anthropologist;
an expert in the anthropology of shelter; an architect with expertise in
disasters and housing; a seismologist; and a social geographer. They
will combine their skills and perspectives to examine the resistance of
houses and settlements in the Karakoram region to earthquakes and other
natural hazards. Their method of analysis will include some of the items
listed above but not include any attempt to modify the existing houses
within the expedition. The basic presuppositions of this group include:-

(i)   the recognition that the local community may have other far
        more pressing priorities than adjustment of their house forms
        or settlements to hazards;

(ii)   the awareness that the local population are the key resource
        if any reduction can be made in the vulnerability of their
        homes and settlements;

(iii)   the knowledge from the examples cited above that only marginal changes in traditional ways of building are feasible.

The overall issue that will be addressed in the study is to ask whether modification of housing is feasible. To answer this the following question will have to be answered: Given the risks facing local communities, and given the local awareness of these risks, and given local and regional resources, and given local priorities is there sufficient evidence to suggest a programme of housing modification?   If the answer is positive then a further detailed set of studies is necessary, such as what hazards to design against and how to do this given local building materials and skills.

This process of analysis requires a genuine sensitivity to local traditions including housing, but not extracting it from its context, as well as a deep understanding of the activities and needs of the given local community.   In the past both attitudes have frequently been lacking in officials, outside experts, and external agencies.

Finally, the process of analysis must be regarded as essentially a two-way reciprocal process.   Although outside experts can contribute, if they do so in the spirit referred to above, they can also receive local insights and wisdom from people who continually cope with and adjust to multiple hazards.   This knowledge is a vital ingredient in any significant advance in this emerging subject.   In certain instances local attitudes have placed a question mark against the entire concept of modifying traditional housing, this has occurred when there have been rising expectations away from traditional houses towards a 'superior type' of house probably derived from urban images.   In such situations where this has arisen from within a culture, any attempt to introduce a modification programme has been greeted with deaf ears.   This has contributed to the lack of enthusiasm for the ARTIC housing in India and is probably one of the most significant unresolved problems in this field.   (26)

Given the scale of disasters noted at the outset of this paper, the present level of concern is paltry and will remain so until far more academic groups, funding bodies, and governmental officials start taking a serious interest in this topic, it is hoped that this conference and the Karakoram project will accelerate this process.

REFERENCES

(1)   COMMITTEE ON THE IMPLICATIONS OF DISASTER ASSISTANCE (CIDA), (1978).   'The United States Government Foreign Disaster Assistance Programme';   National Academy of Sciences, Washington, D.C.

(2)   BURTON, I., KATES, R. W. and WHITE, G. F., (1978).   'The Environment as Hazard'; Oxford University Press, pp 1 - 18.

(3)   WISNER, B., WESTGATE, K. and O'KEEFE, P., (1976).   'Poverty and Disasters', New Society, 9 September 1976.

(4)   OLIVER, P., (1978). The Cultural Context of Shelter Provision', Disasters, Vol. 2, No. 2/3, pp 125 - 128.

(5)   RAZANI, R., (1978).   'Seismic Protection of Unreinforced Masonry and Adobe Low-Cost Housing in Less Developed Countries: Policy

Issues and Design Criteria'; Disasters, Vol. 2, No. 2/3, pp 137 – 147.

(6) SINNAMON, I. T., (1976). 'Natural Disasters and Educational Building Design'. An Introductory Review and Annotated Bibliography for the Asian Region. UNESCO, Regional Office for Education in Asia, Bangkok.

(7) WESTGATE, K., (1979). 'Land-Use Planning Vulnerability and the Low-income Dwelling'; Disasters, Vol. 3, No. 3, pp 244 – 248.

(8) DAVIS, I., LEWIS, S. and WINCHESTER, P., (1979). 'The Modification of Unsafe Housing Following Disasters'; Architectural Design, No. 7, pp 193 – 198.

(9) MABOGUNJE, A. L., HARDOY, J. E. and MISRA, R. P., (1978). 'Shelter Provision in Developing Countries'; John Wiley and Sons, p 3.

(10) NATIONAL BUREAU OF STANDARDS, (1977). 'Building to Resist the Effect of Wind'; (5 volume set), United States Department of Commerce, National Bureau of Standards, Washington, D.C.

(11) DALDY, A. F., (1972). 'Small Buildings in Earthquake Areas'; Building Research Establishment, Garston, Herts.

(12) AMBRASEYS, N. N., (1976). 'Earthquakes in History'; UNESCO Courier, May 1976, pp 24 – 29.

(13) PORPHYRIOS, D. T. G., (1971). 'Traditional Earthquake-Resistant Construction on a Greek Island'; Society of Architectural Historians Journal, Vol. 30, No. 1, March 1971, pp 31 – 39.

(14) MOINFAR, A. A., (1972). '4th European Proceedings: Symposium on Earthquake Engineering'. Imperial College, London.

(15) DAVIS, I., (1977). 'Housing and Shelter Provision Following the Guatemalan Earthquakes of February 4 and 6, 1976'; Disasters, Vol. 1, No. 2, pp 82 – 90.

(16) OXFAM – (America), (1977). 'The Oxfam/World Neighbours Housing Reconstruction Programme, Guatemala'; Oxfam America, Boston,

(17) MACKAY, M., (1978). 'The Oxfam/World Neighbours Housing Education Programme in Guatemala'; Disasters, Vol. 2, No. 2/3, pp 152 – 157.

(18) BOWERS, J., (1980). 'Some Thoughts on Communication'; Disasters, Vol. 4, No. 1, pp 22 – 26.

(19) HOLT, J., (1980). 'Some Observations on Communication with Non-Literate Communities'; Disasters, Vol, 4, No. 1, pp 19 – 21.

(20) FUGLESANG, A., (1973). 'Communication for Developing Countries'; Dag Hammarskjold Foundation.

(21) CARNEGIE-MELLON UNIVERSITY PRINCIPAL INVESTIGATOR: HARTKOPF, V., (1980). 'Indigenous Building Techniques of Peru

and their Potential for Improvement to better withstand Earthquakes'; Agency for International Development, Washington, D.C., and Oficina de Investigacion y Normalizacion Ministerio de Vivienda y Construccion, Lima, Peru.

(22) CUNY, F. C., (1979). 'Scenario for a Housing Improvement Program in Disaster Prone Areas'; Disasters, Vol. 3, No. 3, pp 253 – 257.

(23) CUNY, F. C., (1979). 'Analysis of the Potential for Housing Improvement in High Risk, Vulnerable Areas of Peru, INTERTECT, Box 1025, Dallas, Texas.

(24) CUNY, F. C., (1979). 'Analysis of the Potential for Introduction of Stabilised Adobe in Peru'; INTERTECT, Box 1025, Dallas, Texas.

(25) CUNY, F. C., (1979). 'Improvement of Adobe Houses in Peru: A Guide to Agencies'; INTERTECT, Box 1025, Dallas, Texas.

(26) DAVIS, I. R., (1979). 'View over the Fence' (Guest Editorial); Disasters, Vol. 3, No. 2, p 121.

APPENDIX I

THE VULNERABILITY OF SETTLEMENTS

TYPICAL LOCATION OF SETTLEMENT	POSSIBLE REASONS FOR CHOICE OF SITE OR HOUSE FORM	POSSIBLE HAZARDS
1. Peasant farmers living on a steep slope (Eastern Turkey: Lice)	1.1 Wind protection building on the lee-side of a hill or mountain. 1.2 In cold climates a south facing slope will maximise warmth from the sun. 1.3 Good drainage. 1.4 Views - particularly over fields to see if anyone is interfering with crops or cattle, or if animals are straying. 1.5 Cut and fill foundation digging is probably easier on hillside sites. 1.6 To build on less fertile soils thus keeping flat, fertile valley bottoms for crops or grazing land. 1.7 For families keeping animals in their house, a hillside site is a sensible location since it allows access for animals at the lower level and families at the upper level, without the need for building complicated staircases.	1.1 Rockfalls and land-slides - either seismically induced or from normal frost action or erosion. 1.2 Flash-floods that can wash down slopes. 1.3 In Alpine regions, avalanches. 1.4 Earthquakes (particularly if the house construction is vulnerable).

APPENDIX I (cont'd)

TYPICAL LOCATION OF SETTLEMENT	POSSIBLE REASONS FOR CHOICE OF SITE OR HOUSE FORM	POSSIBLE HAZARDS
2. Fishermen living in a flood plain delta region or river estuary (Divi Seema Region of Andhra Pradesh delta region)	2.1 Access to water for their livelihood. 2.2 A need to work very long hours in order secure a subsistance income. Therefore, any travel from home to work-place will have very serious implications. 2.3 In crowded cities, this may be the only land available within his price range.	2.1 Flooding. 2.2 In earthquake areas, Tsunamis. 2.3 Tidal surges induced by cyclone winds. 2.4 High winds (maximum wind forces in coastal areas - diminishing in intensity inland).
3. Urban squatters living on steep ravines in close proximity to city centre (Guatemala city)	3.1 The need for quick access to the city centre where there are always 'unpopular' jobs available (i.e. porterage; street cleaning; clothes washing; general labouring). 3.2 The only land available at the income level of recent migrants to the city from outlying rural areas.	3.1 Rockfalls and land-slides, either seis-mically induced or from normal erosion caused by rains. 3.2 Flash floods washing down slopes can remove these flimsy structures. 3.3 Earthquakes.

APPENDIX I (cont'd)

TYPICAL LOCATION SETTLEMENT	POSSIBLE REASONS FOR CHOICE OF SITE OR HOUSE FORM	POSSIBLE HAZARDS
4. Farmer living in an unreinforced Adobe house with heavy tiled roof (Guatemalan Highlands area).	4.1 He builds in this manner since it is the established building tradition, (ever since the Spanish Conquest, Adobe houses with Spanish tiled roofs have denoted a certain status over and above thatch roof housing on cornstalk infill walls in timber frame). 4.2 He cannot afford to buy bricks or concrete blocks for the walls or lightweight roofing. 4.3 There are good climatic reasons (in an area that has a wide seasonal range) in having thick walls and roof.	4.1 The shape of the house may place it at risk. 4.2 If the house is closely packed to the next house this may constitute a further risk. 4.3 If the adobe is unreinforced, then walls may fall in if an earthquake occurs. 4.4 The heavy tiled roof constitutes a risk due to its massive weight, and very frequently a badly tied timber roof structure.

Settlements and buildings as the physical expression of a culture: a case study in Nigeria

J.C. Moughtin

Institute of Planning Studies, University of Nottingham

ABSTRACT

Settlements and their buildings are facets of man's physical and spiritual culture and can therefore be considered as the expression of such factors as the level of technological development, and the social, economic, political and religious structures of the society which gave rise to them. Settlements can also be considered as an expression of a society's response to its environmental setting. It is possible therefore that settlement form can be affected by external factors such as climate soils, vegetation and water table. The subdivision of factors affecting the distribution, size and shape of settlements into those which are "internal" or cultural and those which are "external" or environmental is however misleading. It is common knowledge that the environment both shapes man's activities and is in turn shaped by them. There is also evidence to show that similar environmental conditions can elicit widely different cultural responses. With this important proviso it is proposed to analyse the settlements in Northern Nigeria in terms of some of the main internal and external factors which have affected their form and distribution.

LOCATION AND ENVIRONMENT OR HAUSALAND

Hausaland, <u>Kasar Hausa</u>, is in the northwestern part of the Federal Republic of Nigeria and lies between the confluence of the Niger and Benue Rivers (1). It extends approximately from latitude 3.5° east to latitude 11.0° east – or 525 miles (844 km) and from longitude 10.5° north to longitude 14.0° north – or 250 miles (400 km). Hausaland consists of mature or old-age plains upon which the forces of erosion operate slowly. Weathering has taken place to a great depth and to such an extent that the rocks have been decomposed, and it is not possible to recognise their original type. Although there is a wide variety of underlying rock structure, large areas of Hausaland are covered by a common end product, laterite, with limited exposures of fresh rock (2). From the viewpoint of traditional building operations in the area, laterite is the single most important material in use.

As R. J. H. Church points out, "Nigeria has a greater variety of climate than any other West African country; the Cameroon, Equatorial, Semi-seasonal Equatorial, Seasonal Equatorial, Southern Savannah, Jos Plateau, Savannah and Sahel types of climate being found." (3). Hausaland has a climate with pronounced wet and dry seasons. The rainfall drops from the south to the north of Hausaland and varies from a savannah type to a Sahel climate in the extreme northern districts. Hausaland's climate is influenced by two principal wind systems: a hot dry north-easterly wind blowing from the Sahara and a south-westerly, warm, moist air current from the southern Atlantic Ocean. The harmattan or north-easterly wind brings dust with it from the Sahara; when it blows from November to April, day temperatures are high in the afternoon and low at night and in the morning (4). During this time of year, humidity is low; it is frequently hazy with visibility reduced - sometimes to less than half a mile. During the remainder of the year, the south-westerly monsoon winds bring moist, warm air, most of the rainfall, and high humidity. Surface winds tend to be generally light; the only exception is the line squalls which are at the beginning and the end of the rainy season and can gust up to 90 m.p.h.

Climatic conditions in Hausaland affect building operations in two important ways. Shelter has to be designed for two distinctive climatic conditions, a warm, humid climate where cross ventilation is important and for a hot, dry climate where it is important to keep the interior cool and enclosed during daylight hours. In simplistic terms, the first condition requires light walls with many openings screened by a large roof, and the second requires thick walls with few and small openings. Unfortunately, laterite, the chief building material in Hausaland, is highly unstable in rainy conditions; this imposes a whole series of problems for the builder to solve. The measures adopted to solve these problems give Hausa architecture some of its distinctive features.

There are two main vegetation zones in Hausaland: the Northern Guinea zone and the Sudan zone. A zone of transition between the vegetation types runs broadly east-west through Northern Zaria province. The climax vegetation of the Northern Guinea zone is a western extension of the great miombo woodlands, the Isoberlinia-Brachystegia of east, central and southern tropical Africa. It has been suggested that the climax vegetation of the Sudan zone was probably similar to the mutemwa vegetation of Zambia which consists of large emergent trees with open canopy, an under-story of deciduous shrubs, and sparse or no grass (5). In both zones the natural vegetation has been completely modified as a result of human occupation because the bush has been cleared and burnt for cultivation, hunting and cattle grazing.

In the areas near large settlements such as Kano and Katsina, natural bush vegetation is almost completely absent. Trees such as Acacia albida Tammarrind indica and Butyro spermum parkii have been planted in settle-ments for shade and fruit, but timber itself is rarely used in building operations. An important exception is the dumb palm, Hyphanene thebaica, which occurs in the Sudan zone either singly or in dense groves. Azara, the wood from this tree, is a fibrous material which is free from attack by white ants. Consequently, it makes excellent roofing joists and may also be used as reinforcement for mud buildings. In addition to azara, other vegetable material used in building includes small branches for light framing, grass thatching, grass matting of various types, and vegetable additives to mud for waterproofing purposes.

313

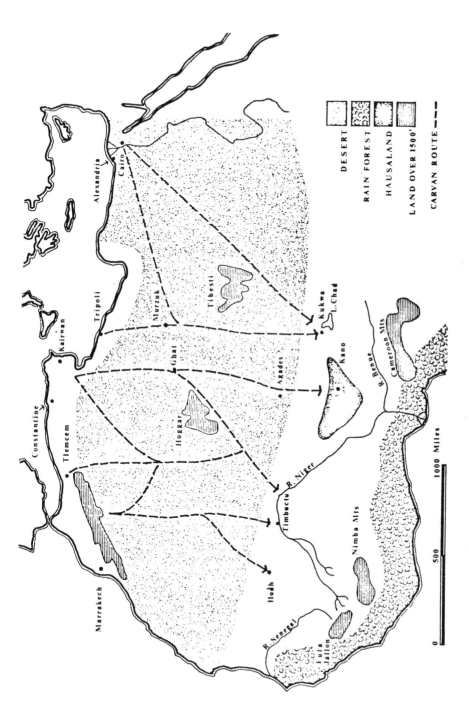

Fig. 1. Location of Hausaland. Nigeria. (Ref. 10)

314

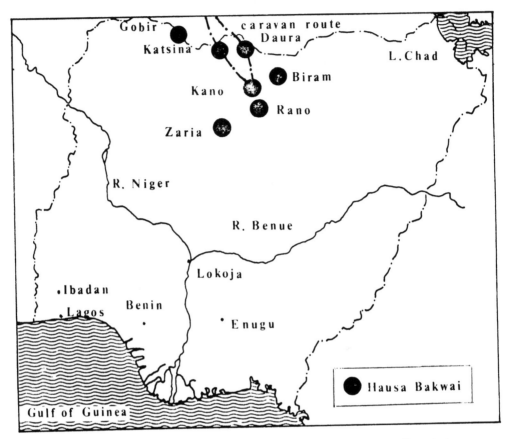

Fig. 2. Nigeria, showing the location of the seven city states.

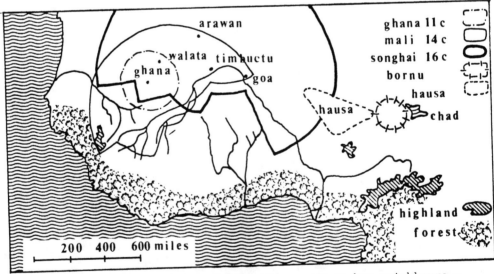

Fig. 3. Location of Hausaland with respect to her neighbours –
a historical perspective. (Ref.7)

THE DEVELOPMENT OF HAUSA SETTLEMENTS

The built forms evolved by different cultural groups to solve similar environmental problems are many and diverse and are dependent almost entirely upon the perception of those problems (6). Such perceptions can vary from almost totally ignoring environmental problems and placing greater emphasis on socioeconomic or symbolic considerations at one extreme, to an attitude which places a premium on design for optimum internal environmental conditions in which built form is a climatic control mechanism. For this reason it is important to understand the cultural history of the Hausa in order to appreciate the attitude of these people towards the development of the architectural programme and to understand the values attached by the Hausa to the design factors governing traditional built forms.

HISTORICAL ASPECTS

Until the coming of the Europeans, Hausaland was cut off from the coast by an impenetrable barrier of tropical rain forest; all contact with the outside world was by way of the Sahara. For many centuries the great bullock trains, and later the camel trains, which crossed the Sahara were the commercial lifeline that connected Hausaland to north Africa. Along these routes travelled the wealth of west Africa, gold and slaves in return for salt and Mediterranean produce (7). Up to the first millenium A.D., the areas just south of the Sahara were undergoing a change in the structure of their vegetation; in part this was due to world-wide climatic changes, but later man's agricultural activity brought about great changes to this region. There was, therefore, in what is now the Sahel vegetation zone, a consequent increase of population pressure which produced a southward movement of people in Hausaland and other areas of west Africa (8). This great southward movement of peoples is recorded by the Hausa and other Nigerian peoples in their legends which generally place their origin in the north (9).

In the savannah lands between the great northern bend of the Niger and Lake Chad was founded a group of seven small related city states: Daura, Kano, Zaria, Gobir, Katsina, Rano and Biram, called the Hausa Bakwai or true Hausa. The origin of the Hausa states is obscured by legend, but it is probable that they owe their origins to a series of related invasions of peoples from the western shores of Lake Chad in the first millenium A.D.

Beginning in the ninth century A.D., Hausaland was polarized between the powerful states of Ghana, Mali, and Songhai to the west and Kanem-Bornu to the east. Hausaland occupied the no-man's-land between these two power blocks and acted as a buffer state, alternating under the domination of either west or east. Although in constant subjection to foreign rulers and despite having their farm lands laid waste and their trade dislocated by the occasional attacks of their more powerful neighbours, the Hausa states managed to retain their identity and their local autonomy which enabled them to develop an elaborate social organization.

Because the Hausa states were located at the terminus of the central Sharan trade route, they were able to develop as a series of high density urban centres with highly structured political and administrative systems of government and with a prosperous economic base. Continual tribal warfare before the British occupation (1900 to 1906) of northern Nigeria added impetus to the establishment of a pattern of densely occupied, walled settlements in areas of unoccupied savannah land. In addition to the major administrative cities such as Kano and Zaria, there are also many smaller walled settlements ringed by bush hamlets (10).

CULTURAL ELEMENTS

Islam has had a considerable effect on many aspects of Hausa culture. Building techniques and architectural style are no exception. Imported ideas of what constitutes a good building have changed the indigenous architecture from a savannah style to one that is more suited to an arid Saharan climate. The early contact of the Hausa states with Islam were by way of the western Sudan and the Maghrib; tradition says that Islam was brought to Hausaland by merchants from Mali in the fourteenth century. For many centuries Timbuctu remained a renowned centre of Muslim scholarship, and it was from this source that the Hausa received their introduction to Islam. In the sixteenth century the Hausa city of Katsina became an important centre of Islamic learning with its own special quarter in the town for students (11).

ZARIA

City Structure

It is difficult to generalize about the layout of the major Emirate cities. Most of them, however, have well-defined quarters separated from the old city. The layout pattern of Zaria, in particular, emphasizes the separation of the various quarters. The barriers between the quarters are the railway track, the river, or the city wall. The settlement resembles a cluster of independent urban centres rather than a single unit, and the pattern it forms reflects religious, racial and cultural differences overlaid by the recent need to accommodate the motorcar, modern educational establishments, or administrative and commercial institutions (12).

The one-time "European" or residential quarter is laid out with large avenues of madaci and neem trees. At its centre is the Zaria club and race course, and in this area are found administrative offices, law courts, and the post office. The Sabon Gari (new town) is occupied chiefly by southern Nigerians who work in the industries, commercial firms, railway, and administrative organizations. The street pattern in the Sabon Gari is set out in the form of a rigid grid. The buildings are mainly concrete block walls with tin roofs totally unsuited to the environmental conditions. At its centre is its market, and everywhere there are small commercial firms, bars and night clubs. In fact, it is a shanty town, full of brash character, noise and colour, but tremendously overcrowded with polluted open drains and a primitive night soil collection system.

Tudan Wada, another distinct quarter just outside the walls of the old city, is inhabited chiefly by Muslim outsiders. This area, like the Sabon Gari, is laid out in a grid street pattern conforming with health and byelaw regulations introduced by the British earlier in the century. Here, however, many of the buildings are regularized versions of traditional Hausa buildings, although the ubiquitous tin roof is fast replacing the mud roof, and concrete block construction is also in evidence.

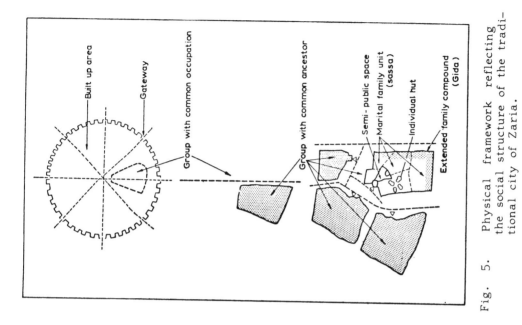

Fig. 5. Physical framework reflecting the social structure of the tradi-tional city of Zaria.

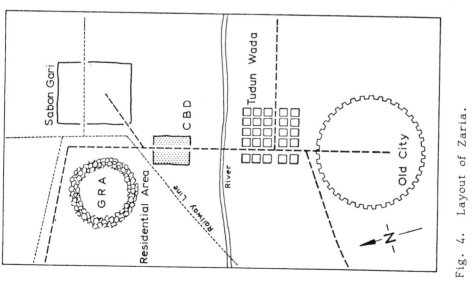

Fig. 4. Layout of Zaria.

GRA = Government Reserve Area
CBD = Central Business District

318

Fig. 6. Daura, an outdoor praying area

Fig. 7. Street scene, Zaria, showing modern geometrical patterning.

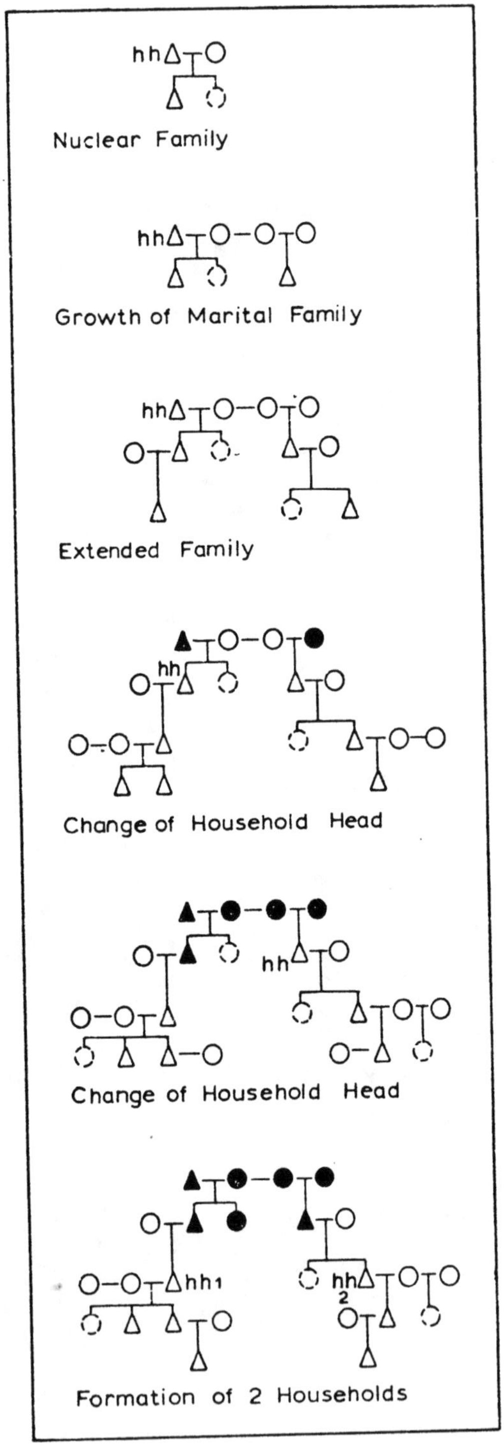

Fig. 8.   Chart indicating Hausa co-residential kin groups.

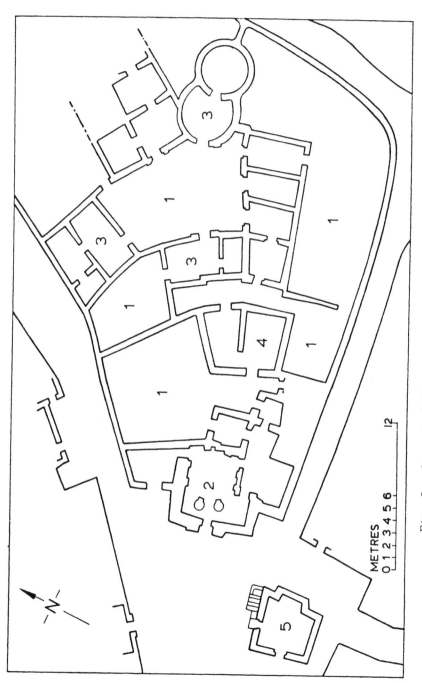

Fig. 9. Layout of the house of Babban Gwani, Zaria.(Ref. 19)

Key: (1) Courtyard;   (4) Young unmarried men
        (2) Entrance hut (zaure); (5) Mosque
        (3) Wives' huts;

A large residential area in <u>Samaru</u> has been established in the last 15 years to serve the Ahamadu <u>Bello</u> University, which is some five miles from the city. This modern area of <u>Samaru</u> resembles the <u>Sabon Gari</u> in form, density, and social structure and exhibits all the environmental problems that can be found there. Between the university and the older parts of Zaria, there is ribbon residential development consisting of modern suburban areas for senior administrative, commercial, or university staff. There are also traditional expanded Hausa villages following closely patterns of development in the ancient city and low-cost housing schemes built by government agencies.

Traditional City of Zaria

Within the apparently formless complex of mud dwellings that exist inside the walls of the ancient city, there is a basic physical framework which reflects its social structure. The residential area in the old city is subdivided into zones usually associated with one of the main gateways. Each zone is occupied by groups having certain similarities, common occupations (dyeing, weaving, potting, etc.) or membership in one of the royal families. These zones are further subdivided into areas of a few street blocks occupied by families having a recent common ancestor. The basic element within this complex pattern, however, is the street block, irregular in shape and enclosed by a high mud wall. The street block contains the main social and economic unit of the extended family. The family compounds are linked by pedestrian routes which connect the market place, the Friday Mosque, and the gateways. These routes also connect outdoor praying places, shaded sitting and teaching areas, and the large formless borrow pits and take the form of a sequence of spaces, some of which are narrow passages barely wide enough for a fully laden donkey.

FAMILY STRUCTURE AND HOUSE FORM

Hausa families are of three types: individual families; a married man, his married sons and their dependents; or a group of collateral agnates and their dependents. These varieties of Hausa co-residential kin groups are not formal alternatives but manifestations of the same rhythmic and dynamic cycles (13). The effect of this household cycle is most evident in the organic nature of the settlement pattern. New family units are constantly being formed, maturing and breaking up. During the dry season each year, new homes are built; and during the rains the unused parts of decaying compounds are reduced, first to rubble, then to simple mounds of laterite. This process of growth and decay is assisted by the impermanent nature of the building materials. Once vacated, a building soon disappears, either naturally or by being demolished; its materials are then reused for building on another site.

A typical house plan follows the traditional African pattern with rooms arranged within or around a courtyard. The compound within a wall is also an important feature of traditional African housing and contains the main economic unit of the extended family which works the same fields, shares the same grain store and eats from the same pot. The extended family is broken down within the compound into marital family units or sassa, each of which occupies a separate part of the house delineated by low walls of mud or matting made from guinea corn stalks (14). Figure 9 shows the plan of a typical house in Zaria which provides privacy for the womenfolk. Entrance to the house (<u>gida</u>) is through the entrance hut (<u>zaure</u>). Circulation from the entrance hut to the family

Fig. 10. Repair of well-weathered perimeter wall of large
family home, Zaria. (Ref. 15)

part of the house (cikin gida) is through one or more courtyards. In these courtyards are huts for unmarried youths and male guests, and around them are further huts which are screened entrances to the various quarters of the individual marital groups. Within each sassa each wife has one or two huts which she decorates with her dowry and other belongings and where she sleeps with her children. In addition, there may also be a hut for the husband and huts for his other dependent relatives (15).

Changes in the family structure result in either the subdivision of the compound between the new compound heads, that is, the male inheritors of the estate; or the group may break up completely, and one or more sections build new cells on a new site. If the compound is subdivided, the divisions between sassa become harder, and each new compound has its own perimeter wall and external gateway. Field investigations indicate that the man who sets up a new compound on a virgin site builds first his perimeter wall and then an entrance hut and huts for his wife and himself. As his family grows, he adds huts where and when they become necessary; later still, walls are built to subdivide the compound into sassa. When the family shrinks, through death or the loss of a breakaway sassa, the land soon returns to agricultural use. Like the settlement, the house is dynamic during a man's lifetime; it is constantly adapted to suit his changing needs with the unit of growth and decay being the hut (16).

In traditional Hausa society, land and the buildings erected upon it have no market value, the ownership of land being usufructory only (17). The fabric of the house acts mainly as a shelter and is not regarded as a long-term investment. Until recently Hausa builders were not preoccupied with the durability of building materials. However, even in the rural areas of Hausaland and over large areas of the Emirate cities, there is evidence of experimentation with new forms of material and with new systems of construction. For example, concrete block walls are used with mud walls, and corrugated iron roofs are used with either wall system. Mono-pitched iron roofs are quite common in Kano with gable walls taken above the roof line. The traditional outlook, however, of the Hausa builder has been a concern to prolong the life of the building to suit the needs only of the current occupants.

The great ingenuity of the Hausa builders has produced many and varied techniques of weatherproofing for the critical surfaces of the building. Figure 10 shows Hausa men at work repairing a well-weathered structure. The result of the system of construction which has been developed is a perfect match between the changing needs during the life-cycle of the family and the organic nature of the houseform. Experiments, therefore, with methods of extending the life-span of buildings either with new construction methods or with new additives to the existing materials are not entirely appropriate unless confined to those areas of the home such as the zaure (entrance hut), which often is expected to outlast the lifespan of the occupants.

Climatic Influence on House Structure and Layout

As indicated earlier it is difficult to design for comfort in both types of climate in Hausaland when traditional building techniques are used. To find a solution to this problem in the form of a single building type is expensive and can be achieved only by using a modern structure with a mechanical support system. The Hausa, therefore, have adopted a com-

Fig. 11. Tubali construction techniques in Hausaland: Four layers left to dry after construction (foreground); a completed wall with mud plaster on the outside (background). (Ref. 15)

promise using different structural types for different times of the day or year.

The house normally consists of structures with different properties; rooms built entirely from mud; simple structures consisting of a light frame and a large grass roof with shade trees and large walls providing additional shaded areas. The all-mud building (soro) has a heavy mud roof and thick walls with few openings which give it good properties of thermal insulation. Within a building of this type, it is possible to trap the cool night air and prevent the interior from being warmed by the sun or external air; conversely, during cold nights the warm air is retained inside the building. In this way, the extremes of external temperature associated with the dry season are reduced, and a more even internal climate is achieved. In contrast, the areas beneath the shade trees and under the grass-roofed open structures are ideal in the humid rainy season when air movement through the building is important to keep skin temperature cool.

In discussing the adaptation of built form to climatic conditions in the case of the Hausa, it is important to consider the complete range of buildings and spaces to be found in the family compound. It becomes clear that the Hausa system of house planning provides comfortable living conditions at all times of the year. It is also clear that climatic control cannot be as effectively achieved using cheap modern materials particularly on a tight urban site.

Construction Techniques

The construction process for a mud building is a long one, and careful preparation of the materials is important. The mud walls are made up of regular courses of unbaked bricks (tubali) laid in rough courses and set in mud mortar. The tubali are coneshaped, and in Zaria they are usually about 15.2 cm in diameter (18). It is considered good practice to lay only two or three courses of mud bricks a day; when the wall reaches the height of the door lintel, the work is suspended for 24 hours so that the walls are thoroughly dry before finishing the top courses

The normal method of building is for the builder to sit astride the top of the unfinished wall. Mud bricks and pats of mortar are thrown up to him; as he finishes the day's courses which are within his reach, he moves away backwards over that part of the wall which was built the previous day. Scaffolding and ladders are then needed only for the construction of the larger buildings, such as the Friday Mosque in Zaria. Heavy roofs require the construction of thick walls; for example, those of the Friday Mosque in Zaria measure 1.2 m at the base. All such walls are further strengthened with azara laid transversely across the width of the wall, on top of which are placed layers running longitudinally, at about 0.6 m to 1.0 m above the ground and again just above the height of the door. Piers and columns are also reinforced by using groups of bound azara bonded together with mud mortar and surrounded by a thick coating of mud.

The most basic type of mud roof is formed simply by spanning a space 1.8 m wide between supporting mud walls. Spaces which are larger than this economic span for azara can be made by forming a series of mud corbels at the tops of the walls. These mud corbels are reinforced with several layers of azara and project 45 cm from the face of the wall.

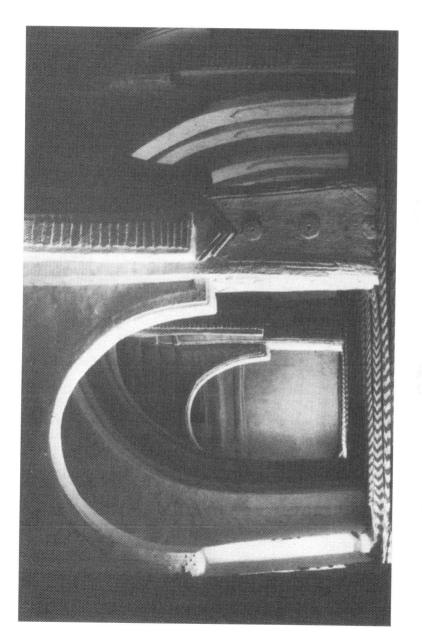

Fig. 12. Main Hall, Friday Mosque, Zaria. (Ref. 19)

328

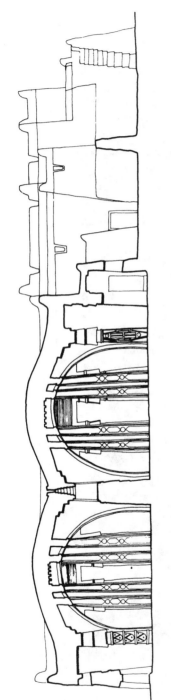

Section B-B

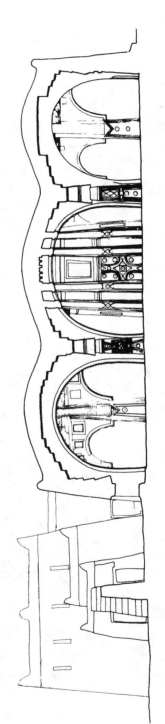

Section C-C.

Fig. 13.   Cross sections, Friday Mosque, Zaria.   (Ref. 19)

They are usually at about 2.25cm on centre, and the space between them is spanned by a beam made up of several layers of azara. From this beam, azara joists span across to the other wall in the usual way. Using this construction system, the room width can be increased to 2.7m. In order to increase the room size to about 3.45 , azara must be placed diagonally across the corners of the room. For the construction of very large rooms, the roof may be supported on a series of columns connected by beams. The mud column is usually surmounted by a simple capital, consisting of two or four azara corbels which are used in order to increase the spacing between the columns to about 2.4 or 2.7m. The spaces between columns are spanned by mud beams reinforced in the usual way with azara; the roof joists then span between the beams. Figure 16 is a drawing of the roof plan of the Mosque at Kazaure which illustrates the roof form of a building with a simple trabeated construction and with some use of corbelling techniques.

Spanning spaces larger than 3.5m^2 (as, for example, in the Friday Mosque depicted in Figure 12) requires the use of the mud arch; and with this structural system, it is possible to construct rooms 8.0m^2. The mud arch is essentially a series of reinforced mud corbels placed one on top of the other. Figure 17 is a section through a mud arch showing a typical arrangement of reinforcement. In better class construction, the layers of azara reinforcement should not project more than about 70cm, nor should the change in angle between succeeding azara be too great.

For these reasons the mud arches should start quite low down, and they are normally constructed in the following manner. Each layer of azara reinforcement (kafi) is tied back to the preceding one; when the gap between the two halves of the arch is small enough, horizontal azara (biko) are used to complete the arch. Each corbel is allowed to dry overnight before the next one is constructed, and in this way the arch can be built without the use of centring or scaffolding. Additional lengths of azara are used for reinforcement and are placed at right angles to the wall and project into the body of the arch. Lengths of azara (Mashim fidi) are also placed in the wall at the back of the arch and at right angles to it so that cracking is prevented by distributing any thrusts through a large area of the wall.

In a rectangular room where the shortest side is less than about 4.5m the room is usually divided into bays of 2.1m; and simple arches span across the room parallel to the shortest side. The area between the arches, which is usually about 1.8m, is covered in the usual way with azara rafters. When the shortest side of a room without internal piers exceeds 4.5m it is usually spanned by two or more arches built at right angles to each other. When this system is used and the arches have been built, diagonal lengths of azara are laid across the corners of the room; from the triangular platforms so formed, azara beams are carried over the backs of the arches. The arches and these beams form a rigid framework on which the azara rafters run from the apex of the arch over to the wall beams, presumably in order to throw most of the load directly on to the wall and reduce the loading on the arches.

Mud is an extremely unstable building material when used in a climate such as that of northern Nigeria where rainfall is high and frequent. The Hausa, however, have developed ingenious methods of adapting mud construction to the climate. Rainwater is directed away from the roof in one of two ways, either by long rainwater spouts which project just over half a metre from the face of the wall, or down deeply incised

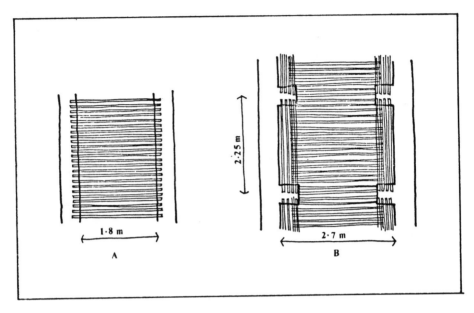

Fig. 14. Mud roof construction: (A) basic span 1.8m wide between supporting mud walls; (B) increased span allowing room width of 2.7 m by means of corbelling and a beam. (Ref. 19)

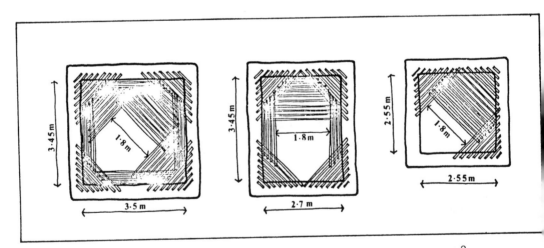

Fig. 15. Method of constructing mud roof for room 3.45m^2. (Ref. 19)

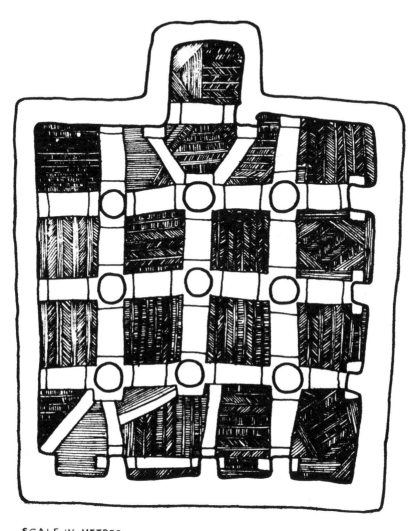

SCALE IN METRES

0   10   20   30   40   50   60   70   80   90   100   110   120m

Fig. 16.   Mud roof plan, Friday Mosque, Kazaure, using
columns and beams in addition to the corbelling and diagonal azara used for smaller rooms.
(Ref. 19)

332

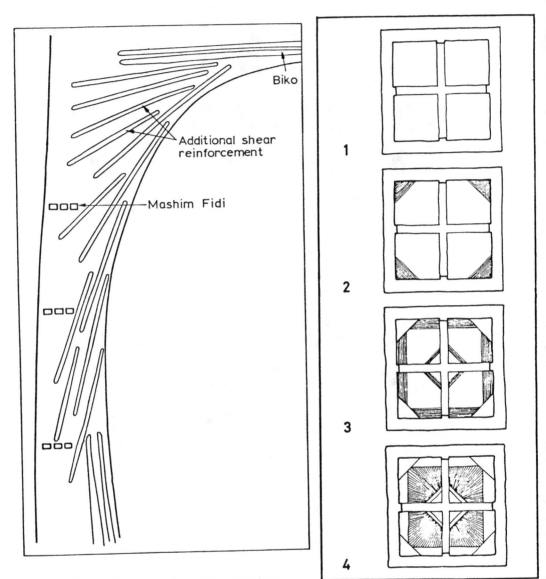

Fig. 17. Schematic diagram, construction of mud arch. (Ref. 19)

Fig. 18. Schematic representation, mud roof construction for room without internal piers and more than 4.5 m on its shortest side. (Ref. 19)

vertical channels cut into the face of the wall and lined with a waterproof finish. The base of the wall is usually protected by a plinth projecting about half a metre. The Hausa, too, have developed many types of waterproof finishes which cannot be described in detail here, but which prolong the life of the building (20). Each finish has a recommended use and a recognized lifespan. For example, laso which is made from Katsi, is a byeproduct of the dyeing trade. It is used for the most exposed surfaces, such as roofs, pinnacles (zankwaye), and rainwater channels; it has a life expectancy of five to six years when used for walls. The best and most expensive finish for walls is cafe which is reputed to last for many years without maintenance. It usually consists of black earth mixed with the solution of the pounded seeds of the bagaruwa tree.

Hausa building techniques are highly developed and extremely complex; it is difficult to conceive improvements to the system without destroying its integrity. The extension or continuation of its use will require that building codes will allow its use along with other modes of construction so that it can be confined to those areas and for those problems for which it is best suited. In such circumstances, it will simply involve the management of the basic and necessary resources for the construction process, ensuring that palm plantations are renewed, that the building craft remains attractive to young men, and that proposed developments are properly related to mud borrow pits (21).

Limitations of the House System

Most technological systems have strict limitations on their use, and the Hausa system for providing shelter is no exception. For example, the Hausa system is thought to contribute to the spread of epidemics such as meningitis. For part of the dry season, the harmattan blows from the desert, bringing with it fine particles of suffocating dust. The only method the Hausa know to reduce the effect of the harmattan is to retire within sealed rooms; it is at this time that dust-borne epidemics occur. Doctors believe that these epidemics are caused by the confinement of many people in overcrowded and stuffy rooms where the chances of contact with affected cases increase. This failure of the system in terms of the control of the spread of epidemics may be due in part to the growth in population and its concentration in large urban complexes such as Kano; as such, it may represent a fundamental limitation of the system's ability to cope with the problems associated with rapid urbanization. It must be pointed out, however, that epidemics are just as prevalent in areas such as Sabon Gari and Samaru where the physical designs are quite different; therefore, eradication of the disease may require more fundamental treatment than a simple change of built form.

Yet another problem associated with scale and density of development relates to sewage disposal and water supply. It is the Hausa custom to provide both a water supply from a well and also a pit latrine within each compound, and it is not known under what conditions this aspect of the system will constitute a health hazard. Therefore, rather than concentrating on research, experimenting with climatic control durability, etc. it may be wiser to investigate the limits of traditional systems such as this one in order to determine the critical factors and to suggest methods for extending the boundaries of the system.

CONCLUSION

This paper has attempted to explain the form and distribution of the Hausa settlements of northern Nigeria in terms of political, social, economic technological and environmental factors. The present regional location of the Hausa settlements can be traced to ecological and demographic changes in the area south of the Sahara which occurred up until the first millennium A.D. The regional location was later confirmed by Hausaland's role as a political buffer between the great African empires to its east and west. The distribution of discrete nucleated and highly defensive settlement form adopted by the Hausa reflected the generally unstable conditions prevailing in the area right up until the end of the 19th century and the colonisation of Nigeria. The internal structure of the city has been shown to reflect the ethnic and social divisions within the community and the built form as a product of both technological developments, reactions to the environment and a reflection of religious and social requirements. This paper has emphasised only a few of the factors involved primarily because of the reason of brevity. It is, therefore, in many respects an over-simplification and if time and space had permitted other themes such as the availability of good agricultural land, potable water supply and the system of marketing could well have been developed.

REFERENCES AND NOTES

1) A better translation of Kasar Hausa is "the land where Hausa is spoken."

2) BUCHANAN, K. M. and PUGH, D. Land and People in Nigeria, (London: University of London Press, 1955), p 2.

3) CHURCH, R. J. H. West Africa, 6th ed. (London: Longmans, 1957), pp 448 - 449.

4) UDO, R. K. Geographical Regions of Nigeria, (London: Heinemann, 1970), p 2.

5) KEAY, R. W. J. An Outline of Nigerian Vegetation, (Lagos: Federal Government Printer, 1959), pp 22 - 25.

6) RAPOPORT, Amos. House Form and Culture, (London: Prentice Hall International, Inc., 1969), p 47.

7) FAGE, J. D. An Introduction to the History of West Africa, (Cambridge: Cambridge University Press, 1962), p 9. Also see HOGBEN, S. J. and KIRK-GREENE, A. H. M. The Emirates of Northern Nigeria, (London: Oxford University Press, 1966), pp 30, 31.

8) SMITH, Abdullah. "The Early States of Central Sudan", in History of West Africa, ed. London: Longman (AJAYI, J. F. A. and CROWDER, Michael, 1978), 1, pp 158 - 201.

9) ARNETT, E. J. trans. "A Hausa Chronicle", Journal of the African Society, 9, No. 34 (1910), pp 161 - 167.

10) MOUGHTIN, J. C. "The Traditional Settlements of the Hausa People", The Town Planning Review, 35, No. 1. (April 1964), pp 21 - 34.

11) HOGBEN and KIRK-GREENE, op. cit., pp 163 - 166.

12) MOUGHTIN. "The Traditional Settlements of the Hausa People", op. cit., p 23.

13) SMITH, M. G. Introduction to Mary F. Smith, Baba of Karo: a Woman of the Moslem Hausa, (London: Faber and Faber Ltd., 1954), pp 21 - 22.

14) MOUGHTIN, J. C. "New Homes for Old Societies", in Proceedings of the Town and Country Planning Summer School, University of Manchester, 1968, pp 74 - 75.

15) MOUGHTIN. "The Traditional Settlements of the Hausa People", op. cit., p 26.

16) MOUGHTIN. "New Homes for Old Societies", op. cit., pp 74 - 75.

17) In some districts of the older cities, houses are bought and sold, but in general, this procedure does not conform to the traditional system of land tenure.

18) DALDY, A. F. Temporary Buildings in Northern Nigeria, technical paper No. 10, (Lagos: Public Works Department, 1945).

19) MOUGHTIN, J. C. "The Friday Mosque, Zaria City", Savanna, 1, No. 2 (1972), op. cit., pp 143 - 163.

20) TAYLOR, F. W. and WEBB, A. G. G., trans. Labarun Al'Adun Hausawa Da Zantatukansu, Accounts and conversations describing certain customs of the Hausas (London: Oxford University Press, 1932), pp 168 - 191.

21) MOUGHTIN, J. C. "Building Materials in the Third World: a Case Study in Northern Nigeria", in Proceedings of the Town and Country Planning Summer School, University of Nottingham, 1976, p 75.

The analysis of local building materials and building techniques

R.E. Hughes

Ove Arup & Partners, London

ABSTRACT

There are many factors that affect the vernacular building's ability to withstand an earthquake. These not only include features of the structure's 'type' but also its construction quality, the environment, condition, history of deformation, and history of stresses and stability. It is shown, illustrated with British examples, how these aspects form the survey techniques for the study of Karakoram buildings.

INTRODUCTION

Sadly, decay of our buildings is an inevitable fact of life with which we have to live (1). Although it may take hundreds of years to have a total effect, material and structural decay starts off slowly and speeds up, and the result is always the same - a rotting mass of vegetation- a pile of rubble. This is incidentally what my colleagues would call a valuable archaeological site.

However, the rate of decay can be modified, if not controlled, and here we see the skill of the architect, engineer or builder, in providing numerous protective features (2). Perhaps massive hard stone walls need minimum design since they may lose less than 1% of their mass in 100 years, so extending the building wall beyond its estimated life expectancy. Soil and some soft stones can totally erode within just a few years and need special designed features. In England we have a saying that "all a cob house wants is a good top hat and a pair of shoes" (hat = roof, and shoes = foundations) (3). To ensure against the effect of earthquakes, floods and rock falls, the most dramatic forms of decay and regularly experienced in Karakoram, needs even more specialised skills.

During construction, occupation and subsequent alteration, a building starts to decay and 'maintenance' keeps the structure in good order. A badly decayed building can be brought back into active use by 'renovation' and 'restoration' and an important monument of ruin can be 'conserved', 'stabilised' or even 're-built'. These secondary building

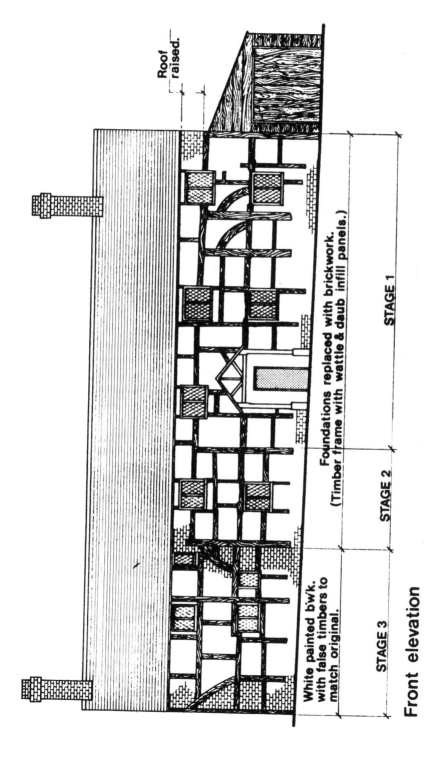

Roof raised.

Foundations replaced with brickwork.
(Timber frame with wattle & daub infill panels.)

STAGE 1

STAGE 2

STAGE 3

White painted b'wk.
with false timbers to
match original.

Front elevation

FIG. 1. Initial Survey of Hazelwick Grange front elevation.

activities are never quite invisible and indeed are often in a totally
different style. Over the years we therefore see a building slowly developing a unique character and the appearance of age is one main quality
we value.

It is the purpose of my three month study programme to locate and
examine the clues to the character of selected buildings and accurately
record them (4). Those features designed and adopted to resist the
harsh Karakoram environment are to be analysed in great detail. By
necessity, much of my own work will be historical detection and can
be compared with the historian who analyses the content and meaning
of an old document. This detection work is generally called "structural
archaeology".

My approach is to be quite distinct from that of the interested visitor
to the Karakoram who has noted buildings with 'snap shots', sketches
and perhaps outline plans. Basically, the recording of a building must
revolve around a study of its history regarding its construction, environment, materials, physical and chemical condition, deformation stresses
and stability.

For the general interest of my audience I will illustrate the approach
to my study with examples from projects I have participated in, in the
United Kingdom, but clearly this could also be done with activities abroad.
This work has been undertaken for Ove Arup and Partners, an internationally known firm of consulting civil engineers and for the
International Council of Monuments and Sites. This has been recognised
by UNESCO since 1966 as being the principal non-governmental organisation
in the field of architectural heritage.

HISTORY OF CONSTRUCTION

In order to understand the structural interaction between the various
parts of a building a detailed history of construction is required (5).
This involves recording the sequence and extent of each construction
operation by plans, rectified and stereoscopic photography (6) and written
description (see Fig. 1)

(i) Where possible records should be made of previous buildings
on the site. Often man's activities can detrimentally affect
the ground conditions and foundations can't be re-used. The
location and type of nearby previous land use can affect
the existing building's structural integrity (see Fig. 2)

(ii) It is essential to have a complete history of the building
including the extent, type and dates of the original building,
subsequent alterations, additions, types of foundations, periods
of neglect or disuse and any strengthening or significant
repair work.

HISTORY OF THE ENVIRONMENT

A building is constructed into a continuously changing environment
and also the building itself can affect the environment. It is therefore
important to appreciate the environment at the time of the building's

FIG. 2. Excavated Archaeological Remains Below a Proposed
Redevelopment. Lower Thames Street, London.

340

FIG. 3. Because of decayed roof timbers, Christ Church near
Marble Arch, London, has now been demolished.

construction as well as at the time of survey. The factors of chief concern are:

(i) Subsoil: previous buildings and vegetation can improve or detrimentally affect the soil properties. Geotechnical investigations can usually determine whether the soil strength properties are sufficient for the building. Existing nearby trees can extract water and mineral nutrients from the ground and in the case of clayey soils this results in soil shrinkage (7).

(ii) Water table: changes in the groundwater regime can seriously affect the integrity of foundations. For example, a falling water level can cause timber foundations to decay and peaty soils to consolidate. A rising water table, perhaps next to a new canal, can cause a soil to soften or can induce capillary water to rise up a wall.

(iii) Climate: temperature, rain, wind and the 'aspect' of the building on both a micro and macro scale can seriously cause structural instability of walls and the decay of building materials.

(iv) Air pollution: the production of slightly acidic rain and air moisture, the transportation of dirt and the colonisation of bacteria and simple organisms all cause materials to decay. It is also worth noting that smoke from domestic fires can preserve organic materials.

HISTORY OF PHYSICAL AND CHEMICAL CONDITION

It is essential to recognise that the decay of the structure and materials are inherent from the day of construction. It is firstly important to understand the composition and durability of the materials and whether they were chosen to resist certain decay mechanisms. A study should then be made to see how the materials have changed or deteriorated with time and the effect of the prevailing climatic conditions. Decay of the fabric must then be recorded, for example: the erosion of masonry facings, joint weathering, dry and wet rot in timbers (see Fig. 3), salt efflorescing and surface spalling, animal attack, root penetration, soil slumping and foundation undermining (8). (see Fig. 4A and B).

HISTORY OF DEFORMATION

By how much a building has differentially moved is a measure of the skill of the original builder as well as a product of its decay history (9). Since we do not have plans of the original construction of most buildings, the structure should be surveyed particularly noting the phases of construction and estimating defects in the original building procedure. The introduction of modern and 'alien' materials should be noted because they can often cause new stresses and strains to develop. It is important to supplement this with measurements of the building's existing deformation (10). This is to include differential settlement, verticality of the walls, position orientation and width of cracks. It is important to determine whether the building is currently being deformed, and here a measurement recording system is installed for future

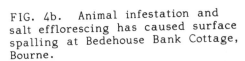

FIG. 4a. The high water table
at Theobald's Park Farm barn
has caused the soil walls to
slump.

FIG. 4b. Animal infestation and
salt efflorescing has caused surface
spalling at Bedehouse Bank Cottage,
Bourne.

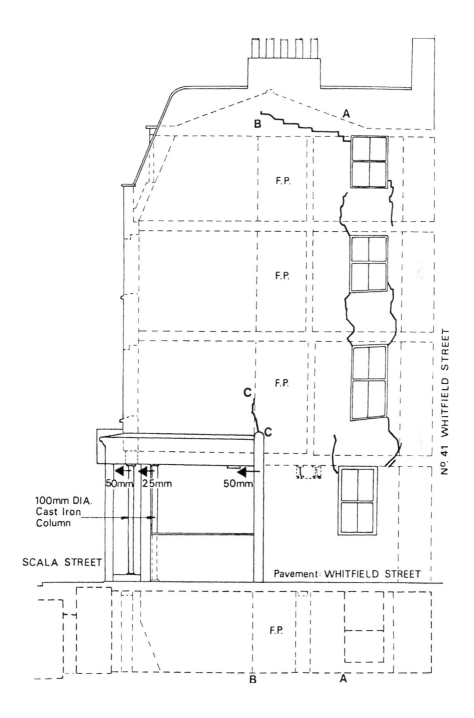

FIG. 5. Simple crack analysis at Pollock's Toy Museum, London.

FIG. 6. Structural deformation at Eagle House Annexe, Clapham Park.

readings (see Fig. 5).

HISTORY OF STRESSES AND STABILITY

The deformation of a building provides essential data for determining the stability and stresses in a building. This can somewhat be checked and supplemented by calculations on a 'model' of the original building and then on the building in its existing form. This helps to ascertain the safety of the building against overall collapse (see Fig. 6).

CONCLUSION

It is this methodical approach that is to be adopted during the following three month study. It is hoped that this will result in a thorough and detailed survey of an extremely interesting group of structures that exist in one of the most physically harsh environments in the world. In particular, the outcome will produce an appreciation of building techniques developed by empirical knowledge to resist catastrophic events such as earthquakes. Much of the information will be collated for use by other members of the research team. It is also predicted that this information will benefit other societies where such sophisticated technology is not yet available. It is also hoped that the research will provide a corpus of data relevant to the development of a conservation programme (11).

REFERENCES

1) FAULKNER, P. A. A Philosophy for the Preservation of our Historical Heritage. Three Bossam Lectures. Royal Society of Arts, Journal CXXVI, July 1978, pp 452 - 480.

2) DAVEY, N. A History of Building Materials. Phoenix House, London 1961.

3) WILLIAMS-ELLIS, C. Building in Cob, Plate and Stabilised Earth. Country Life Ltd., 1919.

4) BRUNSKILL, R. W. Vernacular Architecture. Faber & Faber, 1971.

5) SINN, B. H. (Editor). The Role of Concrete in Conservation of Historic Buildings. CEMBUREAU, Paris 1976.

6) DALLAS, R. Surveying with a Camera - Photogrammetry. Architects Journal, 30th January, 1980, pp 249 - 252.

7) WARD, W. H. (1947). The effects of fast growing trees and shrubs on shallow foundations. Journal Inst. Landscape Architects, 11: pp 7 - 16.

346

8) HOLMSTROM, I. and SANDSTROM, C.   Maintenance of Old Buildings. National Swedish Building Research Bulletin, B10, 1972.

9) HUTCHINSON, B. D. and BARTON, J.   Maintenance and Repair of Buildings.   Newness & Butterworth, 1975.

10) BECKMANN, P.   Measurement of Crack Movements.   Ove Arup and Partners, News Letter, 115, July/August 1979.

11) LEWCOCK, R.   Working Paper presented at the Meeting on the Conservation of the Architectural Heritage of the Islamic Period. Lahore, Pakistan.   April 1980.

Geomorphology

Sedimentological analysis of glacial and proglacial debris: a framework for the study of Karakoram glaciers

E. Derbyshire

Dept. of Geography, University of Keele

ABSTRACT

Sediments of mixed grain sizes exhibiting multimodal particle size curves (diamictons) are widely distributed. Processes producing them include glacial deposition, periglacial solifluction, debris slide and flow, mudflow, lahar flow (volcanic), and lacustrine and marine density currents (turbidites). Glaciers and ice sheets are a major source, deposits of this origin, mostly of Pleistocene age, being found over about one third of the land area of the earth.

Diamictons of glacial, periglacial and mud and debris flow origin are abundant in glaciated terrain, and they have considerable significance in the study of both pure and applied geomorphology.

This paper presents a framework for the study of Karakoram glacial and proglacial debris based on knowledge of the processes involved and the properties of sediments.

FACTORS INFLUENCING SEDIMENT PROPERTIES

The sedimentological and geotechnical properties of glacial sediments are dictated by a wide range of factors, some of the most important being the lithology of the glacier bed, the activity and regime of the glacier, the dominant mode of glacier movement, the type, incidence and rate of degradation of subaerial slopes above the glacier, the ice debris facies and debris flowpaths, the processes acting on sediments along the flowpath, the mode of primary deposition and the type, incidence and rate of secondary processing of the deposited material.

Variability in the lithology of the glacier sole is azonal but not random. Glacial sediment properties respond within 1 km to changes in lithology of the glacier bed and, on a regional scale, this is exemplified by the tills left behind by the Pleistocene ice sheets of Great Britain (1) and (2). Rocks of geosynclinal facies which have been stressed by severe folding, faulting and brecciation, together with some igneous and medium to high grade metamorphics, are characteristic of the Tertiary Alpine orogenic belts, parts of which are undergoing the greatest rates of current uplift.   These include the Karakoram and Himalaya, as well as the American cordilleras and the New Zealand Alps, for example.

Glacier regime and activity (throughput) are substantially controlled by prevailing climate.   Montane glaciers in temperate climates with high snowfall input and high losses arising from warm summer temperatures tend to have high activity indices (3).   This enhances their potential for erosion (especially abrasion) and transportation of debris.   In areas of active uplift, such high activity glaciers generate steep-sided glacier valleys and may carry abundant supraglacial debris often leading to large terminal zones of stagnation.   The third factor (subaerial degradation) plays a role in this and is influenced by both climate and tectonism; e.g. see (4, 5 and 6).

The relative importance of subglacial and supraglacial sediment input in alpine glaciers is influenced by glacier regimen and dominant flow mode and in turn, by the relative size of the ablation and accumulation zones, a further product of climate.   With input from the bed, sediment flowpaths will be in the traction zone of maximum comminution.   Sediment incorporated and inter-bedded with snow above the equilibrium line will suffer relatively little comminution in the englacial transport path, and supraglacial debris will again be passively transported (7) but may suffer comminution by geli-faction and sorting by supraglacial rills and streams and by deflation of fines, see Fig. 1.

Variation in sediment properties arising from primary depositional mode has been much studied over the past decade, e.g. (7, 9, 10, 11, 12, 13, 14, 15, 16, 17, 18, 19, 20 and 21).   The dominant depositional mode is greatly influenced by glaciological regime which, in turn, is dictated by climate.   Tills deposited by the primary processes of lodgement and meltout and by the secondary processes of sliding and flow have distinctive sedimentological and geotechnical properties, including particle size (17) particle shape (7) and total sediment fabric (macro-, meso- and micro-fabric) (18, 21 and 22).   It has been tentatively suggested (23) that the mesofabric (pebble fabric) of till laid down by surging glaciers differs from the established relationship for non-surge deposits, but this requires further testing in glaciers over a wide climatic range, including the Karakoram.

Tills of different origin vary in their moisture content, index pro-perties, bulk density, void ratio, coefficient of consolidation and compressive and shear strengths (e.g. 24, 25, 26, 27 and 28).   Some general relation-ships between genetic mode and sedimentological and geotechnical properties of mid-latitude tills are shown in Fig. 2.   Some tills, including those from valley glaciers in climates with warm summers, behave as collapsing soils and some have a marked strength anisotropy (22).

FACIES DISCRIMINATION

Diamictons of different origin (alluvial fan, glacial outwash, mudflow, till, glacimarine sediment) have been successfully discriminated using selected particle size descriptors.   Passega (29) used plots of the coarsest one - percentile against the median grain size to discriminate sediments transported by tractive currents from those produced by turbidity currents. Buller and McManus (30) differentiated Pleistocene tills into three grain size populations using plots of quartile deviations and median grain size. Landim and Frakes (31) used several graphic measures including standard

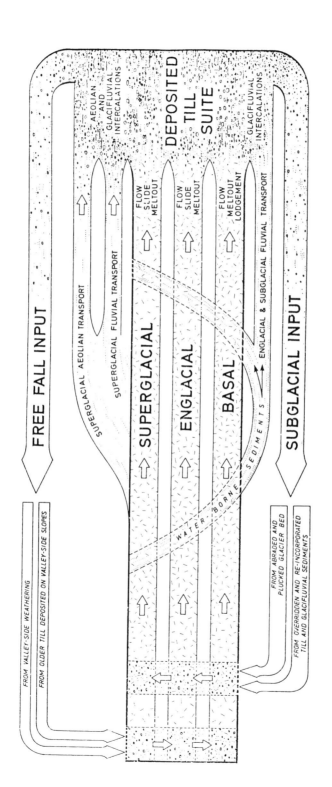

FIG. 1.   THE GLACIAL SEDIMENT SYSTEM (8).

TILL PROCESS CHARACTERISTICS			DOMINANT SOIL FRACTION	FABRIC FEATURES	RELATIVE DENSITY	OVERCONSOLIDATION RATIO	PERMEABILITY
FORMATION	TRANSPORT	DEPOSITION					
	SUPERGLACIAL	FLOW	G	MACRO: Interlayering with glacifluvials common. Segregation, contortions, layering and fissuring in upper section and nose of flow.	3		7
			W	MESO: Aligned low angle orientation of clasts conforming to flow direction rather than ice direction.	4	1 – 2	4
			Mg	MICRO: Rather compact parallel arrangement of fines related to flow rather than ice flow direction	5		6
			Mc		5		3
COMMINUTION	ENGLACIAL	MELTOUT	G	MACRO: Occasional interlayering with glacifluvials. Preferred orientation of clasts often retained from englacial state.	2 – 4		7 – 8
			W	MESO: Moderate to high degree of preservation of preferred orientation of clasts from englacial state, especially in subglacial type.	2 – 6	1 – 2	4 – 5
			Mg	MICRO: Very open to moderately closed arrangements with many englacial arrangements retained, especially in subglacial type.	2 – 6		5 – 7
			Mc		2 – 7		3 – 4
	BASAL	LODGEMENT	G	MACRO: Interlayering of glacifluvials, joints, fissures, contortions. Consistent preferred clast orientation.	4 – 7		5 – 6
			W	MESO: Fissuring. Contortion. Moderate to very high consistency of preferred orientation of clasts.	5 – 8	2 – 5	2 – 3
			Mg	MICRO: Moderate to high degree of parallelism of fines in sympathy with clast surfaces.	6 – 8		4 – 5
			Mc		6 – 8		2
	WATERLAIN		G	Not documented			
			W	Range probably includes cardinal features of lodgement, meltout and flow tills.		← NOT KNOWN →	
			Mg				
			Mc				

Qualitative Scale 1 -10

FIG. 2. SOME RELATIONSHIPS BETWEEN GLACIAL SEDIMENT PROCESSING, FABRIC AND SELECTED GEOTECHNICAL PROPERTIES (22).

deviation (as a measure of degree of sorting), mean, skewness and kurtosis in discriminating tills, outwash, alluvial fans and mudflows. Frakes and Crowel (32) segregated terrestrial tills, and two types of glacially derived diamictons by means of standard deviation/mean grain size plots.

Environmental discrimination within the very variable glacial facies is relatively poorly developed. Mills (33) examined the textural properties of several representative North American glaciers and, on the basis of progressive increase in both mean particle size and degree of sorting, discriminated basal, recessional moraine, end moraine, ablation and stratified drift sub-facies. The percentage of silt plus clay was found to decrease in this series, a characteristic now widely recognised in the differentiation of basal tills and those of supraglacial provenance (Fig. 3). Apart from

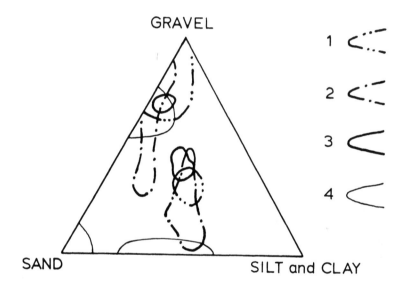

Key to ornament: 1 = North American Cordilleran glaciers (33);
                      2 = Matanuska Glacier, Alaska (20);
                      3 = Breidamerkurjokull, S. E. Iceland (19);
                      4 = Hooker Glacier, New Zealand.

FIG. 3. TERNARY DIAGRAM SHOWING PERCENTAGE GRAVEL, SAND, SILT AND CLAY IN SUPRAGLACIAL (PERIPHERAL DISTRIBUTION) AND BASAL DEBRIS (CENTRAL DISTRIBUTION) FROM SELECTED MODERN GLACIERS.

basal tills which Mills found disposed symmetrically about the zero skew line, the other sub-facies were all positively skewed, but no correlation was evident between kurtosis and depositional mode. No correlation was found between the selected statistical parameters and bedrock lithology, thus throwing some doubt on Slatt's (34) positive correlations, although Mills recognised a positive correlation in the fine till fractions (clay plus silt). Boulton (7) was critical of Mills' approach which discriminated between grain size characteristics produced by secondary (depositional) rather than primary processes. He emphasized the constrasting characteristics of debris processed in transit on and within the glacier (supraglacial and englacial) on the one hand, and debris which is processed in the basal zone (tractional debris on the other). Different flow paths produce sediment populations with different statistical groupings on standard deviation/mean grain size and standard deviation/skewness plots. Some limited data recently published e.g. (35, 36 and 37), plot clearly within one or both of these two flow path zones.

The limited data subjected to this type of analysis have come from two main sources: mid-latitude ice-sheet tills of Pleistocene age and deposits of isothermal valley glaciers. Little attention appears to have been paid to valley glaciers in climates with hot summers yet, in view of the fundamental influence of climate on glacier activity and sediment entrainment, transport and deposition, such environments (ranging from warm temperate to montane equatorial) merit investigation with respect to the range of their sub-facies.

The glaciers on the eastern side of the Southern Alps of New Zealand lend some support to this view. Many display large stagnant zones with widespread supraglacial debris. Till is being deposited by lodgement, but meltout is dominant and deposition by supraglacial sliding is also important. Deposition of flow till is rare. All the tills of the lower Hooker Glacier, for example, (38) including lodgement and meltout types, are deficient in fines (Fig. 4) as a result of general mechanical eluviation by percolating

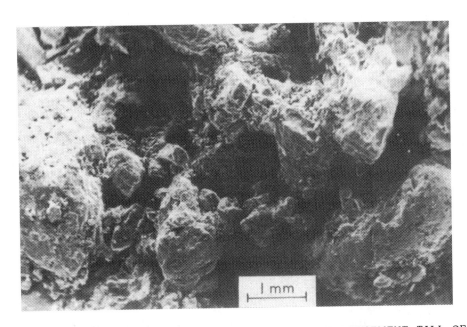

FIG. 4.   SCANNING ELECTRON PHOTOMICROGRAPH OF LODGEMENT TILL OF THE HOOKER GLACIER, NEW ZEALAND.

Till is impoverished in clay and silt and has a (metastable) silt matrix as a result of post-depositional mechanical eluviation. Clast fabric is unaffected.

meltwaters and also by the wind during deposition by sliding (cf 37, 38 and 40). Selected grain size parameters of the tills of this glacier are shown as envelopes on Figs. 5, 6 and 7, together with envelopes for a wide range of glacial sediments, modern and Pleistocene.

In Fig. 5 it can be seen that Pleistocene tills from North America, Britain (above and below present sea level) and New Zealand are distinctly more poorly sorted than the debris of modern glaciers, regardless of deposit-ional type or flow path. Modern sediments tend to be positively skewed while the Pleistocene tills range only from −0.25 to +0.32 with the mode close to zero skew. The Pleistocene tills also tend to have a much finer mean grain size (Fig. 6). Supraglacial debris from the New Zealand glacier plots as two major zones on Fig. 6: a fine, moderately sorted group trans-ported by supraglacial streams and a coarse, very poorly sorted group of supraglacial debris with the general appearance in the field of a granular till. Between these two groups lie poorly sorted and very poorly sorted zones representing recently deposited kame terrace sediments which are clearly transitional in granulometry. It will be noted that the supra-glacial debris from the New Zealand glacier overlaps the 'high level debris' of Boulton (7) and the 'ablation drift' of the Cordilleran glaciers studied by Mills (33). The distributional range is very much wider, however, and clearly reflects the sub-facies recognised on the glacial surface during field work. The relationship between the Alpine, Cordilleran and New Zealand 'high level debris' sediments is equally distinctive in Fig. 5, the latter lying peripheral to the other two envelopes. The range of kurtosis is not facies-specific on an inter-glacier basis (Fig. 7), but it may be specific when plotted against mean grain size for one particular glacier.

Several possible explanations might be advanced to explain these differences, including differences in the density and frequency of sampling. However, experience of European valley and plateau outlet glaciers suggests that the differences between the data sets are not merely a product of sampling. The supraglacial debris in the high temperature summer con-ditions on the eastern side of the Southern Alps suffers a great deal of pro-cessing during transit and lowering by the glacier. Fluvial, meltout and slide facies are clearly to be seen on the glacier surface. Photographs of Himalayan and Karakoram glaciers suggest certain similarities: they will provide a test of the hypothesis suggested here.

One of the criteria used by Boulton and Eyles (19) for distinguishing basal and supraglacial sediment sub-facies is a lack of preferred particle organisation. On several glaciers studied in New Zealand, deposition of till by a combination of slide and flow of supraglacial debris produces a fabric anisotropy as great as those found in many tills derived from the basal flowpath (Figs. 8 and 9). Post deposition mechanical eluviation of fines does not necessarily produce significant alteration of the clast (pebble) fabric (cf 37 and 39).

Well marked clast anisotropy (high anisotropy number: Fig. 8) with clasts lying parallel to the distal slopes of lateral moraines has been attributed to sliding of supraglacial debris from the flanks of the Athabasca Glacier (41) and the Bethartoli Glacier, Garwhal Himalaya (42).

APPLICATION TO THE KARAKORAM: A FRAMEWORK

Field description, mapping and sampling of sediments upon, within and adjacent to the Karakoram glaciers is necessary in order to establish

354

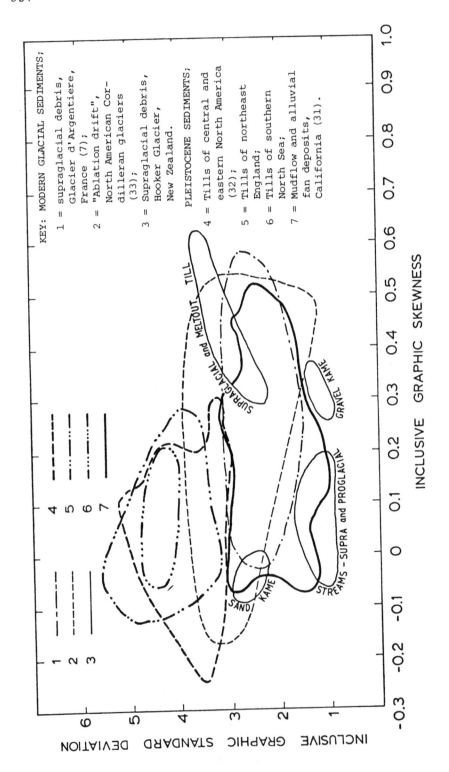

FIG. 5. INCLUSIVE GRAPHIC STANDARD DEVIATION AND INCLUSIVE GRAPHIC SKEWNESS, ALL IN PHI UNITS, FOR SELECTED PLEISTOCENE SEDIMENTS AND DEBRIS FROM SOME MODERN GLACIERS.

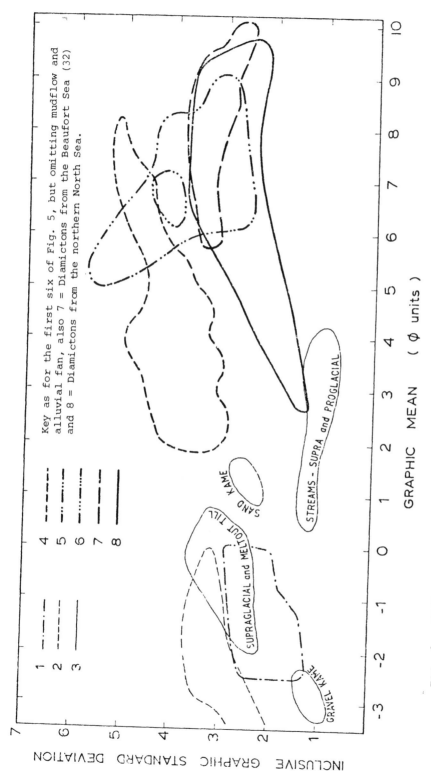

FIG. 6. INCLUSIVE GRAPHIC STANDARD DEVIATION AND GRAPHIC MEAN, IN PHI UNITS, FOR SELECTED PLEISTOCENE SEDIMENTS AND DEBRIS FROM SOME MODERN GLACIERS.

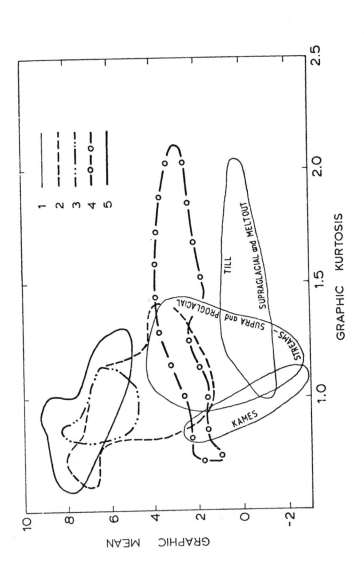

FIG. 7. GRAPHIC MEAN AND GRAPHIC KURTOSIS, IN PHI UNITS, FOR SELECTED PLEISTOCENE SEDIMENTS AND DEBRIS FROM SOME MODERN GLACIERS.

Key: MODERN GLACIAL SEDIMENTS; 1 = Hooker Glacier, New Zealand:

PLEISTOCENE SEDIMENTS;

2 = Tills of central and eastern North America (32);
3 = Tills of northeast England;
4 = Outwash, New York State (31);
5 = Diamictons from northern North Sea.

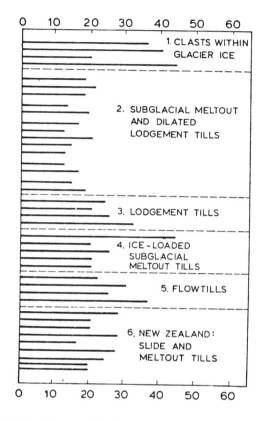

FIG. 8. ANISOTROPY NUMBER (PERCENTAGE OF CLASTS WITH SHORT AXES LYING WITHIN ANY ONE PER-CENT OF AN EQUAL AREA HEMI-SPHERICAL PLOT) FOR PARTICLES GREATER THAN 2 MM.

Categories 1 - 5 based on samples from Iceland, Norway, Antarctica and England.

genetically critical properties of different types of diamicton and the relative importance of each.

This should involve at least three main components as follows:

(i)   the provenance, transport history, and glacial facies contribut-ing to deposition in the valleys;

(ii)   variations in the sedimentological and geotechnical properties of diamictons arising from different modes of primary deposition; and

(iii)   variations in sedimentological and geotechnical properties arising from natural re-moulding, re-sedimentation and in situ post-depositional changes.

Ideally, work of this nature should be integrated with studies of glacier fluctuations, and with studies of processes at high altitudes so that provenance and environment of glacial debris supply and sediment flowpaths and in transit processes can be determined.

The study of systematic variability in the behavioural properties of glacial diamictons and the application of this knowledge in civil engineer-ing is a relatively recent development and is still gaining momentum. Most work has been concerned with the deposits left behind by the Pleistocene

358

FIG. 9. CONTOURED EQUAL AREA GRATICULE
PLOT OF SHORT AXIS AZIMUTH AND DIP OF CLASTS
IN A RECENT LATERAL MORAINE OF THE HOOKER
GLACIER, NEW ZEALAND.

Anisotropy number is 30. Clast dip mode is
diametrically opposed to the surface ('ice
contact') moraine slope, indicating accumulation by sliding of debris off the lateral slope
of the glacier.

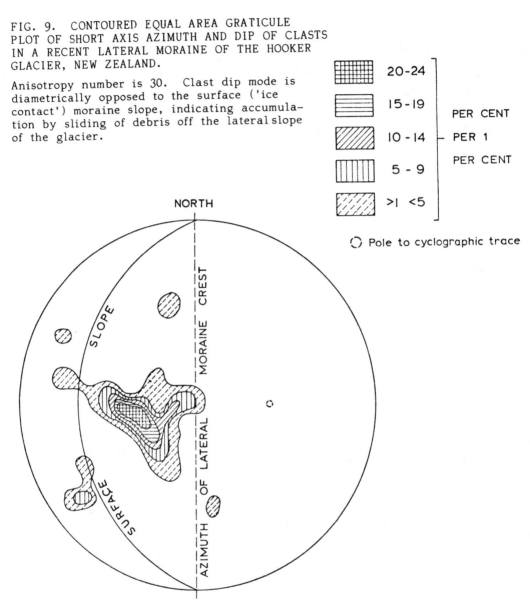

continental ice sheets in North America and N. W. Europe, and standard
methods in site investigations are only now beginning to emerge (e.g. 21,
28 and 43).

Subtropical glaciers, including those of the Karakoram seem likely to
provide valuable data on systematic variations in the properties of high-
activity ice masses on a scale not yet studied. The findings should be
capable of application and development in other glaciated montane areas of
the third world including the Himalaya and the Andes.

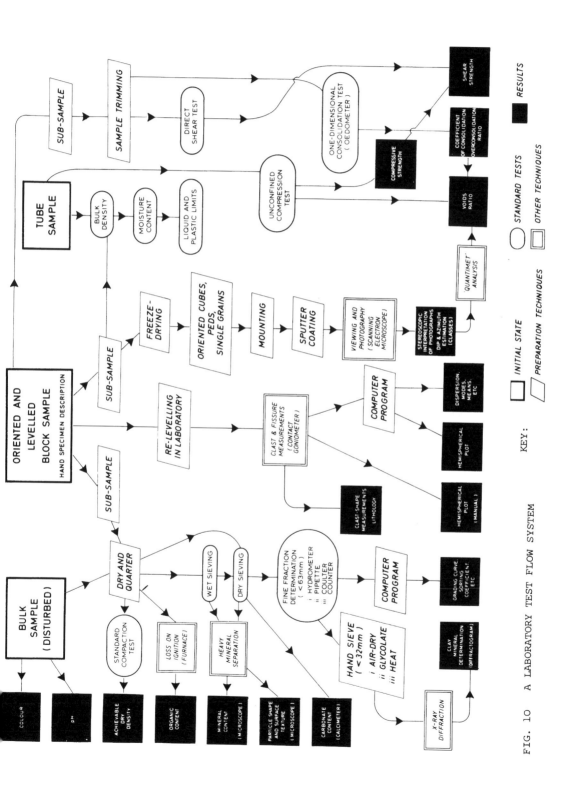

FIG. 10 A LABORATORY TEST FLOW SYSTEM

REFERENCES

(1) DERBYSHIRE, E., (1975). 'Distribution of glacial soils in Great Britain. In Engineering Behaviour of Glacial Materials, Midland Soil Mechanics and Foundation Engineering Society, pp 6 - 17.

(2) DREIMANIS, A. and VAGNERS, U., (1972). 'The effect of lithology upon texture of till'. In Research Methods in Pleistocene Geology (ed. E Yatsu and A. Falconer), pp 66 - 82.

(3) ANDREWS, J., (1972). 'Glacier power, mass balances, velocities and erosional potential. Z. Geomorph., 13, pp 1 - 17.

(4) POST, A., (1967). ;'Effects of the March 1964 Alaska earthquake on glaciers'. U. S. Geol. Surv. Prof. Pap., 544D, 42pp.

(5) POST, A., (1968). 'Effects on Glaciers'. The Great Alaska Earthquake of 1964. Hydrology. Pub. 1603, Nat. Acad. of Sciences, Washington, D.C., pp 266 - 308.

(6) FIELD, W. O., (1968). 'The effect of previous earthquakes on glaciers'. The Great Alaska Earthquake of 1964. Hydrology. Pub. 1603, Nat. Acad. of Sciences, Washington, D.C., pp 252 - 265.

(7) BOULTON, G. S., (1978). 'Boulder shapes and grain-size distributions of debris as indicators of transport paths through a glacier and till genesis'. Sedimentology, 25, pp 773 - 799.

(8) DERBYSHIRE, E., GREGORY, K. J. and HAILS, J. R., (1979). Geomorphological Processes, Dawson - Westview, 312 pp.

(9) HEWITT, K., (1967). 'Ice-front deposition and the seasonal effect: a Himalayan example'. Inst. Brit. Geogr. Trans., 42, pp 93 - 106.

(10) BOULTON, G. S., (1968). 'Flow tills and related deposits on some Vestspitsbergen glaciers'. J. Glaciol., 7, pp 391 - 412.

(11) BOULTON, G. S., (1970). 'On the deposition of subglacial and meltout tills at the margins of certain Svalbard glaciers'. J. Glaciol., 9, pp 231 - 245.

(12) BOULTON, G. S., (1976). 'The development of geotechnical properties in glacial tills'. In Glacial Till (ed. R. F. Leggett), Roy. Soc. Canada, Spec. Publ., No. 12, pp 292 - 303.

(13) MICKELSON, D. M., (1973). 'Nature and rate of basal till deposition in a stagnating ice mass, Burroughs Glacier, Alaska'. Arctic and Alpine Research, 5, pp 17 - 27.

(14) REHEIS, M. H., (1975).   Source, transportation and deposition of debris on Arapaho Glacier, Front Range, Colarado, U. S. A.'. J. Glaciol., 14, pp 407 – 420.

(15) LINDSAY, J. F., (1970).   'Clast fabric of till and its development'. J. Sed. Petrology., 40, pp 629 – 641.

(16) DRAKE, L., (1971).   'Evidence for ablation and basal till in east central New Hampshire'. In Till: a symposium (ed. R. P. Goldthwait), pp 73 – 91.

(17) McGOWN, A., (1971).   'The classification for engineering purposes of tills from moraines and associated landforms'. Q. J. Engng. Geol., 4, pp 115 – 130.

(18) DERBYSHIRE, E., McGOWN, A. and RADWAN, A., (1976).   '"Total" fabric of some till landforms'. Earth Surface Processes, 1, pp 17–26.

(19) BOULTON, G. S. and EYLES, N., (1979).   'Sedimentation by valley glaciers; a model and genetic classification'. In Moraines and Varves, (ed. Ch. Schluchter), pp 11 – 23.

(20) LAWSON, E. E., (1979).   'Sedimentological analysis of the western terminus region of the Matanuska Glacier, Alaska'. U. S. Corps of Engineers, CRREL Report, 79 – 9, 122 pp.

(21) DERBYSHIRE, E., (1980).   'The relationship between depositional mode and fabric strength in tills: schema and test from two temperate glaciers'. Geografia (Poznan), 20, pp 41 – 48.

(22) McGOWN, A. and DERBYSHIRE, E., (1977).   'Genetic influences on the properties of tills'. Quart. J. Engng. Geol, 10, pp 389 – 410.

(23) RUTTER, N. W., (1969).   'Comparison of moraines formed by surging and normal glaciers'. Can. J. Earth Sciences, 6, pp 991 – 999.

(24) EASTERBROOK, D. J., (1964).   'Void ratios and bulk densities as means of identifying Pleistocene till'. Geol. Soc. America Bull., 75, pp 745 – 750.

(25) BOULTON, G. S. and PAUL, M. A., (1976).   'The influence of genetic processes on some geotechnical properties of glacial tills'. Quart. J. Engng. Geol., 9, pp 159 – 194.

(26) MILLIGAN, V., (1976).   'Geotechnical aspects of glacial tills'. In Glacial Till (ed. R. F. Leggett). pp 269 – 291.

(27) QUIGLEY, R. M. and OGUNBADEJO, T. A., (1976).   'Till geology, mineralogy and geotechnical behaviour, Sarnia, Ontario'. In Glacial Till (ed. R. F. Leggett), pp 336 – 354.

(28) McGOWN, A., MARSLAND, A. and DERBYSHIRE, E., (1980).   'Soil profile mapping in relation to site evaluation for foundations and earthworks'. Intnl. Assoc. Engng. Geol, 21, pp 139 – 155.

(29) PASSEGA, R., (1964).   'Grain size representation by CM patterns as a geological tool'. J. Sed. Petrology, 43, 168 – 187.

(30) BULLER, A. T. and McMANUS, J., (1973). 'The quartile-deviation/ median diameter relationships of glacial deposits'. Sed. Geology, 10, pp 135 - 146.

(31) LANDIM, P. B. and FRAKES, L. A., (1968). 'Distinction between tills and other diamictons based on textural characteristics'. J. Sed. Petrology, 38, pp 1213 - 1223.

(32) FRAKES, L. A. and CROWELL, J. C., (1975). In Gondwana Geology (ed. K. S. W. Campbell 374).

(33) MILLS, H. H., (1977). 'Textural characteristics of drift from some representative Cordilleran glaciers'. Geol. Soc. America Bull., 88, pp 1135 - 1143.

(34) SLATT, R. M., (1971). 'Texture of ice-cored deposits from ten Alaskan valley glaciers'. J. Sed. Petrol., 41, pp 828 - 834.

(35) GERMAN, R., MADER, M. and KILGER, B., (1979). 'Glacigenic and glaciofluvial sediments, typification and sediment parameters'. In Moraines and Varves (ed. Ch. Schluchter), pp 127 - 143.

(36) RABASSA, J. and ALIOTTA, G., (1979). 'Sedimentology of two super- imposed tills in the Bariloche Moraine (Nahuel Huapi Drift, Late Glacial) Rio Negro, Argentina'. In Moraines and Varves, (ed. Ch. Schluchter), 81 - 92.

(37) SCHUBERT, C., (1979). 'Glacial sediments in the Venezuelan Andes'. In Moraines and Varves (ed. Ch. Schluchter), pp 43 - 49.

(38) DERBYSHIRE, E. and BARRETT, P. J., (in litt). 'Recent deposits of the Hooker Glacier, South Island, New Zealand.

(39) SPALLETTI, L. A. and GUTIERREZ, R., (1976). 'Estudio granulo- metrico de sedimentos glaciales, fluviales y lacustres de la region del Monte de San Lorenzo, Provincia de Santa Cruz'. Assoc. Geol. Argent. Rev., 31, pp 95 - 117.

(40) NAKAWO, M., (1979). 'Supraglacial debris of G2 glacier in Hidden Valley, Mukut Himal, Nepal'. J. Glaciol, 22, pp 273 - 283.

(41) MILLS, H. H., (1977a). 'Differentiation of glacier environments by sediment characteristics: Athabasca Glacier, Alberta, Canada'. J. Sed. Petrol., 47, pp 728 - 737.

(42) OSBORN, G. D., (1978). 'Fabric and origin of lateral moraines, Bethartoli Glacier, Garwhal Himalaya, India'. J. Glaciol., 20, 547 - 553.

(43) DERBYSHIRE, E., (1979). 'The influence of modes of deposition on the fabric and other properties of glacial soils'. Proc. Symp. on Offshore Geotechnical Research, Nov. 1978, Building Research Establishment, Department of Energy Technology Reports Centre, No. QT-R-7937, pp 15 - 16.

(44) DERBYSHIRE, E., (1978). 'A pilot study of till microfabrics using the scanning electron microscope'. In Scanning Electron Microscopy in the Study of Sediments (ed. B. W. Whalley), pp 41 - 59.

APPENDIX

DATA COLLECTION: TECHNIQUES AND EQUIPMENT

Techniques used to discriminate diamictons of different origins include the following:

(i)   Observation and recording and detailed photography of measured sections of englacial and proglacial material. Many of the locations can be fixed on large scale maps produced under expedition conditions, but the construction of detailed local basemaps by using a lightweight plane table and alidade is a well established geomorphological field technique. Specific sites are recorded by stereoscopic field photography.

(ii)   Particle shape measurements provide the basic data in facies discrimination of supraglacial and proglacial debris, and sediments studied along selected sediment flow paths. Within- and between-site variation can be considerable. Standardised photography of pebble sets for later two-dimensional analysis is a technique which is becoming more widely used.

(iii)   Particle size analysis of large rock fragments is effected by axial measurement: finer grades are determined by dry sieving. In addition, wet-sieving and sedimentation methods are applied to determine clay and silt percentages.

(iv)   Macrofabric and mesofabric properties are determined with contact goniometers and, less precisely, with compass clino-meters.

(v)   Selected oriented and levelled samples must be collected (ideally using a portable freeze-drying unit in the case of clay rich sediments), for detailed analysis of mesofabric and microfabric variation using contact goniometer and scanning electron micro-scope (44).

(vi)   Sediment colour is recorded as a routine field procedure but pH is best done in the field laboratory, total carbonate content and clay mineralogy being determined in a sedimentological laboratory using sealed disturbed samples.

(vii)   Bulk density, natural moisture content, shear strength and unconfined compressive strength can be determined, in part at least, under field conditions.

Sealed samples are usually collected for subsequent laboratory determination of voids ratios, plasticity and liquidity indices, coefficients of consolidation and shear strengths of selected diamictons.

EQUIPMENT

The following equipment is required for investigations of this kind.

A - FIELD

Survey

1. Standard levelling and linear measuring equipment.
2. Lightweight plane table, alidade, level and 30 m tape.
3. A Four-inch compass and tripod.

Data Recording

1. Scaled photography: 35 mm cameras and tripod.
2. Particle measurement:

   (a) calipers
   (b) compass - clinometer
   (c) pocket penetrometer
   (d) selected shear vanes

Sediment Colour

Standard soil colour charts

Sediment Sampling

1. Spade and trowel
2. Bubble level and compass
3. Plaster of Paris
4. 1.1/2 inch sampling tubes (unconfined compressive strength)
5. 100 ml and 500 ml sampling tubes (bulk density)
6. Tube sampler and ram
7. Wax and saucepan
8. Portable 'gaz' stove
9. Moisture content tins
10. Portable freeze - dry apparatus

B - FIELD LABORATORY

1. Particle size analysis:

   (a) Set of single $\emptyset$ 200 m sieves
   (b) Electrical sieve-shaker

2. Balance (accurate to 0.01 g)
3. 2 kg beam balance
4. 1.1/2 inch tube sample extruder
5. Drying oven (small)
6. Drying oven (large, fan assisted)
7. Unconfined compression machine
8. Proctor compaction apparatus
9. Contact goniometer
10. Low power binocular microscope

Several other techniques used to determine sedimentological and geotechnical properties of diamictons require a fully equipped laboratory. One example of a laboratory flow system in current use is shown in Fig. 10.

High altitude rock weathering processes

W.B. Whalley
Department of Geography
The Queen's University, Belfast

ABSTRACT

High altitude rock weathering is poorly understood both in terms of processes involved and rates of activity. As well as freeze–thaw activity which produces small blocks, larger scale rockfalls appear to be common. In addition, chemical processes are probably more important at high altitude than has been generally realized. This chemical weathering can take place in either cracks or on the surface to give sand and silt–sized particles (grus). Rock surface rinds, as a form of desert varnish, appear to form where special conditions operate, although these environmental parameters are not yet elucidated. Other places show grain by grain disintegration due to weathering in micro–cracks. The varnish rinds are of complex chemical composition and show preferential enrichment of Mn and Fe over the parent rock. A variety of techniques have been used to investigate these crusts. Examples are shown of scanning electron micrographs and Mössbauer spectroscopy which suggest formation of haematite and a gel form of $Fe(OH)_3$. The form of Mn concentration is not yet known but appears to be in small nodules (1 – 2μm diameter) on the varnish surface.

INTRODUCTION

The high mountains of the world have rarely been investigated from the point of view of weathering processes actually taking place in such regions. With denudation rates, for example, inferences are usually made from river sampling well outside the mountains and which incorporate weathering products from many processes. This paper is primarily concerned with the examination of some micro–scale rock weathering processes which contribute to the denudation of a high mountain area. Work to date has been primarily from the Alps and the Hindu Kush of Afghanistan, but the Karakoram should provide comparable conditions. The work of Hewitt (1) has indicated the importance of frost and chemical activity in this region. However, virtually no detailed investigations have been made at high altitude in such a dry area. The studies reported here are a first step into some specific investigations of weathering processes.

Fig. 1 Mountains of the Hindu Kush showing typical gullied geometry of rock faces and glaciers heavily charged with debris (G). Small block breakdown can be seen front and left and desert varnish and rocks rounded by chemical weathering on the right (V).

Fig. 2 Granular disintegration of a granite rock surface with some desert varnish centre bottom.

THE SCALE OF WEATHERING PRODUCTS

Rock weathering at high altitude appears to take two basic forms. One gives blocky products a few cm in size up to several metres, and the other is the production of finer material of sand or silt size (often known as grus). It is commonly thought that all weathering processes which produce these sizes of detritus in mountains are essentially physical processes of freeze-thaw activity. Some authors (2) have suggested that chemical weathering may play an important part however, and the complexity of the process referred to as "freeze-thaw" has been reviewed by McGreevy (3). It is likely that a complex of factors is in operation and that physico-chemical agencies may be responsible for production of the complete size range of weathering products mentioned above. Some, essentially chemical, factors will be discussed separately below.

The problems associated with using the mechanical action of freezing water as an explanation of the means whereby large blocks can be split off mountains have been discussed by Whalley (4). By using some of the ideas of Gerber and Scheidegger (5) he also (6) has suggested the importance of endogenic and long term climatic factors. The endogenic agencies are those of creep induced by mountain geometry (e.g. Fig. 1) and stress relief. The rather characteristic buttress and gully shape of mountains has been attributed to this (5) but the importance of melt-water at depth in a cliff (7) is frequently a means of promoting a cliff fall from tens to millions of metres cubed (4). Such processes, although they may be infrequent, are particularly important in mountains, especially if the debris falls onto glaciers which can expedite their removal (Fig. 1). It is also possible that blocks a few tens of cm can also be produced in this way as well as by "freeze-thaw" processes. The crack-forming processes can, however, be particularly prone to chemical action at their tip by stress-corrosion processes. Such ideas to help explain rock weathering are still in their infancy (8).

The small size fraction is most probably produced by weathering of certain minerals, feldspars for example, perhaps aided by crack-forming processes. Figure 2 shows the grain-by-grain removal of a granite surface at high altitude. Even though winter and night-time temperatures may be very low, the day rock temperatures can easily reach 40°C. With water supplied from melting snow it is likely that considerable chemical activity can occur on warm rock surfaces and Reuslatten and Jørgensen (9) have shown changes in cation concentrations across bedrock even at temperatures just above zero. Such reactions deserve more detailed study as do more careful measurements of rock temperature changes in relation to meteorological conditions.

HIGH ALTITUDE DESERT VARNISH

One feature of rock surfaces even up to 6000m in the Hindu Kush was found to be a surface crust or rind similar to desert varnish (Fig. 3). This is a thin (<10 μm) coat of much darker material on top of unweathered rock. Under some circumstances this had flaked off and revealed a new surface undergoing alteration (Fig. 4). Some relative dating on surfaces produced at different times may be possible if note is taken of the degree of surface colouration.

Some recent studies in southwestern USA (10) have suggested that desert varnish there is produced by the "plastering-on" of previously weathered day minerals. This appears to be an unlikely origin at high altitude

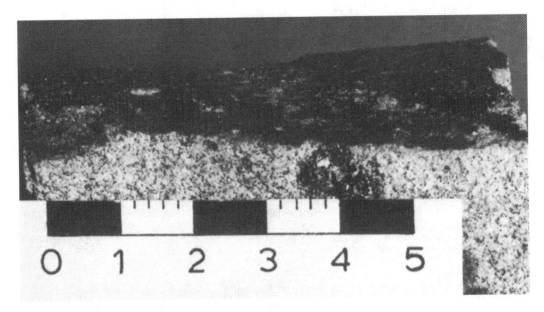

Fig 3   Desert varnish top
and fresh surface
below.   Scale in cm

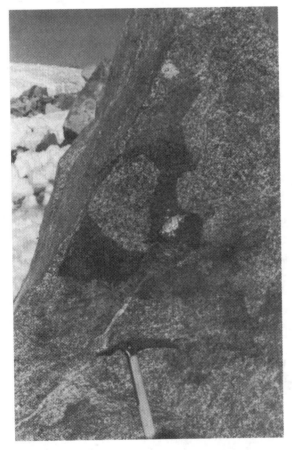

Fig. 4   Desert varnish next
to a glacier at 4000m
showing a flaked off
removal.

because a source area of suitable clay minerals is difficult to envisage and different sides of rocks are uniformly covered. Rather, an in situ (autochthonous) origin is favoured similar to that proposed by Hooke et al (11) in which weathering products from the minerals are concentrated at the surface. Several techniques have been used to investigate these rinds including x-ray diffraction, thermo-analytical methods and infra-red spectroscopy. Some recent findings on the character of the varnish from scanning electron microscope (SEM) and Mössbauer spectroscopy (MS) will be reported here.

SCANNING ELECTRON MICROSCOPE EXAMINATION

To facilitate detailed examination of the thin crusts, a Cambridge Instruments Stereoscan 180 was used to examine them in the secondary-emissive mode at 10-20 kV. The specimens were glued to a mounting stub and sputter coated with a conducting gold coating; carbon coating was used for EDS analysis.

The extreme thinness of the crust can be seen in Fig. 5 and the way it may sometimes flake off from the barely altered rock below. This does not necessarily indicate an allochthonous origin such as Potter and Rossman (10) suggest because the weathering products are formed at the surface, probably in solution, and are in effect precipitated there. This differs from the ideas of Hooke et al (11) as it is not a diffusion-controlled process within the rock but is rather a translocation of material.

Figure 6 shows two important features of the coat which are common. One is the wrinkled appearance which is suggestive of the contraction of a gel upon dehydration. This gel probably consists of complexed ions of Al, Fe, Mn, and Si in the main. The more soluble cations Mg, Ca, K and Na are less likely. This idea is supported by the findings of area scans of surfaces with an energy-dispersive x-ray analyser (EDS) fitted to the SEM (Table 1).

Table 1 Energy Dispersive Spectrometer semi-quantitative analyses of major cations for various areas of varnish coat.

Zone		Area μm	Si	Al	Ca	Mg	K	Na	Fe	Mn
Surface	1	50x50	60	15	4	1	3	1	10	3
	2	10x10	45	14	0	0	4	5	15	8
	3	10x10	63	21	0	0	0	0	5	2
'Nodules'	1		48	38	0	0	0	0	3	0
	2		50	25	5	2	0	0	12	12
	3		80	3	0	2	0	0	4	0
	4		48	21	0	4	0	0	15	10

370

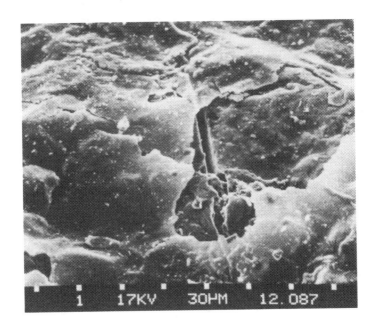

Fig. 5 Scanning electron micrograph showing thin layer
of varnish coat. Scale between bars is 30 μm.

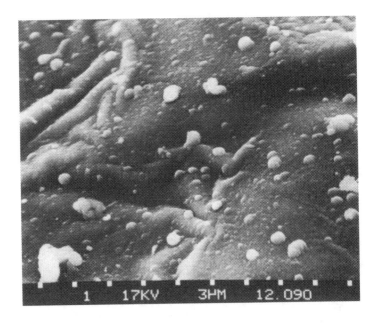

Fig. 6 Close up of wrinkled varnish surface with small
nodules. Scale between bars is 3 μm

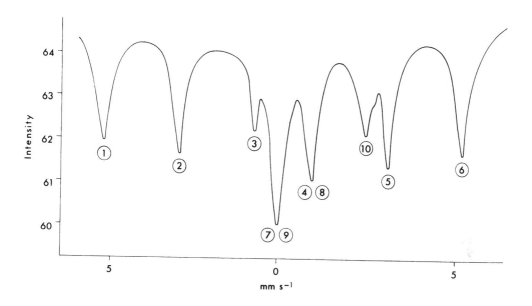

Fig. 7   300K Mössbauer spectrum of surface coat.
The numbers are referred to in the text.

Peaks	Type	δ mm s^{-1}	ΔE mm s^{-1}	Possible Interpretation
1,2,3,4,5,6	Ferromagnetic			haematite
7,8	High Spin Fe^{III}	0.34	1.04	Fe(OH)$_3$ gel
9,10	High Spin Fe^{II}	1.14	2.61	weathered biotite

Table 2   Peaks and parameters for Mössbauer Spectrum

Also seen in Fig. 6 are small nodules, usually 1-2 μm in diameter. These appear to be on top of the skin or coat but are assumed to be part of the overall varnish structure. The composition of the coat shows concentrations of Mn and Fe and these nodules are mostly rich in these elements too (Table 1). The reason for the formation and composition of these nodules is now known.

Cross sections of fracture surfaces of the varnish often show a layered structure where the skin is more than a few micrometres thick. No distinct clay minerals have been seen to have formed and x-ray diffraction has given indeterminate results to date.

MÖSSBAUER SPECTROSCOPY

This technique (12) has been used to try to characterize the iron component in the varnish. Material for the specimen was obtained by carefully scraping some of the crust and concentrating this in the specimen holder. A typical spectrum is shown in Fig. 7. This was obtained at 300K; low temperature spectra were much less distinct.

The interpretation is still not clear but tentatively the peaks can be grouped into three components with parameters δ (isomer shift) and ΔE (quadrupole splitting) shown together with possible interpretations in Table 2.

The origin of the ferromagnetic material which has the form of the iron spectrum may be a form of haematite. The gel structure, also found by Herzenberg and Toms (13) is also in line with the idea mentioned above that the weathering of the rock is complex and produces a gel which may contain Fe, Al, Mn and Si. Whalley (14) has shown that a rapidly drying solution containing $Si(OH)_4$ produces a gel-like precipitate which can show a contraction-wrinkled surface. Further work on the elucidation of the varnish structure is in progress.

DESERT VARNISH AND WEATHERING

The varnish formation outlined above presents some restriction to overall weathering as the material is barely moved from the rock where it is produced. However, it does show the importance of chemical as well as mechanical processes acting at high altitudes. Furthermore, the varnish coat does tend to splinter off (Fig. 4) and the process can be repeated. It is possible that granular disintegration as shown in Fig. 2 may be dependent upon the size and mineralogy of the rock concerned, or it may be controlled largely by environmental factors such as precipitation.

Clearly there is much yet to learn about high altitude rock weathering processes of all kinds, but the importance of chemical effects is undeniable.

ACKNOWLEDGEMENTS

I thank Mr. R. Reed of Queen's University Electron Microscopy Laboratory for his assistance and Dr. M. Nelson of the Department of Chemistry, Queen's University for running the Mössbauer spectra. I should also like to thank members and sponsors of the Cambridge University 1976 Hindu Kush Expedition for their considerable field assistance.

REFERENCES

1) HEWITT, K., (1969). 'Geomorphology of Mountain Regions of the Upper Indus Basin', Ph.D. Thesis, London University.

2) REYNOLDS, R. C. and JOHNSON, N. M., (1972). 'Chemical Weathering in the Temperate Glacial Environment of the Northern Cascade Mountains', Geochem. Cosmochim. Acta., 35, pp 537 – 554.

3) McGREEVY, J. P., (1982). 'Some field and laboratory investigations of rock weathering', Ph.D. Thesis, Queen's University, Belfast, p 486.

4) WHALLEY, W. B. (In Press). 'Rockfalls'. Chapter 8 in Mass Movement Processes on Slopes, (ed.) D. Brunsden and D. B. Prior. Wiley.

5) GERBER, E. and SCHEIDEGGER, A. E., (1969). 'Stress-induced weathering of rock masses', Ecl. Geol. Helv. 62, pp 401 – 416.

6) WHALLEY, W. B., (1974). 'The Mechanics of High Magnitude – Low Frequency Rock Failure and its Importance in Mountainous Areas', Geogr. Papers (Reading) No. 27, pp 1 – 48.

7) MULLER, L., (1964). 'Application of Rock Mechanics in the Design of Rock Slopes'. In: W. R. Judd (ed.) The State of Stress in the Earth's Crust. Elsevier, New York, pp 575 – 598.

8) WHALLEY, W. B., DOUGLAS, G. R. and McGREEVY, J. P., (1982). 'Crack propagation and associated weathering in igneous rocks', Zeit. Geomorph. 26, pp 33 – 54.

9) RUESLATTEN, H. G. and JØRGENSEN, P., (1978). 'Interaction between Bedrock and Precipitation at Temperatures close to $0^{\circ}C$', Nordic Hydrol. 9, pp 1 – 6.

10) POTTER, R. M. and ROSSMAN, G. R., (1977). 'Desert Varnish: The Importance of Clay Minerals', Science 196, pp 1446 – 1448.

11) HOOKE, R. LeB., YANG, H. and WEIBLEN, P. W., (1969). 'Desert Varnish: an Electron Probe Study', J. Geol. 77, pp 275 – 288.

12) BANCROFT, G. M., (1973). Mössbauer Spectroscopy. McGraw Hill, London.

13) HERZENBERG, C. L. and TOMS, D., (1966). 'Mössbauer Absorption Measurements in Iron-containing Minerals', J. Geophys. Res. 71, pp 2661 – 2677.

14) WHALLEY, W. B., (1978). 'Scanning Electron Microscopic Study of Laboratory-Simulated Silcrete'. In: W. B. Whalley (ed.) Scanning Electron Microscopy in the Study of Sediments. Geo. Abstracts, Norwich, pp 399 – 405.

Techniques for investigating meltwater runoff and erosion

R.I. Ferguson*
D.N. Collins**
W.B. Whalley***

*Stirling University, **Manchester University, ***Queen's University, Belfast

ABSTRACT

Meltwater runoff from glaciers and snow fields varies seasonally, with day to day hydrometeorological conditions, and diurnally. Suspended sediment transport is high in meltwater rivers but does not have a simple relationship with discharge, and solute concentrations depend on the englacial or subglacial path of the meltwater as well as on rock type. Quantitative understanding of meltwater runoff, solid load and hydro-chemistry in the Alps, Scandinavia, and North America has been improved by continuous monitoring of river conductivity and turbidity as well as stage, automatic water sampling at frequent intervals, and use of atomic absorption spectroscopy and ion-selective electrodes.

The sediment removed by meltwater runoff may come from beneath, above or downstream from the glacier. Sources of suspended and dissolved sediment can be inferred crudely from changes in turbidity and conductivity along meltwater routes. They can also be investigated at the microscale by laboratory analysis of rock samples and weathered debris from different altitudes using scanning electron microscopy, X-ray diffraction, microprobe analysis and other techniques already applied to rock debris from the Alps and Hindu Kush. Rockfall contributions can be assessed by time-lapse photography, and bedload transport in meltwater rivers may be predictable from hydraulic properties obtained during gauging by the velocity-area or dilution methods.

INTRODUCTION

Each year the Indus River carries some 500 million tonnes of solid and dissolved rock to the ocean (1). This huge load, 2 - 3% of the total erosion of the continents, is derived very largely from the Karakoram, Himalaya, and Hindu Kush mountains via major tributaries from the Kabul River in the northwest frontier to the Chenab in the Punjab. The highest recorded erosion rate of all is for the Hunza River tributary to the Indus in the western Karakoram, whose suspended sediment concentrations (2) indicate annual erosion of c. 5000 tonnes km^{-2} which is equivalent to an average lowering of the entire landscape by c. 2 m every 1000 years.

This exceptionally rapid erosion is almost entirely restricted to the summer months (June-September) when river discharge reaches 20 – 50 times, and silt concentration 500 – 1000 times, the winter level (2). In this northern extremity of Pakistan the source of summer runoff, and thus erosion, is not so much the monsoon rains as rapid melting of winter snow and glacier ice. The meltwater floods are one of Pakistan's major natural resources (irrigation water, hydro-electric power) but also pose serious problems (flood hazard, silting of reservoirs, washing away of roads and cultivated land).

Fuller understanding of the sources of meltwater runoff and its solid and dissolved load should be of value in managing Pakistan's water resources. Our work in the Karakoram, following on as it does from similar investigations in other glacierised mountain ranges, will we hope assist this understanding while also revealing scientifically important contrasts with the lower, less geomorphically active, mountains of higher altitudes. We summarise below some modern scientific techniques for the investigations of meltwater runoff and erosion, under four headings: runoff measurement, sediment load measurement, runoff sources and sediment sources.

RUNOFF MEASUREMENT

Rivers fed by meltwater are characterised by high peak discharges and cyclic variations in flow, especially seasonally but also diurnally, since melting rates of snow and ice depend on air temperature and other hydrometeorological conditions. Study of meltwater hydrology thus requires frequent monitoring of stream levels, or preferably continuous measurement using a float mechanism enclosed in a stilling well and linked to the pen of a clockwork chart recorder. Quasi-continuous recording can also be achieved by digital data logging of the signal from a float-driven rotary potentiometer or from a pressure transducer on the stream bed. Conversion of continuous records of water level to discharge requires the estimation of a discharge rating curve by measuring discharge at different water levels. At permanent gauging stations this is normally done by means of a cable-suspended current meter, and where road bridges are available this is the best technique for fairly large meltwater rivers. Small melt streams may be wadeable so that velocities can be measured by floats (3) or hand-held current meter, but in both intermediate-sized and small streams, including supraglacial ones (4), gauging is best done by the dilution method (5, 6) using either common salt (NaCl) or rhodamine-WT dye as the tracer, and a conductivity meter or fluorimeter respectively to measure the degree of dilution of either a constant-rate or sudden injection of tracer. The high turbulent diffusion in steep mountain streams (7) reduces the necessary mixing length but makes its correct choice critical if tracer concentration is not to drop below the detectable limit; and most Karakoram rivers have such high discharges that impossibly bulky and/or expensive quantities of tracer would be required. Gauging here is dependent on the availability of winch-supported current meters and of bridges from which to use them.

SEDIMENT LOAD MEASUREMENT

The mass transport rate of suspended or dissolved sediment in a river is found as the product of water discharge and sediment concentration. Since suspended sediment concentration tends to increase greatly with discharge the long-term solid load of a river can be far greater than the product of average discharge and average sediment concentration; and the opposite is true of dissolved load. In each case a more accurate picture can only be pieced together by intensive water sampling and analysis, especially during flood periods.

Collins (8, 9) has successfully used automatic vacuum samplers to investigate the diurnal variation in suspended load in Swiss meltwater rivers. These devices (Fig. 1) obtain twenty-four successive samples by suction, using a clockwork or battery-powered timer to release at a constant interval (e.g., hourly) the clips on tubes leading to successive sample bottles. Increasing the sampling interval allows the collection of less frequent samples over a longer total unattended period. Once all twenty-four samples have been collected they are pressure filtered through preweighed filter papers which are sealed in numbered plastic bags for subsequent weighing in the laboratory. The chief problem in small meltwater streams is the coarse nature of the suspended load, which can clog up the suction tubes and also leads to considerable random sampling variation since the loss or capture of individual large grains may have a major effect on total sediment mass in the sample.

Total dissolved sediment (TDS) concentrations can be determined from automatically-collected samples by evaporation and weighing, and concentrations of individual dissolved ions may be determined by atomic absorption spectroscopy. These techniques require laboratory facilities and therefore the storage and transport of water samples. To minimise alteration of ionic balance and solute loss by absorption onto suspended solids it is essential to filter samples in the field and store them under refrigeration. An alternative analytic technique that avoids this problem is the use of ion-selective electrodes and a suitably calibrated pH meter (10), but field results using battery-powered instruments are questionable because of systematic instrumental drift. A third possibility for field analysis is the range of titration and calorimetric methods described in standard handbooks (11).

A final and very useful field technique is to obtain a continuous chart record of the electrical conductivity of a meltwater stream, which is linearly proportional to TDS assuming constant temperature and ionic balance. Battery-powered conductivity meters and chart recorders (Fig. 1) have proved very reliable in Switzerland and Canada (9, 12) and as explained below the continuous record gives insight into runoff sources as well as sediment loads.

RUNOFF SOURCES

Streamflow from heavily glacierised catchments arises from various sources in proportions that change over time in absolute and relative importance. Extraglacial rainfall inputs can readily be assessed using a simple temporary rain-gauge network, but the extraglacial snowmelt contribution in the Karakoram will remain uncertain in the absence of year-round precipitation measurements and snowpack surveys.

Available climatological information suggests that the bulk of Karakoram runoff is derived from glacier ablation, but the streams emerging from glacier snouts may also contain water initially derived from rainfall, valley-side snowmelt, or groundwater springs. In typical temperate valley glaciers water from all these sources passes along an organised network of progressively bigger subglacial conduits (13). The nature of this

KEY: A. Automatic water sampler
      B. Crate of sampled bottles for A
      C. Meter for conductivity measurements
      D. Recorder for C
      E. Conductivity probe
FIG. 1. TYPICAL FIELD INSTALLATION FOR SEDIMENT AND
              SOLUTE MONITORING.

internal drainage system is thought to have a bearing on both short-term variations in glacier sliding (14) and long-term periodic surging (15), and meltwater floods may be related to subglacial drainage of ice-dammed lakes (13) or sudden shifts in the position of subglacial channels as in 1972 under the Batura glacier (16). Compared to the alpine glaciers hitherto investigated, the Karakoram glaciers have greater summer precipitation and avalanche nourishment and are much longer. Their hydrological behaviour may well be different in detail but the same techniques of investigation are thought appropriate.

The seasonal cycle of snow and ice melt depends primarily on gross solar radiation but weather variations (especially cloud cover or fresh snowfall) may substantially reduce net insolation and thus river discharge. Basic meteorological observations are thus an essential prerequisite for interpreting short-term fluctuations in runoff, and rugged lightweight automatic weather stations scanned frequently by a digital data logger (17) are ideal. Correlation coefficients calculated between mean or maximum daily discharge and temperature or radiation at different lags can indicate the delays involved in meltwater flow from source areas to proglacial rivers: 1 - 3 days in Switzerland (18) but probably more in the very long (>50 km) valley glaciers of the Karakoram.

The small temperate valley glaciers studied in other parts of the world invariably show marked diurnal runoff cycles whose afternoon or evening peaks are interpreted as the day's meltwater from the ablation zone. Flow does not cease entirely at night; instead there is a seasonally varying background contribution which could arise either from the greater delay experienced by meltwater conveyed subglacially from distant parts of the glacier or from delays due to storage in firn, crevasses, or at the glacier bed away from conduits.

The relative importance of rapid and delayed runoff may be assessed by conventional hydrograph separation (e.g. 19) or by hydrochemical analysis as pioneered by Collins (20). Variations in electrical conductivity (and thus TDS) in glacier-fed rivers reflect the mixing, in changing proportion through time, of almost pure surface-derived meltwater and chemically enriched water that has been in prolonged contact with the glacier bed either as diffuse sheets or in small, slow subglacial streams. The relative contributions of baseflow and rapid melt runoff can be quantified using a simple mass balance model. This technique has been applied successfully in Switzerland (21) and Canada (12) and will now be tried on the much larger Karakoram glaciers.

SEDIMENT SOURCES

It is clear from the measured sediment concentrations in the Indus and its tributaries that the Karakoram and adjacent mountains are undergoing rapid erosion, but far less clear whether the main sediment sources are high and remote (e.g. mountain sides and glacier beds) or lower and more likely to affect the local inhabitants (as with erosion of river channels or cultivated valley lands). Of course these are not mutually exclusive alternatives: debris eroded from mountain sides, for example by rockfalls or avalanches, may be deposited and stored for long periods before once again being eroded, perhaps this time by rain or river action. Sediment stores such as lateral moraines, ice-cored terminal moraines, and river terraces act as buffers between the hillslope and river sediment transport systems and make it very difficult to assess the long-term importance of different sediment sources from only short-term measurements. Nonetheless much can be learned by piecing together qualitative and quantitative evidence at a variety of spatial scales.

A first way of comparing erosion rates in different local areas, and thus beginning to pinpoint major sediment sources, is to measure river flow and sediment concentration (solid and dissolved) at not just one but several sites. Measurements on two different tributaries of a river may enable a comparison of erosion from glacier-covered and glacier-free terrain, or on two different types of rock experiencing the same geomorphic processes. Measurements at different sites along the same river allow estimation of the net erosion (or deposition) along a stretch of valley. A carefully located set of sampling sites can thus reveal much about the geographic variability of erosion within the mountain area. Moreover, intensive measurements at each site may indicate whether erosion in the contributing area is limited by the availability of weathered sediment or by the capacity of the rivers: if the latter, the same discharge should be accompanied by the same suspended and dissolved sediment concentrations on every occasion, whereas differences in concentration for a given discharge suggest a variable supply of sediment. For example, the silt concentrations of the Hunza River appear to be higher in June than September despite similar discharges (2) suggesting flushing out by the first meltwaters of loose material accumulated over the winter; and the wide range of concentrations observed at similar discharges in a Swiss proglacial river led Collins (8) to suggest that subglacial channels periodically migrate and tap fresh pockets of till.

The river measurements described so far neglect the transport of sediment as bedload. In meltwater streams this may comprise up to half the total load (22) but is almost impossible to measure without elaborate permanent structures. The size and amount of material moved can however be estimated from hydraulic properties of the river using the well-known Shields criterion and any of several bedload equations. Formulae applicable to coarse-bed meltwater rivers are reviewed by Church and Gilbert (23); one promising addition is that of Bagnold (24). The necessary information on river discharge, depth, width, and slope can be obtained by stream gauging using the velocity-area or dilution methods, plus elementary surveying using an automatic level or less sophisticated equipment.

Hewitt (25) considered rockfalls to be a major sediment source in the Karakoram, not surprisingly in view of the precipitous unvegetated slopes and intense frost-shattering prevalent in the glacier valleys. Techniques for assessing rockfall magnitude and frequency, including the use of time-lapse photography, are discussed by Whalley (26) and Brunsden and Jones (this volume).

Complementary to all these techniques for estimating mean erosion rates over more or less extensive areas is the investigation by field and laboratory techniques of microscale weathering processes and products. In the arid high mountains of the Hindu Kush, Whalley (27) has found desert-varnish weathering rinds that are possibly promoted by the very high ultraviolet radiation flux at altitudes above 5000 m. Assessment of this, and other more familiar weathering processes such as frost-shattering, salt crystallisation and hydration, requires measurement of diurnal fluctuations in such microclimatic variables as surface temperature, wetness, and UV and shortwave radiation flux, also of temperatures at different depths within rocks. Many of these can be recorded automatically at hourly or shorter intervals using a battery-powered multichannel magnetic cassette data logger.

Modern techniques of laboratory analysis can reveal much about the weathering history of rock and debris samples from different altitudes and microenvironments. One useful technique is scanning electron microscopy, which has revealed characteristic differences in surface textures of quartz grains from different glacial and fluvial environments in the Alps and Hindu Kush (28, 29) and throws much light on the origin of the silt load in high-mountain rivers (30). Clues to the nature of chemical weathering processes are given by the elemental composition of surface weathering crusts, which can be determined in the laboratory by microprobe analysis, and the presence of different clay minerals, determined by X-ray diffraction and thermal methods. Complementary to this is the chemical analysis of snowmelt waters from the sites at which weathered rock samples are collected. The techniques here are essentially the same as for river water chemistry analysis except that pH may also be of interest.

CONCLUSION

It will be apparent from this review that the scientific study of melt-water runoff and erosion in high mountain areas is fraught with difficulties quite apart from the intrinsic hazards and inaccessibility. It is impossible to observe what is happening on river beds or beneath glaciers, and the high variability in runoff and geomorphic processes necessitates intensive measurements. Modern field techniques developed by geographers and other earth scientists are of great value in allowing unattended sampling and monitoring of environmental properties, and laboratory analysis of water and rock samples provides very significant additional information.

REFERENCES

1) MEYBECK, M., (1976). 'Total mineral dissolved transport by world major rivers', Hydrol. Sci. Bull. 21, pp 265 - 284.

2) WATER AND POWER DEVELOPMENT AGENCY. 'Annual Report 1973', Vol. 1, Lahore.

3) NAKAWO, M., FUJII, Y. and SHRESTHA, M. L., (1976). 'Water discharge of Rikha Samba Khola in Hidden Valley, Mukut Himal', Seppyo 38, pp 27 - 30.

4) FERGUSON, R. I., (1973). 'Sinuosity of supraglacial streams', Geol. Soc. Amer. Bull. 84, pp 251 - 256.

5) ØSTREM, G., (1964). 'A method of measuring water discharge in turbulent streams', Geogr. Bull. 21, pp 21 - 43.

6) CHURCH, M., (1974). 'Electrochemical and fluorimetric tracer techniques for streamflow measurements', Brit. Geomorph. Res. Gp. Tech. Bull. 12.

7) DAY, T. J., (1977). 'Observed mixing lengths in mountain streams', J. Hydrol. 35, pp 125 - 136.

8) COLLINS, D. N., (1979). 'Sediment concentration in meltwaters as an indicator of erosion processes beneath an Alpine glacier', Jour.

Glac. 23, pp 247 - 257.

9) COLLINS, D. N., (1979). 'Hydrochemistry of meltwaters draining from an Alpine Glacier', Arctic and Alpine Research 11, p 3.

10) REYNOLDS, R. C., (1971). 'Analysis of alpine waters by ion-electrode methods', Water Resources Research 7, pp 1333 - 1337.

11) GOLTERMAN, H. L., (1978). 'Methods for physical and chemical analysis of fresh waters', 2nd ed. Blackwell, Oxford.

12) COLLINS, D. N. and YOUNG, G. J., (1979). 'Hydrochemical separation of components of discharge in alpine catchments', Proceedings of Western Snow Conference 47, pp 1 - 9.

13) RÖTHLISBERGER, H. 'Water pressure in intra- and subglacial channels', J. Glac. 11, pp 177 - 204.

14) IKEN, A., (1978). 'Variations of surface velocities of some Alpine glaciers', Zeitschrift für Gletscherkunde und Glazialgeologie 13, pp 23 - 35.

15) ENGELHARDT, H., KAMB, B., RAYMOND, C. F. and HARRISON, W. D., (1979). 'Observation of basal sliding of Variegated Glacier, Alaska'. J. Glac. 23, pp 406 - 407.

16) BATURA GLACIER INVESTIGATION GROUP, (1976). 'Investigation report on the Batura Glacier in the Karakoram mountains, the Islamic Republic of Pakistan (1974 - 1975)'. 55 pp. Batura Glacier Investigation Group of Karakoram Highway Engineering Headquarters. The People's Republic of China, Peking.

17) STRANGEWAYS, I. C., (1972). 'The reliability of unattended remotely-sited electronic systems for hydrological data collection', Int. Assoc. Hydrol. Sci. 97, Vol. 2, pp 438 - 446.

18) LANG, H., (1973). 'Variations in the relation between glacier discharge and meteorological elements', I.A.S.H. 95, pp 85 - 94.

19) ELLISTON, G. R., (1973). 'Water movement through Gornergletscher', I.A.S.H. 95, pp 79 - 84.

20) COLLINS, D. N., (1977). 'Hydrology of an Alpine glacier as indicated by the chemical composition of meltwater', Zeitschrift für Gletscher-kunde und Glazialgeologie 13, pp 219 - 238.

21) COLLINS, D. N., (1979). 'Quantitative determination of the subglacial hydrology of two Alpine glaciers', J. Glac. 23, pp 347 - 362.

22) ØSTREM, G., (1975). 'Sediment transport in glacial meltwater streams', Soc. Econ. Paleont. and Min. Sp. Publ. 23, pp 101 - 122.

23) CHURCH, M. and GILBERT, R., (1975). 'Proglacial fluvial and lacustrine environments', Soc. Econ. Paleont. and Min. Sp. Publ. 23, pp 22 - 100.

24) BAGNOLD, R. A., (1977). 'Bedload transport by natural rivers', Water Resources Research 13, pp 303 - 312.

382

25) HEWITT, K., (1967). 'Studies in the geomorphology of the mountain regions of the upper Indus Basin'. Unpublished Ph.D Thesis, King's College, London University.

26) WHALLEY, W. B. In press. Rockfalls. Chapter 8 in 'Mass Movement Processes'. ed. D. Brunsden and D. B. Prior. Wiley.

27) WHALLEY, W. B. In preparation. 'High Altitude Desert Varnish' for Nature.

28) WHALLEY, W. B. and KRINSLEY, D. H., (1974). 'A scanning electron microscope study of surface textures of quartz grains from glacial environments', Sedimentology 21, pp 87 - 105.

29) WHALLEY, W. B., (1978). 'An SEM examination of quartz grains from subglacial and associated environments and some methods for their characterization', Scanning Electron Microscopy 1978:I, pp 353 - 360.

30) WHALLEY, W. B., (1979). 'Quartz silt production and grain surface textures from fluvial and glacial environments', Scanning Electron Microscopy 1979:I, pp 547 - 554.

The geomorphology of high magnitude – low frequency events in the Karakoram mountains

D. Brunsden

University of London, King's College

D.K.C. Jones

London School of Economics

ABSTRACT

The conceptual basis of high magnitude–low frequency event studies in high mountains is described and the use of a Transient Form Ratio to isolate the stability state of the landform evolution pattern is proposed. Methods of study and the use of the Karakoram Project (1980) for Hazard surveys are outlined.

INTRODUCTION

Recent reviews of the geomorphology of high mountains (1) (2) (3) have emphasised that high magnitude–low frequency events should be considered as the formative events of medium and long term landscape evolution. During these events, the amount of work accomplished, and the way in which slope and channel units are intimately linked in a sediment transfer system, suggests that gross valley and slope forms are best considered as high magnitude response forms. Intervening periods are characterised by only slow adjustments of slope form, redistribution of debris, channel incision, revegetation or degradational smoothing of morphological discontinuities (4) (5) (6) (7).

If progress is to be made in understanding such landscapes, it is essential that research programmes are devised to assemble information on (a) the frequency and mean arrival times of events of different magnitude, (b) an analysis of those events which actually do most work in initiating characteristic or persistent form; the formative events, (c) the length of time required for a return to the previous or to a new system state. This information is required to assess the occurrence of characteristic and transient landform patterns for the region as defined by the Transient Form Ratio, (TFR).

$$TFR = \frac{mean \ recovery \ time}{mean \ arrival \ time \ of \ formative \ events}$$

when TFR > 1.0   the landform may be regarded as stable, and

when TFR < 1.0   the landform may be regarded as unstable.

DISCUSSION

Currently available data, see Table 1, suggest that the mean arrival times of formative events on hillslopes in high mountains are approximately 5 to 20 years compared with 1 to 30-100 years in New Zealand and 1 to 500 years for arid zone areas and 1 to greater than 1000 years for northern Europe. For a further discussion, see references (8) and (9).

TABLE 1   Formative Event Frequencies, Relaxation Times and Transient Form Ratios for some Landform Systems

Landform system	Formative Event Frequency	Relaxation Time	Transient Form Ratio
	Years	Years	
Cliffs			
Clay sea cliffs UK	9	100–500	11–55
Sand cliffs UK	40	100	2.5
Clay hillslopes UK	1000–5000	10,000	2–10
Gravel moraines (Athabasca)	1000–5000	10	0.01–0.003
Silt cliffs, Louisiana	>1000–5000	150	0.15
Granite, Dartmoor	>50,000	1,000,000	25
Landslides			
Appalachians	>100	> 25	0.25
Auckland, N.Z.	30	> 25–30	1.0
Himalaya	10–20	10	1.0–0.5
Japan	5–10	25	2.5–5.0
Hawaii	1	3–5	3–5
Tanzania	10	2–10	0.2–1.0
Channels and Floods			
S. Fork R. USA	50–100	5	0.1–0.05
Mad R. USA	50–100	10	0.2–0.1
Patuxet R.	100–200	15	0.075
Appalachians	100	10	0.1

Recovery times given in Table 1 range from 10^{-1} to 10^{-4} years for soft rocks to 10^{-6} years for hard rocks with obvious variations for environment and geomorphological sub-systems. These figures indicate that high mountains have rapid recovery times but a very high frequency of events capable of causing landscape change. They may well, therefore, be characterised by a form or suite of forms typical of frequent disturbance, continuous adjustment and rapid transmission times for debris. The process balance on hillslopes is dependent on the rate of weathering and that of sediment slopes (screes) or valley floors dependent on transport rates and storage times. Equivalent figures for channel systems (Table 1) indicate that characteristic, more stable cross sectional and downslope forms can exist, for the recovery to a new state is possible before the next "formative" flood event arrives.

In the field study area of the Royal Geographical Society, Karakoram Project 1980 and especially in the Hunza Valley above Gilgit, these concepts are of particular importance; Fig. 1. The enormous absolute elevation, relative relief, slope steepness, climatic vigour and the external controls exerted by earthquake, intense precipitation and snow melt combine to form an environmental domain dominated by extremes.

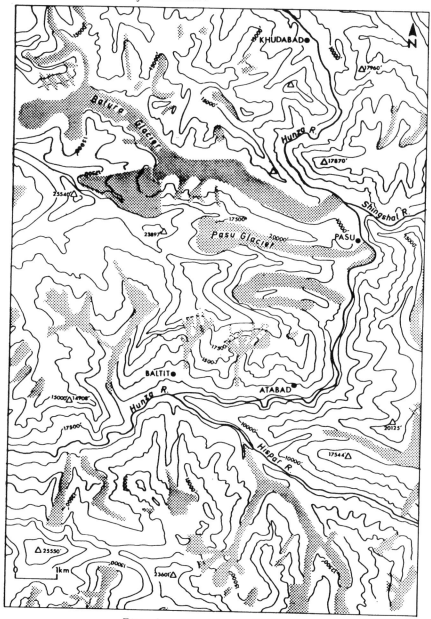

Fig. 1. The Hunza Valley

TABLE **2** Outline of hazard assessment techniques

Approach	Type	Scale	Description
Mapping	Direct (Field)	1:200 – 1:10,000 Site scale	Direct field mapping of hazard forms
Mapping	Indirect	1:10,000 – 1:100,000 Regional Scale	Mapping of secondary or predictive variables. Slope geometry, material types, drainage performance, vegetation, anthropogenic.
	Other	Site-Region	Isopleth map, hazard scoring techniques.
Non Mapping		Site	Proformas and checklists for semi-quantitative data collections
		Site	Instrumentation
		Site	Statistical risk and stability analysis.

In these circumstances it is necessary to design a field programme which records the location, date, pattern and magnitude of events and which can be used as a sampling framework for the elucidation of recovery times and persistence times of such events in the landscape.

An essential element in the programme is to obtain a realistic picture of frequency based upon reliable dates, and all common dating methods are employed including relative dating by stratigraphical and morphological methods, the use of historical sources, oral tradition and folk memory, dendrochronology and lichenometry, and pollen analysis or radio-carbon dating on soils or organic materials buried by the catastrophic debris deposits. Of particular importance is lichenometry using Rhizocarpon geographicum, Acarosporam Chloropana and Lecidea lactea (10) which are common in the area (for a convenient summary see Hewitt (11)) and several fine records on geomorphological change (12). The 1980 Karakoram project will provide one more dateline point to assist future studies.

A major aim of the research will be to use the field maps and the parameter values of the controlling environmental variables to generate a hazard assessment of the valley for planning and engineering purposes. The Hunza Valley must represent one of the most awesome and intractable landscapes on earth for sophisticated engineering designs. Existing projects such as the Karakoram Highway, as well as future ideas for river flood control, irrigation, bridge construction and access, can only benefit from a survey of possible natural hazards.

The main hazards already known for the area, see Table 3 , include flooding, glacier surge, rockfall, major landslide and catastrophic disintegrating debris flows caused by fluidisation and natural dam bursts. These phenomena will be studied using a variety of hazard mapping techniques (Table 2) and will supplement the studies of the earthquake resistant building survey to produce a framework for future development projects. Site

Table **3** Information on major hazards

Location	Hazard	Date or Frequency	Source
1. 80 miles North of Gilgit	Landslide	1 to 2 months	Far East Ec. Rev. 14.10.79
2. 45 miles North of Gilgit	Rockfall	Winter thaw	Far East Ec. Rev. 14.10.79
3. {120 km NNE of Tarbela Dam {140 km SW of Gilgit	Pattan Earthquake (2000 dead)	28.12.74	Far East Ec. Rev. 14.10.79
4. Balthar Glacier (Sheeshkat)	Glacier surge & Lake 27 km long.	1976	Far East Ec. Rev. 14.10.79
5. Sheeshkat	43m Friendship Bridge flooded	1976 to 13.7.78	Far East Ec. Rev. 14.10.79
6. Hunza Valley	Frozen culverts	1978	Far East Ec. Rev. 14.10.79
7. Yongutsa Glacier	Glacier surge	1901	Mason Rec. Geol. Surv. Ind. 1930
8. Minapin Glacier	Glacier surge?	1889-1892	"
9. Hasanabad Glacier	Glacier surge	1903	"
10. Hunza R. Ghammesar Landslide, Atabad & Pasu	Landslide & lake	1858	"
11. Shimshal Valley	Glacier surge & river dams on various glaciers	1884 1893 1905 1906 1927	
12. Batura Glacier & Hunza	Flood-glacial melt. 30m bridge damaged	1972	Chinese Invest. Team 1976
13. Batura Glacier & Hunza	Flood 8m bridge	1973	"
14. Batura Glacier & Hunza	Surge	Approx 1770	"
15. Batura Glacier	Surge?	1925? (and between (1885-1930)	"
	Surge(517.5m/yr)	1974	"
16. Hunza R. (Batura) (30 year flood?)	Rapid river migration of Batura outlet	1973	"
17. Hunza River (Batura)	Mud-rock flow (10.3mm rain in 2 days) Blocked river	Aug. 14, 1972 and 1975	" " "

investigations for slope stability, flood frequency, hydrology, material properties and available water and aggregate properties will benefit from a study of these hazard maps.

Conclusion

The landscape of Hunza, described by the Hon. George Nathaniel Curzon as "This great workshop of primaeval forces" where "nature seems to exert her supremest energy" and the Hunza river where John Keay

388

was so impressed by "this thundering discharge of mud and rock ...
it is a revolting sight" presents a massive challenge to future geomorphological research and to the engineers of the Frontier Works Organization
of the Pakistan Army.   Our present state of knowledge reveals that high
magnitude events are normal for the region and that any engineering
projects must be designed to overcome very severe conditions of slope
stability and river discharge.   Of primary importance is the danger
of landslide dams which may threaten reservoir developments and the
effect of silting on reservoir life.   Highway, industrial and housing
development must also consider the danger, from landslide and mudflow
activity.   In these respects the hazard mapping project should provide
valuable planning documents and locational guidelines.

REFERENCES

(1)   Starkel, L. (1972) "The role of catastrophic rainfall in the shaping
         of the relief of the Lower Himalaya (Darjeeling Hills)" Geographica
         Polonica 21, 103–143.

(2)   Starkel, L. (1976) "The role of extreme (catastrophic) meteoro-
         logical events in contemporary evolution of slopes". Geomorphology
         and Climate. ed. E. Derbyshire (Wiley, London) 203–246.

(3)   Brunsden, D., Jones, D.K.C., Martin, R.P., Doornkamp, J.C.
         (1980) "The geomorphological character of part of the low
         Himalayas of eastern Nepal" Zeits. fur Geomorphologie. (in
         press).

(4)   Welch, D.M. (1970) "Substitution of space for time in a study
         of slope development" Journ. Geol. 78, 234–238.

(5)   Brunsden, D. and Thornes, J.B., (1979) "Landscape sensitivity and
         change". Trans Inst. Brit. Geogrs. Vol 4, No. 4

(6)   Schick, A.P. (1968) "The storm of March 11, 1966 in the southern
         Negev and its geomorphic significance". Studies Geog. Is. 6, 20–52.

(7)   Wolman, M.G. and Gerson, R. (1978) "Relative scales of time and
         effectiveness of climate in watershed geomorphology". Earth Surface
         Processes, 3, 189–208.

(8)   Selby, M.J. (1974) "Dominant geomorphic events in landform evol-
         ution" Bull. Int. Assoc. of Engineering Geol. 9, 85–89

(9)   Selby, M.J. (1979) "Slopes and Weathering" in Gregory, K.J. and
         Walling, D.E. eds 'Man and Environmental Processes'. Dawson.
         Folkstone. 105–122.

(10)   Nikonov, A.A. and Shebalina, T. Yu, (1979) "Lichenometry and
         earthquake age determination in central Asia". Nature 280, 675–677

(11)   Hewitt, K. (1969) "Geomorphology of mountain regions of the Upper
         Indus Basin" (PhD dissertation. University of London).

(12)   Chinese Investigation Team (1976) "Investigation report on the
         Batura Glacier, in the Karakoram Mountains, the Islamic Republic
         of Pakistan".

New perspectives on the Pleistocene and Holocene sequences of the Potwar plateau and adjacent areas of Northern Pakistan

H.M. Rendell
Dept. of Geography, University of Sussex

ABSTRACT

The Pleistocene and Holocene sequence of the Potwar Plateau requires re-examination, particularly in the light of the development of new dating techniques. Some preliminary results of work on the post-Siwalik deposits of the Soan valley are presented. A revision of the relative chronology of the Pleistocene deposits is suggested, and the possibilities of the construction of an absolute chronology are discussed.

INTRODUCTION

The Potwar Plateau has long been recognised as a key area for both the Siwalik series and the later Pleistocene and Palaeolithic sequences of northern Pakistan and India (1, 2, 3). In recent years, detailed studies of the middle and upper Siwalik sequences have been undertaken on an inter-disciplinary basis (4, 5, 6, 7) and have included the establishment of a magnetic polarity stratigraphy for the upper Siwalik subgroup (7). Given the facies changes that characterise the upper part of the Siwalik sequence, the establishment of a chronology that is independent of the vertebrate faunal assemblages represents a most important breakthrough. However, the post-Siwalik deposits of the Potwar Plateau have been largely neglected since the pioneering work of the Yale-Cambridge expedition in the 1930's (8, 9). The sequence deduced by De Terra and Paterson (see Table 1) has been widely adopted by archaeologists (10, 11, 12, 13) and has recently been correlated with the sequence of Pleistocene deposits in the adjacent Peshawar Basin (14) (see Fig. 1). Although De Terra and Paterson were undoubtedly aware of the complexity of sequences that might result from an interaction of climatic and tectonic agencies, their correlation of the evidence of Himalayan glacials with that of the Alps has perhaps served to obscure the Potwar sequence rather than illuminate it. Certainly their sequence is ripe for re-examination, and this paper presents some preliminary observations based on fieldwork undertaken as part of the first season's work of the Cambridge University Archaeological Mission to Pakistan in the Potwar area.

TABLE 1   Relative chronology of the Pleistocene deposits of the Potwar
          according to De Terra and Paterson

4th glaciation	T4 pink loam, silt, gravel	fluvial sedimentation
3rd interglacial	T3 thin loam	erosion/warping
3rd glaciation	T2 Potwar loessic silt and gravel	aeolian, fluvial, lacustrine
2nd interglacial	T1 upper terrace gravel	
	------- erosion/tilting/folding --------------------------	
2nd glaciation	Boulder Conglomerate	fluvial and fluvio-glacial sedimentation
1st interglacial	----------------- erosion/tilting/folding ----------------	
	Pinjor Beds conglomerates, sands, clays	
1st glaciation	Tatrot Beds conglomerates and sands	

PLEISTOCENE SEQUENCE OF THE MIDDLE SOAN VALLEY

The Middle Soan Valley (Fig. 2) has provided a focus for Palaeolithic studies within the Potwar. The Siwalik and older rocks were first mapped by Wynne (15) in the 1870's, and later by Cotter (16) and Wadia (3). Part of the area was subsequently remapped by Gill (17). The Yale-Cambridge exped- ition (8,9) concentrated on the stratigraphy of upper Siwalik and post-Siwalik deposits and their relationship with various Palaeolithic sites. Some of De Terra's conclusions are however directly contradicted by the findings of Gill (17) and Pilgrim (2) and on the basis of field observations and recently published material, the following observations may be made on the Pleistocene sequence:-

First, the major hiatus between upper Siwalik and post-Siwalik depos- ition, with various en echelon structures achieving surface expression, appears to have commenced between 700,000 and 500,000 BP (7). The post-Siwalik deposits overlie Siwalik and older rocks with a marked angular unconformity while the junction between the middle (Dhok Pathan) and upper (Tatrot) Siwalik beds is a disconformity (18). How- ever, De Terra (9) states that in the Soan valley Tatrot and Pinjor deposits "unconformably overlie the slopes of a planed elevated surface" (p.255). In the opinion of this author, De Terra's contradiction stems from a mapping and interpretation problem that is outlined below.

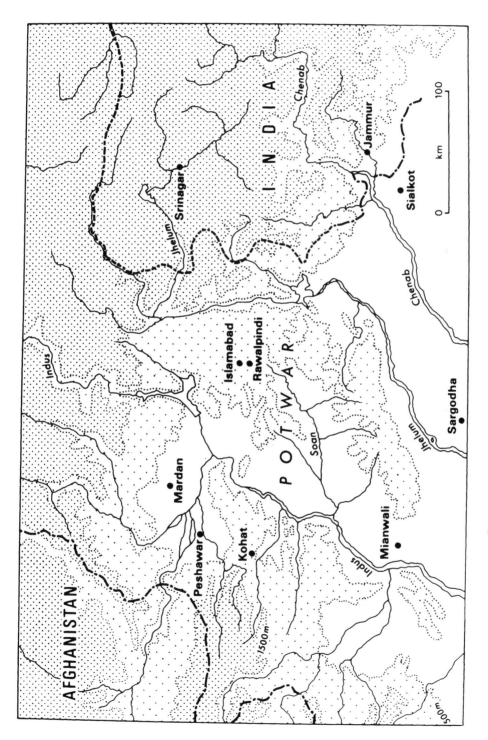

Fig. 1 Northern Pakistan: Relief and Drainage

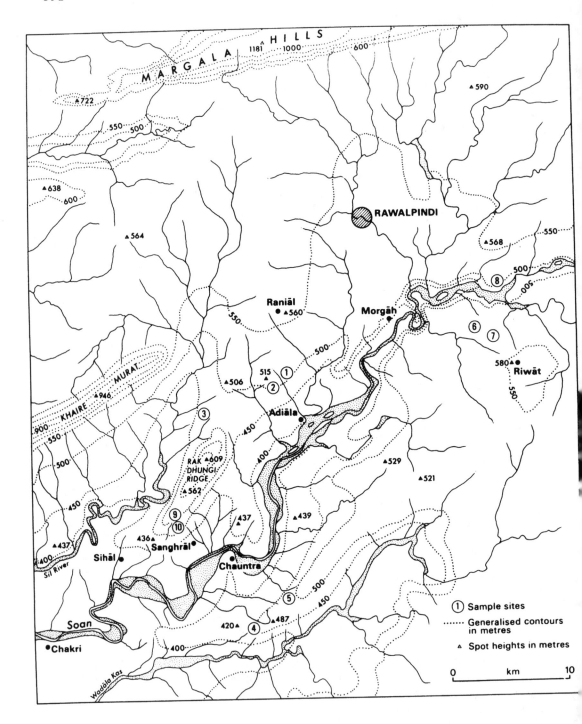

Fig. 2   Middle Soan Valley

Second, the major post-Siwalik formation within the Soan valley is the Lei Conglomerate (17, 19). This formation is a valley-fill deposit consisting of massively bedded conglomerates, predominantly composed of locally-derived limestone pebbles, with subsidiary grey sands and pink and yellow/buff silts, and with occasional palaeosols. This 'conglomerate complex' is currently being incised by the Soan river, revealing sections 60m thick south-east of Rawalpindi. The Lei Conglomerate Complex dominates the middle Soan valley, but the distribution of the conglomerate facies is very variable. Near Adiāla (lat. 33° 28' long. 72° 59' 30") the formation consists of massively bedded pink claystones with one conglomerate horizon near the top of the section, the beds are also slightly tilted, at an angle of 2-3°, towards the main axis of the Soan syncline. The Lei conglomerate may be distinguished from upper Siwalik conglomerates on petrological as well as structural grounds. The upper Siwalik conglomerates are mainly composed of quartzite and, to a lesser extent, sandstone pebbles (18) whereas the Lei conglomerate is predominantly composed of pebbles of locally-derived grey Margala Hill limestone, of Eocene age.

Third, De Terra's use of the term 'Boulder Conglomerate' is misleading. As Gill (17) pointed out, this term was first used by Pilgrim to describe the upper Siwalik conglomerates whereas De Terra uses it as a blanket term for both the upper Siwalik and the post-Siwalik conglomerates. In the absence of palaeontological evidence, De Terra appears to have mapped pink silts underlying his 'Boulder Conglomerate' as Pinjor Beds whereas both the conglomerate and pink silts are merely part of the Lei Conglomerate Complex. Thus De Terra assigns too great an age to a whole series of deposits within the middle Soan valley, and, given that he maps large sections of the finer beds within the Lei Conglomerate as Tatrot and Pinjor zones, it is hardly surprising that he then identifies a major angular unconformity at the base of his 'Tatrot' zone. Typical sections are those south-east of Rawalpindi, north of Adiāla and on the south-eastern flank of the Rak Dhungi ridge north-west of Sanghrāl (lat. 33° 24' long. 72° 53'30"); in each case De Terra has mapped the upper massive conglomerates as 'Boulder Conglomerates' and the underlying finer beds as 'Pinjor' and 'Tatrot'.

Finally, a major period of erosion pre-dated and may also have been partly contemporaneous with the deposition of the Lei Conglomerate Complex, but the conglomerate complex itself suffered a major erosional phase prior to the deposition of a thin cover of loess over the whole area of the middle Soan valley. The Rak Dhungi ridges, north-east of Sihāl, are outliers of the Lei conglomerate complex resting with an angular unconformity on the upturned strata of middle and upper Siwalik beds. These ridges rise to heights of 47 to 93m above the planed-off level of lower and middle Siwalik beds which forms a summit surface at a height of c. 515m (20). The pebble-strewn eroded surfaces of the Lei Conglomerate and of the upper Siwalik conglomerates were exposed in Palaeolithic times and, as Wadia (3) pointed out, "... hundreds of detached chips of flint, fine quartzite and trap (with cores of pebbles bearing signs of intentional fracturing) are met with wherever a suitable type of raw material occurs among the gravel cap covering the surface of the valley" (p.346).

A revised relative chronology for the Pleistocene and Holocene deposits of the middle Soan valley is given in Table 2.

TABLE 2 Modified Relative Chronology of the Siwalik and
Pleistocene Deposits of the Middle Soan Valley

Sequence	Tentative Dates
------------- erosion/deposition ------------	
Loess Deposition	Upper Palaeolithic – Middle Palaeolithic
-------------- erosion/warping -----------	
Lei Conglomerate Complex (valley fill) includes deposition of loess/uplift partly contemporaneous	
-------- uplift/folding/start of erosion -----	– 0.7–0.5 mya
Upper Siwalik Conglomerates	
	– 1.9 mya
Soan Formation Pinjor Beds	
	– 2.5 mya
Tatrot Beds	
----------------- disconformity ---------------	
Dhok Pathan Formation	– 8.6 mya
Nagri Formation	
Chinji Formation	– 10.2 mya

POST-SIWALIK / SIWALIK GROUP

THE 'POTWAR LOESSIC SILT'

The nature of the 'Potwar Loessic Silt' has been hotly debated, with aeolian, fluvial and lacustrine origins suggested. The term appears to have been used by De Terra and Paterson (9) and others (17) to cover at least two distinctly different deposits.

a) Loess

Loess, in the sense of a clastic deposit of predominantly silt–sized particles occurring in wind–laid sheets (21), occurs as a summit–surface deposit over much of the middle Soan valley. It fills the hollows in the surface of the degraded Lei conglomerate and covers spreads of middle Palaeolithic cores, flakes and scrapers. Over much of the area the loess is unstratified and is frequently rich in calcareous nodules; the thickness of the deposit ranges from less than 2m to 8m. Deposits of loess are not confined to the summit surface; loess–like intercalations are present in the sections of Lei conglomerate complex south–east of Rawalpindi and on the south–eastern flank of the Rak Dhungi ridges.

TABLE 3 Analyses of Potwar Loess – Middle Soan Valley Sites

Sample	Median Grain Size (mm)	% Finer than 0.002 mm	% Finer than 0.063 mm	% CaCO$_3$
1	0.015	16.91	91.48	15.50
2	0.020	13.13	77.42	13.75
3	0.0156	13.73	90.31	11.00
4	0.016	18.25	86.37	17.25
5	0.016	15.36	91.83	12.25
6	0.0157	21.25	96.68	11.38
7	0.0165	13.33	90.35	12.00
8	0.02	16.20	94.95	13.25
9	0.024	11.72	83.17	13.63
10	0.019	11.06	92.29	15.38
Mean	0.017	15.09	89.49	13.53

Some preliminary results of the analysis of a series of samples of the loess are given in Table 3. The samples are broadly representative of sections throughout the middle Soan valley (see Fig. 2) and sample numbers 9 and 10 are from the intercalated loess deposits of the Rak Dhungi ridge. It appears that the phases of loess movement into the Potwar area may have been more complex than has hitherto been supposed. However, with the aid of thermoluminescence (TL) dating of the loesses, it should prove possible to construct an absolute chronology for at least part of the post-Siwalik sequence in this area. Although TL dating of loess deposits appears to have developed as a standard technique in the U.S.S.R. during the last decade (22, 23), this dating method is still in its infancy in western Europe and north America (24, 25). The dating technique is highly complex and although many dates have been reported for both loess and other Pleistocene deposits in the Soviet Union, the details of the experimental techniques employed remain obscure (24). Work on the dating of the Soan valley loesses is currently being undertaken in collaboration with the Research Laboratory for Archaeology at Oxford University.

b)   Alluvial Silts

Recent incision has revealed a sequence of thinly-bedded overbank pink silts, buff silts and buff sands, with occasional channel gravels, in the area of the middle Soan valley around Sihāl. These deposits are described as 'Potwar loessic silt' by De Terra and Paterson and appear to post-date the Lei Conglomerate complex. Samples were collected from several localities for analysis and possible dating.

PLEISTOCENE SEQUENCE IN THE PESHAWAR BASIN

A recent study (14) has established the relative sequence of Pleistocene deposits in the Peshawar Basin and has correlated this sequence with that for the Potwar. It has been established that, at one point during the Pleistocene, the Peshawar Basin was dammed to form a large lake into which loess was either blown or carried by flowing water (14, 26). The lacustrine deposits are now exposed as a sequence of thinly-bedded creamy-yellow silt-stones. These siltstones are overlain by torrent gravels and then by a thick deposit of loess. It is envisaged that, with the aid of the TL dating technique, an absolute chronology may be established for these Pleistocene deposits in the course of the programme of work currently being undertaken by the author and others.

CONCLUSION

A preliminary investigation of the upper Siwalik and post-Siwalik deposits of the middle Soan valley has raised serious doubts about the relative chronology established by De Terra and Paterson (9). It is also apparent that the loess deposits of the middle Soan valley represent a far more com-plex sequence than has been hitherto supposed and that the term 'Potwar Loessic Silt' has been applied to alluvial silts as well as to a series of genuine loess deposits. Work is being currently undertaken to date the loess deposits using thermoluminescence.

ACKNOWLEDGEMENTS

This paper represents a preliminary report on one part of the work under-taken by the Cambridge University Archaeological Mission to Pakistan in collaboration with the Government of Pakistan Department of Archaeology and Museums. The author gratefully acknowledges the help of Dr. Ishtiaq Khan and Dr. S. M. Ashfaque of the Department of Archaeology and Museums, Government of Pakistan and of Drs. F. R. and B. Allchin, K. D. Thomas and Mr. J. R. Knox of the Cambridge University Archaeological Mission to Pakistan. The work was undertaken with the aid of a grant from the Arts Research Support Fund, Sussex University.

REFERENCES

(1)   Pilgrim, G. E. (1910) "Preliminary note on a revised classification of the Tertiary freshwater deposits of India". India Geological Survey Records 40 (3), 185–205.

(2)   Pilgrim, G. E. (1944) "The lower limit of the Pleistocene in Europe and Asia". Geological Magazine 81, 28–38.

(3)   Wadia, D. N. (1928) "The geology of Poonch State (Kashmir) and adjacent portions of the northern Punjab". Memoirs, Geological Survey India 51,185-370.

(4)   Pilbeam, D., Barry, J., Meyer, G. E., Ibrahim Shah, S. M.. Pickford, M. H. L., Bishop, W. W., Thomas, H. and Jacobs, L. L. (1977) "Geology and Palaeontology of Neogene strata of Pakistan" Nature 270, 689-695.

(5)   Tauxe, L. (1979) "A new date for Ramapithecus". Nature 282, 399-401.

(6)   Keller, H. M., Tahirkheli, R. A. K., Mirza, M. A., Johnson, G. D., Johnson, N. M. and Opdyke, N. D. (1977) "Magnetic polarity stratigraphy of the Upper Siwalik deposits, Pabbi Hills, Pakistan". Earth Planet. Sci. Lett. 36, 187-201.

(7)   Opdyke, N. D., Lindsay, E., Johnson, G. D., Johnson, N., Tahirkheli, R. A. K. and Mirza, M. A. (1979) "Magnetic polarity stratigraphy and vertebrate palaeontology of the Upper Siwalik subgroup of northern Pakistan". Palaeogeography, Palaeoclimatology, Palaeoecology 27, 1-34.

(8)   De Terra, H. and De Chardin, P. Teilhard (1936) "Observations on the Upper Siwalik formation and later Pleistocene deposits in India" Proc. Amer. Philos. Soc. 76, 791-822.

(9)   De Terra, H. and Paterson, T. T. (1939) "Studies on the Ice Age in India". Carnegie Institute, Washington Publication No. 493.

(10)   Krishnaswamy, V. D. (1947) "Stone Age India" Bulletin Archaeological Survey of India, 3.

(11)   Raza, M. (1972) "Pleistocene Environment and Palaeo-Anthropology in Kashmir". The Geographer (The Aligarth Muslim University Geographical Survey) 19, 47-55.

(12)   Graxiosi, P. (1964) "Prehistoric research in the northwestern Punjab". Scientific Reports of the Italian Expeditions to the Karakoram (K2) and the Hindu Kush, Part V, Vol 1, 7-54.

(13)   Johnson, E. (1973) "Notes on a palaeolithic site survey in Pakistan". Asian Perspectives 15, 60-65.

(14)   Said, M. and Majid, M. (1977) "The Pleistocene history of terrestrial deposits of Bar Daman area, Peshawar Valley". Journal of Science and Technology, Peshawar 1, 39-47.

(15)   Wynne, A. B. (1877) "Notes on the Tertiary zone and underlying rocks in the northwest Punjab". India, Geological Survey Records 10, 107-132.

(16)   Cotter, G. de P. (1933) "The geology of part of Attock District west of longitude 72°45' E". India, Geological Survey Memoirs 55, 665-720.

(17)   Gill, W. D. (1952) "The stratigraphy of the Siwalik series in the northern Potwar, Punjab, Pakistan". Quarterly Journal Geological Society, London 107, 375-394.

(18)   Fatmi, A. N. (eds.) (1973) "Lithostratigraphic Units of the Kohat-Potwar Province, Indus Basin, Pakistan". Memoirs, Geological Survey of Pakistan 10.

(19)   Various Authors (1976) "Stratigraphy of Pakistan". Memoirs, Geological Survey of Pakistan 12.

(20)   Elahi, M. K. and Martin, N. R. (1961) "The physiography of the Potwar, West Pakistan". Geological Bulletin Punjab University 1, 5-14.

(21)   Smalley, I. J. and Vita-Finzi, C. (1968) "The formation of fine particles in sandy deserts and the nature of 'desert' loess". Journal of Sedimentary Petrology 38, 766-774.

(22)   Shelkoplyas, V. N. (1971) "Datirovanie chetvertichnikh otlozhenii termoliuminestsentnim metodom". In: Zubanov, V. A. and Kochegura, V. V. (eds.) "Chronology of the Glacial Age" 115-159 (In Russian).

(23)   Ranov, V. A. and Davis, R. S. (1979) "Toward a new outline of the Soviet Central Asian Palaeolithic". Current Anthropology 20, 249-270.

(24)   Dreimanis, A. Hutt, G., Raukas, A. and Whippey, P. W. (1978) "Dating methods of Pleistocene Deposits and their problems: 1 Thermoluminescence Dating". Geoscience Canada 5, 55-60.

(25)   Wintle, A. G. and Huntley, D. V. (1979) "Thermoluminescence dating of a deep-sea sediment core". Nature 279, 710-712.

(26)   Rendell, H. M. and Thomas, K. D. (1980) "Environmental change during the Pleistocene and Holocene in the Peshawar Basin, Pakistan". Journal of Archaeological Science 7, 69-79.

The loess of Tajik SSR

A.S. Goudie

School of Geography, University of Oxford

H.M. Rendell

Geography Laboratory, University of Sussex

P.A. Bull

Christ Church, Oxford

ABSTRACT

The loess of the Tajik SSR is some of the thickest on Earth and contains many late Pliocene and Pleistocene palaeosols that Soviet workers have dated by palaeomagnetic and thermoluminescence methods. Grain size, chemical and mineralogical data are presented together with analyses of grain characteristics (shape, roundness, surface texture) determined by scanning electron microscopy.

INTRODUCTION

Tajikistan, part of Soviet Central Asia, has a southern frontier with Afghanistan and is bounded on the north by the Ghissar mountains and on the east by the Pamirs. The southern part of Tajikistan has, for much of the Tertiary and Quaternary, been an area of aggradation, and the subaerial aggradation sequence contains alternations of buried soils and loess deposits which provide climato-stratigraphic information extending over the last 1.5 to 2 Ma. Soviet workers have studied these sections intensively and have undertaken palaeomagnetic and thermoluminescence dating of the sequences, together with pollen analysis. In September 1979 observations of these sections were made by two of the authors and representative samples were collected (see Fig. 1 and Table I). Most of the sites lie at about 1400 - 1900 m above sea-level, and much of the land below about 2000 m has a loess cover.

GENERAL DISCUSSION OF THE SEQUENCE

The Tajikistan loesses and associated palaeosols are some of the most important deposits of this type on the earth's surface, in terms of not only thickness but also areal extent. They are also at the forefront of the controversy over "warm" and "cold" origins for loess (1, 2, 3, 4). Whether the loess is derived by deflation of outwash materials from the formerly more extensive glaciers and ice caps of the Pamirs, Ghissar and other mountain ranges in the neighbourhood, or whether the loess was derived from desert plains to the south and east is still a matter of contention.

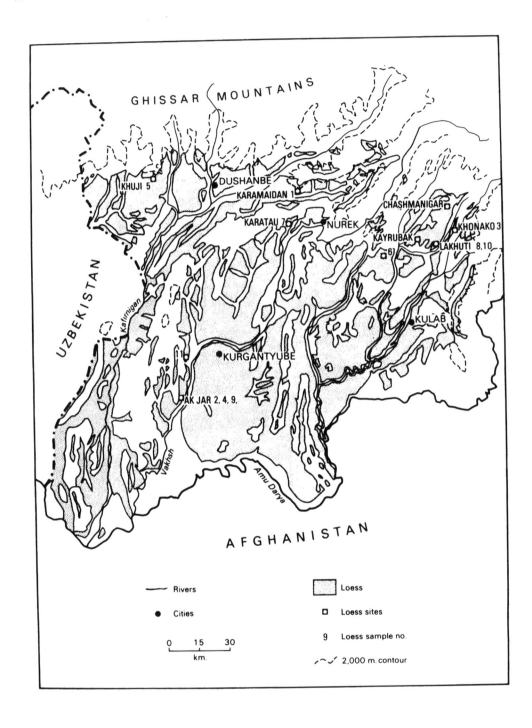

FIG. 1.   DISTRIBUTION OF LOESS IN TAJIKISTAN AND
LOCATION OF LOESS SAMPLES

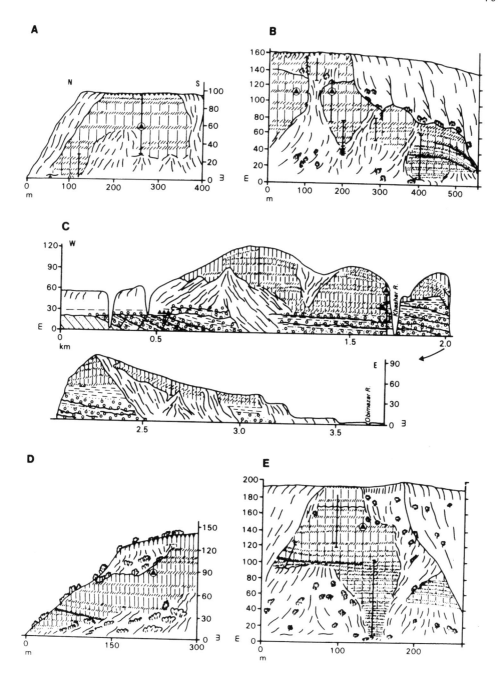

Palaeosols are represented by diagonal shading.

MODIFIED FROM DODONOV et al (10)

FIG. 2.   SELECTED LOESS SECTIONS FROM TAJIKISTAN: A. KHONAKO I,
B. KHONAKO II, C. LAKHUTI, D. KAYBRUBAK, E. CHASMANIGAR.

402

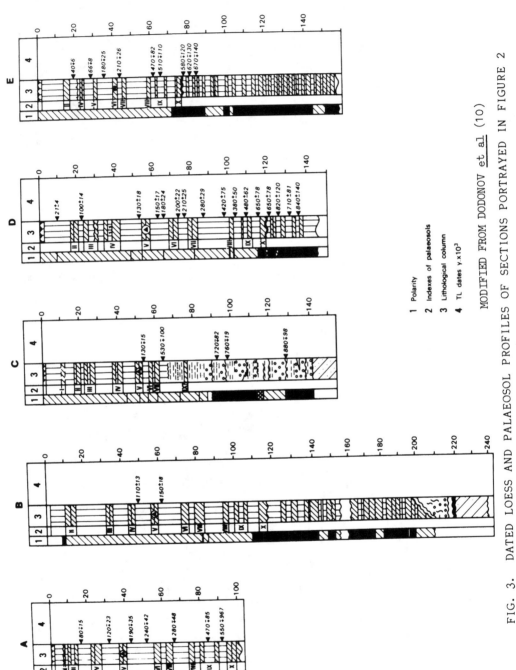

FIG. 3. DATED LOESS AND PALAEOSOL PROFILES OF SECTIONS PORTRAYED IN FIGURE 2

MODIFIED FROM DODONOV et al (10)

1 Polarity
2 Indexes of palaeosols
3 Lithological column
4 TL dates y.×10³

SAMPLE	LOCATION	MEDIAN GRAIN Size (μm)	Percent finer than 2.0 μm	Percent finer than 63 μm	Percent CaCO$_3$	Cation Exchange Capacity*	pH
TJ 79/1	Karamaidan	10.5	22.5	95.40	28.75	43.8	8.25
TJ 79/2	Akjar (Base)	21.0	11.0	94.24	17.50	23.7	8.15
TJ 79/3	Khonako (Upper)	8.6	25.0	99.07	22.75	33.3	8.30
TJ 79/4	Akjar (2 m)	14.0	16.5	97.58	19.50	26.3	8.20
TJ 79/5	Khuji (1.8 m)	11.5	17.5	99.67	20.75	23.7	8.70
TJ 79/6	Kysylsu	12.0	17.0	99.57	22.25	25.5	8.45
TJ 79/7	Karatan (above Palaeosol 5)	11.5	16.5	99.07	10.75	31.5	8.20
TJ 79/8	Lakhuti (above Palaeosol 6)	11.0	23.5	98.7	21.50	28.5	8.60
TJ 79/9	Akjar (8 m)	18.5	21.0	98.45	18.50	27.2	7.85
TJ 79/10	Lakhuti (top)	10.0	23.0	98.9	20.00	34.2	8.60
TJ 79/11	North of Akjar	19.5	8.0	94.3	18.25	23.7	7.75
Mean		13.5	18.32	97.72	20.86	29.2	8.28

* Cation Exchange Capacity in milliequivalents/100 gm.

TABLE I.   CHARACTERISTICS OF TAJIK LOESS SAMPLES

Examination of loess sequences at various locations in Tajikistan shows that they are up to 200 m thick, and are comparable in thickness to the classic loesses of China which have been shown to be 75-175m thick (5, 6). Loess deposits elsewhere in the world appear to have thicknesses of up to 30 m or so (7, 8, 9). A magnetic polarity stratigraphy has been established for the Tajik loess sequences dating back to the late Pliocene and early Pleistocene (10). The Lower Pleistocene formations do not exceed 30 - 35 m in thickness; comparable thicknesses of Middle Pleistocene formations exist whereas the Upper Pleistocene sequence is 50 - 60 m thick.

The sequence preserves evidence of frequent environmental oscillations since the Bruhnes-Matuyama boundary (0.69 Ma) which occurs just above the Xth soil complex from the top. The IInd pedocomplex lies just below the Laschamp event (22,000 years); the age of the Vth soil complex is established by the palaeomagnetic episode recognised above it and correlated with the Blake event (110,000 years). Thermoluminescence (TL) dating (11) has given values of 200,000 and 300,000 years for the VIth and VIIth soil complexes.

Individual loess horizons are generally up to 7 - 10 m in thickness (less frequently 15 m) while the buried soil complexes are generally 5 - 7 m thick. The Upper Pleistocene loess horizons are thicker than the pedocomplexes whereas in the lower Pleistocene the reverse is the case. The density of the loess also increases with age, while its porosity decreases. The fossil soils contain well-developed calcrete hardpan horizons and the associated reddish brown soils have been characterised as of meadow-steppe type. Some geologists judge that the soil formation took place in comparatively warm, dry periods while loess was deposited under relatively cool conditions. On the other hand, on the basis of pollen analysis, Pakhomov (in Ranov and Davis (12)) argues that the soils should be correlated with stadial periods since the interstadials, with their very dry climates and sharp reduction in arboreal vegetation, provided the basis for loess formation.

It is evident from the sections illustrated in Figs. 2 and 3 that there is a relatively consistent sequence within the loess deposits of Tajikistan. This permits tentative correlations with the loess deposits of Central Europe where Kukla (13) has established nine soils and eight loess layers above the Brunhes-Matuyama boundary. In Tajikistan this occurs above the 10th soil complex. This would seem to indicate that there have been at least eight glacial-interglacial cycles in both areas in the last 0.73 million years, though the precision of dating methods does not allow further correlation of individual events since the Matuyama. Some loess horizons occur below the Olduvai-Gilsa palaeomagnetic event (14) and up to 25 cycles have been recognised for the whole Pleistocene. Penkov and Gamov (15) recognise as many as 45 or 46 soils above the Gauss-Matuyama reversal (2.48 Ma). In China the Brunhes-Matuyama boundary is overlain by at least 12 palaeosols (16).

CHARACTERISTICS OF TAJIK LOESS

Loess may be defined as a clastic deposit which consists predominantly of quartz particles 20 - 50 µm in diameter and which occurs in wind-laid sheets (17). It has been strongly argued that glacial action provides the only viable mechanism for the generation of silt- and clay-size quartz particles (1). It is, however, difficult to demonstrate the origin of the quartz particles.   Developments in scanning electron microscopy (SEM) have added greatly to the knowledge of shapes and textures of fine particles and, along with other more general analyses, some results of SEM work on the Tajik loesses are presented below.

a)   Grain Size Characteristics

The results of the granulometric analyses of 11 representative Tajik loess are shown in Table 1.   The loess is predominantly silty with an average median grain size of 13.5 µm.   The mean percentage of clay-size material (2 µm) is 18.32% while there is very little material coarser than 63 µm.   It is evident that the loess of Tajikistan is rather finer than most examples that are cited in the literature (5, 18, 19).

b)   Chemical Composition

The results of general analyses of the loess samples are given in Table 1.   The carbonate content of all samples is high, with a mean value of 20.9 percent by weight.   The partial cementation of most of the samples appears to be the result of recrystallisation of primary carbonate.   The primary nature of the carbonate is attested by the results for sample TJ/11, a completely uncemented loess, with a carbonate value close to the mean.   A partial elemental analysis of the samples was carried out using an elecron probe attachment on a scanning electron microscope.   Relative concentrations of silicon, aluminium and potassium remained fairly constant for all samples with silicon the major component.

c)   Mineralogical Analysis

The mineralogy of the loess samples was determined qualitatively by X-ray diffraction.   Samples were prepared by sedimentation in acetone on to glass slides and were analysed in a Siemens Diffracto-meter with a cobalt source.   The major peaks obtained for all samples were for quartz (4.26 Å, 3.34 Å) with smaller peaks for carbonate (3.03 Å), feldspar (3.19 Å) and minor peaks for kaolinite and montmorillonite.   Peaks for illite and mica were present for most samples, but only TJ/4 possessed a peak for chlorite.   The mineralogy of the samples is therefore fairly uniform.   Apart from the dominance of quartz and carbonate, orthoclase appears to be the major feldspar component and this might prove useful for provenance studies.   Also, although much of the carbonate appears to have undergone secondary crystallization, 'carbonate dust' appears to be an important component (c. 20 percent by weight) of the Tajik loesses, and it would be extremely useful to discover its provenance.

Scanning Electron Microscopy

i)   Grain Shape

The quartz grains to be analysed were puffed onto an electron microscope viewing stub following sample preparation similar to that adopted by Krinsley and Doornkamp (20).   Fifty grains were inspected utilizing the 'adjoining grain examination technique' employed by Bull (21).   Note was taken of the grain dimensions (long, intermediate and short axis), surface textures and roundness and, following the technique of Sneed and Folk

(22), triangular form diagrams were constructed to summarise the characteristics of grain-form in each of the samples (Figs. 4 and 5).

FORM TRIANGLE

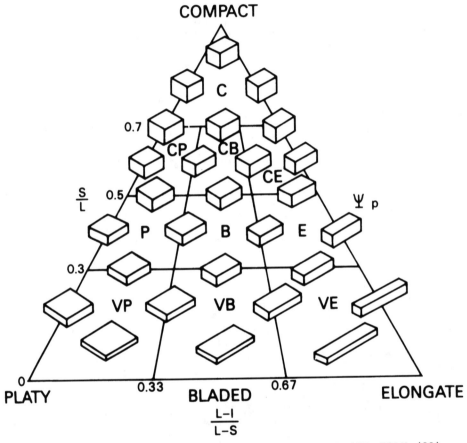

FIG. 4.   TRIANGULAR FORM-DIAGRAM; AFTER SNEED AND FOLK (22)

Although it was considered important to ascertain the shape characteristics of the silt-sized particles per se the underlying aim was to test the hypothesis that small sedimentary quartz particles generally adopted a more platy habit than their larger counterparts (23, 24).   This particle form variation is considered to occur below a critical size limit although there is a certain variation in the estimate of 'critical size' (i.e. 200 μm – 100 μm (20)).   General inspection of the form triangles (Fig. 5) and more particularly of the composite form triangles (Fig. 5; composite) show that neither the critical size limit nor the tendency for particles, flatness seem to apply for the Tajik loess (Fig. 6a).   This anomaly to the generally accepted theory has also been identified from loess-derived

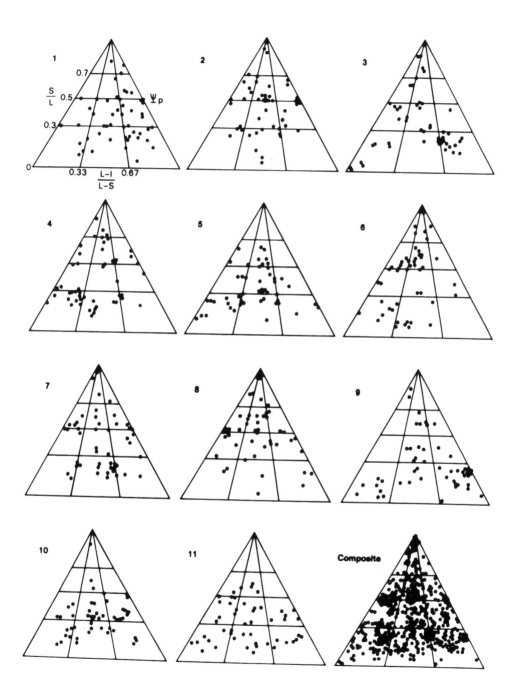

Composite diagram includes all grains measured, n=550.

FIG. 5.   TRIANGULAR FORM-DIAGRAMS FOR TAJIK LOESS SAMPLES

408

Elongate, shard-like particles TJ 79/9

Blocky grains with very few surface features TJ 79/5

General view of TJ 79/9

Edge-rounded grains TJ 79/4

FIG. 6. ELECTRON MICROGRAPHS OF TAJIK LOESS SAMPLES

material currently found in British caves (21) and is supported by general observations undertaken during on-going research (PAB) of loess taken from near Wurzburg, West Germany and Kentucky, U.S.A. Whilst the quartz loess particles adopt this varied form, particles of mica, feldspar and calcite seem to provide the 'traditional' form of the composite loess material. Indeed micas characteristically produced cleavage-controlled plates whilst feldspars, particularly orthoclase, seem responsible for the more elongate particle shapes.

Although there would appear to be the more obvious relation-ships of particle form with mineralogy there is evidently a discrete variation of form of the quartz particles within and between samples. TJ/1, for example (Fig. 5:1) exhibits a generally elongate-compact form with only 18% of the grains showing a platy habit. Similarly, samples TJ/3, TJ/9 and TJ/11 (Figs. 5:3, 5:9, 5:11) also show a strong component of elongate form, although each has its own composite form (Fig. 6B). These residuals of breakdown can be found in samples TJ/1, TJ/4, TJ/7 and TJ/10 in fairly significant proportions. Whilst they may reflect inherent microfracture paucity (25, 26) and hence the nature of the parent material (Metamorphic, igneous etc.) more work needs to be directed into such provenance studies before firm conclusions can be drawn.

(ii) Grain Roundness

Mechanical rounding of silt-sized sedimentary particles has long been shown to be of limited effect (27). However, rounding does occur upon the surfaces of small quartz particles, both by mechanical modification processes (subaqueous and aeolian) and by chemical agencies (precipitation and solution of quartz). The Tajik loess exhibits various degrees of rounding due to processes of both chemical and mechanical alteration (Table II).

Sample No.	Degree of Rounding	Cause of Rounding
TJ 79/1	Sub-Rounded	Mechanical edge rounding
TJ 79/2	Well-Rounded	Initial mechanical rounding, later less extensive chemical rounding
TJ 79/3	Mixed	Largely chemical action
TJ 79/4	Sub-Rounded	Mechanically rounded
TJ 79/5	Angular to Sub-Rounded	Limited mechanical action
TJ 79/6	Angular - Sub-Angular	Limited mechanical action
TJ 79/7	Sub-Rounded	(Chemically obscured surface)
TJ 79/8	Sub-Rounded	Largely chemical action
TJ 79/9	Mixed	Flatter grains more angular: Chemical action
TJ 79/10	Sub-Rounded	Chemical action
TJ 79/11	Sub-Rounded	Mechanical action

TABLE II. ROUNDING EXTENT AND CAUSE OF ROUNDING

Edge-rounding due to mechanically induced abrasion is adjudged to have occurred on six of the eleven samples, whilst silica precipitation (post-depositional?) and hence edge-dulling is evident on the remaining five samples (Fig. 6C). Significantly there is no relationship spatially or stratigraphically between the location of the loess samples and the suggested mechanism of rounding.

(iii)   Grain Surface Textures

Although surface texture studies have been normally concentrated on sand-sized particles, the general concensus is that surface textures are largely absent upon silt-sized grains. It has also been shown, however, (21) that mechanically and chemically derived surface features do exist upon these small grains. Tajik loess samples show surprisingly few 'mechanical' features present upon their surfaces (Fig. 6D). Conchoidal fractures and breakage blocks are present, reflecting the inherent characteristics of quartz but little else has been identified. Although chemically produced precipitation features are relatively abundant (and considered post-deposition) the loess grains exhibit no solutional features at all, perhaps as testimony to post-depositional groundwater conditions. A few solutional hollows were evident upon a number of grains examined but these were considered to be inherent to a previous cycle, probably a first-cycle quartz inclusion subsequently dissolved out. It is, then, the lack of chemical features that is surprising, indicative of quiescent groundwater modification since deposition (Post-depositional chemical etching has been identified by one of us (PAB) upon the surface of loess-derived materials from Germany, Britain and U.S.A.).

CONCLUSIONS

Loess is very extensive in Tajikistan and is of comparable thickness to that of the classic loess terrains of China. It contains multiple layers of palaeosols, many of which contain calcrete horizons, and thermoluminescence and palaeomagnetic dating techniques indicate that the loess has been deposited since the late Pliocene. As in China and Central Europe there have been at least eight or nine cycles of soil formation and loess deposition in the last 0.73 m.y.

The Tajik loess is finer-grained than many examples cited from other parts of the world, has a high carbonate content, and is dominated by quartz. Particle shape studies using the SEM have shown that, contrary to the accepted view, the quartz particles are not characterised by flatness. Shard-like quartz particles were common, and although some rounding had occurred, there was no evidence of solution. The lack of chemical features is perhaps indicative of quiescent ground water conditions since deposition.

ACKNOWLEDGEMENTS

Helen Rendell and Andrew Goudie visited Tajikistan as guests of the Tajik Academy of Science through the exchange agreement between the Royal Society and the Soviet Academy of Sciences. They were introduced to the sites by Dr. V. A. Ranov and his colleagues, notably Dr. A. Penkov. Their kindness and hospitality is here acknowledged as is the company of Dr. Bridget Allchin and Dr. Ken Thomas. Laboratory analyses were made possible through the assistance of Mr. C. B. Jackson, the School of Geography, University of Oxford. Peter Bull wishes to acknowledge financial support from Christ Church, Oxford. The School of Cultural and Community Studies, Sussex University provided funding for analytical work. The figures were kindly drawn by Miss Margaret Loveless.

REFERENCES

1)   SMALLEY, I. J., (1966). "The properties of glacial loess and the formation of loess deposits". Journal of Sedimentary Petrology 36, 669 – 676.

2)   SMALLEY, I. J. and VITA-FINZI, C., (1968). "The formation of fine particles in sandy deserts and the nature of 'desert' loess". Journal of Sedimentary Petrology, 38, 766 – 774.

3)   KEUNEN, Ph. H., (1969). "Origin of quartz silt". Journal of Sedimentary Petrology 39, 1631 – 1633.

4)   SMALLEY, I. J. and KRINSLEY, D. H., (1978). "Loess deposits associated with deserts". Catena 5, 53 – 66.

5)   DERBYSHIRE, E., (1978). "The middle Hwang Ho loess lands". Geographical Journal 144, 191 – 194.

6)   STODDART, D. R., (1978). "Geomorphology in China". Progress in Physical Geography 2, 187 – 236.

7)   SCHULTZ, C. B. and FRYE, J. C. (eds.), (1968). "Loess and related eolian deposits of the world". Lincoln, Nebraska.

8)   SMALLEY, I. J. (ed.), (1975). "Loess: lithology and genesis". Stroudsburg.

9)   MALYCHEFF, V., (1929). "Le Loess". Revue Géomorphologie Dynamique 2, 147 – 183.

10) DODONOV, A. Y., MELAMED, Y. R. and NIKIFOROVA, K. V. (eds.), (1977). "Guidebook, International Symposium on the Neogene-Quaternary Boundary". Moscow.

11) SHELKOPLYAS, V. N., (1977). "Dating of subaerial Pleistocene sediments of Tajikistan by method of thermoluminescence" in Guidebook, International Symposium on the Neogene-Quaternary Boundary". Moscow, 17 – 19.

12) RANOV, V. A. and DAVIS, R. S., (1979). "Toward a new outline of the Soviet Central Asian Palaeolithic". Current Anthropology 20, 249 – 270.

13) KUKLA, G. J., (1977). "Pleistocene land–sea correlations: I – Europe". Earth Science Reviews 13, 307 – 374.

14) LAZARENKO, A. A., (1977). "Loess cover of the Tajik depression

(stratigraphy, lithology, problems of genesis)" in "Guidebook, International Symposium on the Neogene-Quaternary Boundary". Moscow, 35 - 37.

15) PENKOV, A. V. and GAMOV, L. N., (1977). "Palaeomagnetic datums in Pliocene-Quaternary strata of southern Tajikistan" in "Guidebook, International Symposium on the Neogene-Quaternary Boundary". Moscow, 46 - 47.

16) BOWLER, J. M., (1968). "Glacial age aeolian events at high and low latitudes: a southern hemisphere perspective" in Van Zinderen Bakker, E. M. (ed.) "Antarctic Glacial History and World Palaeoenvironments", 149 - 172.

17) FLINT, R. F., (1970). "Glacial and Quaternary Geology".

18) SWINEFORD, A. and FRYE, J. C., (1955). "Petrographic comparison of some loess samples from western Europe with Kansas loess". Journal of Sedimentary Petrology 25, 3 - 23.

19) RUSSELL, R. J., (1944). "Lower Mississippi Valley loess". Bulletin, Geological Society of America 55, 1 - 40.

20) KRINSLEY, D. H. and DOORNKAMP, J. C., (1973). "Atlas of quartz grain surface textures".

21) BULL, P. A., (1978). "Observations on small sedimentary quartz particles analysed by scanning electron microscopy" in Johari, O. (ed.) "Scanning Electron Microscopy 1978" 1, 821 - 828.

22) SNEED, E. D. and FOLK, R. L., (1958). "Pebbles in the lower Colorado River, Texas. A study of particle morphogenesis". Journal of Geology 66, 114 - 150.

23) SMALLEY, I. J., (1974). "Discussion. Fragmentation of granite quartz in water". Sedimentary 21, 633 - 635.

24) HAMMOND, C., MOON, C. F. and SMALLEY, I. J., (1973). "High voltage electron microscopy of quartz particles from postglacial clay soils". Journal of Material Science 8, 509 - 513.

25) MOSS, A. J., WALKER, P. H. and HUTKA, J., (1973). "Fragmentation of granite quartz in water". Sedimentary 20, 489 - 511.

26) MOSS, A. J., WALKER, P. H. and HUTKA, J., (1974). "Fragmentation of granite quartz in water: a reply". Sedimentary 21, 637 - 638.

27) KUENEN, Ph. H., (1956). "Experimental abrasion of pebbles 2. Rolling by current". Journal of Geology 64, 336 - 368.